BIOSENSING

Biosensing

International Research and Development

JEROME SCHULTZ
University of California, Riverside, CA, U.S.A.

MILAN MRKSICH
University of Chicago, IL, U.S.A.

SANGEETA N. BHATIA
University of California, San Diego, CA, U.S.A.

DAVID J. BRADY
Duke University, Durham, NC, U.S.A.

ANTONIO J. RICCO
Stanford University, Los Gatos, CA, U.S.A.

DAVID R. WALT
Tufts University, Medford, MA, U.S.A.

and

CHARLES L. WILKINS
University of Arkansas, Fayetteville, AR, U.S.A.

Springer

A C.I.P. Catalogue record for this book is available from the Library of Congress.

ISBN-10 1-4020-4057-1 (HB)
ISBN-13 978-1-4020-4057-3 (HB)
ISBN-10 1-4020-4058-X (e-book)
ISBN-13 978-1-4020-4058-0 (e-book)

Published by Springer,
P.O. Box 17, 3300 AA Dordrecht, The Netherlands.

www.springer.com

This document was sponsored by the National Science Foundation (NSF) and other agencies of the U.S. Government under awards from the NSF (ENG-0104476) and the Army Research Office (DAAD19-03-1-0067) awarded to the World Technology Evaluation Center, Inc. The government has certain rights in this material. Any opinions, findings, and conclusions or recommendations expressed in this material are those of the authors and do not necessarily reflect the views of the United States Government, the authors' parent institutions, or WTEC, Inc.

Printed on acid-free paper

Printed in the Netherlands.

WTEC Panel on International R&D in Biosensing

Jerome Schultz (chair)
University of California, Riverside, CA, U.S.A.
Milan Mrksich (vice-chair)
The University of Chicago, Chicago, IL, U.S.A.
Sangeeta N. Bhatia
University of California, San Diego, CA, U.S.A.
David J. Brady
University of California, Riverside, CA, U.S.A.
Antonio J. Ricco
Los Gatos, CA, U.S.A.
David R. Walt
Tufts University, Medford, MA, U.S.A.
Charles L. Wilkins
University of Arkansas, Fayetteville, AR, U.S.A.

Abstract

This report reviews international research and development activities in the field of biosensing. Biosensing includes systems that incorporate a variety of means, including electrical, electronic, and photonic devices; biological materials (e.g., tissue, enzymes, nucleic acids, etc.); and chemical analysis to produce detectable signals for the monitoring or identification of biological phenomena. This is distinct from "biosensors" that employ only biological materials or mechanisms for sensing. In a broader sense, the study of biosensing includes any approach to detection of biological elements and the associated software or computer identification technologies (e.g., imaging) that identify biological characteristics. Topics covered include the national initiatives, interactions between industry and universities, technology and manufacturing infrastructure, and emerging applications research. The panel's findings include the following: Europe leads in development and deployment of inexpensive distributed sensing systems. Europe also leads in integration of components and materials in microfabricated systems. Europe and Japan both have much R&D on DNA array technology, but the impact is likely to be only incremental. The United States leads in surface engineering applied to biosensing and in integration of analog-digital systems. Both Europe's and Japan's communication infrastructures are better suited for networked biosensing applications than those of the United States. Integrated biosensing research groups are more common in Europe and Japan. Additional findings are outlined in the panel's executive summary.

WTEC Mission

WTEC provides assessments of foreign research and development in selected technologies under awards from the National Science Foundation, the Office of Naval Research, and other agencies. Formerly part of Loyola College's International Technology Research Institute, WTEC is now a separate non-profit research institute. Michael Reischman, Deputy Assistant Director for Engineering, is NSF Program Director for WTEC. Sponsors interested in international technology assessments and related studies can provide support for the program through NSF or directly through separate grants to WTEC.

WTEC's mission is to inform U.S. scientists, engineers, and policymakers of global trends in science and technology. WTEC assessments cover basic research, advanced development, and applications. Panels of typically six technical experts conduct WTEC assessments. Panelists are leading authorities in their field, technically active, and knowledgeable about U.S. and foreign research programs. As part of the assessment process, panels visit and carry out extensive discussions with foreign scientists and engineers in their labs.

The WTEC staff helps select topics, recruits expert panelists, arranges study visits to foreign laboratories, organizes workshop presentations, and finally, edits and disseminates the final reports.

Executive Editor: Dr. Marvin Cassman, Former Director, National Institute of General Medical Sciences, NIH

Series Editor: Dr. R.D. Shelton, President WTEC

World Technology Evaluation Center, Inc. (WTEC)

FOREWORD

> We have come to know that our ability to survive and grow as a nation to a very large degree depends upon our scientific progress. Moreover, it is not enough simply to keep abreast of the rest of the world in scientific matters. We must maintain our leadership.[1]

President Harry Truman spoke those words in 1950, in the aftermath of World War II and in the midst of the Cold War. Indeed, the scientific and engineering leadership of the United States and its allies in the twentieth century played key roles in the successful outcomes of both World War II and the Cold War, sparing the world the twin horrors of fascism and totalitarian communism, and fueling the economic prosperity that followed. Today, as the United States and its allies once again find themselves at war, President Truman's words ring as true as they did a half-century ago. The goal set out in the Truman Administration of maintaining leadership in science has remained the policy of the U.S. Government to this day: Dr. John Marburger, the Director of the Office of Science and Technology (OSTP) in the Executive Office of the President made remarks to that effect during his confirmation hearings in October 2001.[2]

The United States needs metrics for measuring its success in meeting this goal of maintaining leadership in science and technology. That is one of the reasons that the National Science Foundation (NSF) and many other agencies of the U.S. Government have supported the World Technology Evaluation Center (WTEC) and its predecessor programs for the past 20 years. While other programs have attempted to measure the international competitiveness of U.S. research by comparing funding amounts, publication statistics, or patent activity, WTEC has been the most significant public domain effort in the U.S. Government to use peer review to evaluate the status of U.S. efforts in comparison to those abroad. Since 1983, WTEC has conducted over 50 such assessments in a wide variety of fields, from advanced computing, to nanoscience and technology, to biotechnology.

The results have been extremely useful to NSF and other agencies in evaluating ongoing research programs, and in setting objectives for the

---

[1] Remarks by the President on May 10, 1950, on the occasion of the signing of the law that created the National Science Foundation. *Public Papers of the Presidents* 120: p. 338.

[2] http://www.ostp.gov/html/01_1012.html.

future. WTEC studies also have been important in establishing new lines of communication and identifying opportunities for cooperation between U.S. researchers and their colleagues abroad, thus helping to accelerate the progress of science and technology generally within the international community. WTEC is an excellent example of cooperation and coordination among the many agencies of the U.S. Government that are involved in funding research and development: almost every WTEC study has been supported by a coalition of agencies with interests related to the particular subject at hand.

As President Truman said over 50 years ago, our very survival depends upon continued leadership in science and technology. WTEC plays a key role in determining whether the United States is meeting that challenge, and in promoting that leadership.

Michael Reischman
Deputy Assistant Director for Engineering
National Science Foundation

Table of Contents

APPENDICES

LIST OF FIGURES

List of Tables

PREFACE

This report was prepared by the World Technology Evalutation Center (WTEC), a nonprofit research institute funded by grants and other awards from most of the Federal research agencies. Among other studies, WTEC has provided peer reviews by panels of U.S. experts of international research and development (R&D) in more than 55 fields since 1989. In 2002, WTEC was asked by several agencies to assess international R&D in biosensing. This report is the final product of that study.

We would like to thank our distinguished panel of experts, who are the authors of this report, for all of their efforts to bring this study to a successful conclusion. We also are very grateful to our foreign hosts for their generous hospitality, and to the participants in our preliminary workshop on U.S. biosensing R&D. Of course, this study would not have been possible without encouragement from our sponsor representatives: Bruce Hamilton, Fred Heineken, Elbert Marsh, Deborah Young, Fil Bartoli, and Vijay Jain of the National Science Foundation (NSF); Christine Kelley, Joan Harmon, Dick Swaja, Mollie Sourwine, and Stephen Green of the National Institutes of Health (NIH); John Hines and Steve Davison of the National Aeronautics and Space Administration (NASA); Dan Schmoldt of the U.S. Department of Agriculture (USDA); and Micheline Strand of the U.S. Army Research Office (ARO).

This report covers a broad spectrum of material on the subject, so it may be useful to give a preview here. The Executive Summary was prepared by the chair, Jerome Schultz, with input from all the panelists. The chapters in the body of this report present the panel's findings in an analytical organization by subdiscipline. Appendix A provides the biographies of the panelists. Appendices B and C contain the panel's individual reports on each site visited in Europe and Japan, which form a chronological or geographic organization of much of the material. Appendices D-H present information on U.S. Government sponsored projects in the field. Appendix I presents biosensing program information from the European Union 6th Framework Programme (2002–2006). Appendix J lists recent biosensing-related patents filed by organizations that hosted the panel's site visits in Europe and Japan.

To complement the qualitative assessment by peer review, Appendix K is a quantitative bibliometric study of international biosensors research for 1997-2002. This work was performed by Grant Lewison of the City University, London, for WTEC. Finally, a glossary is provided as Appendix L.

All the products of this project are available at http://www.wtec.org. The full-color electronic version of this report is particularly useful for viewing some of the figures that do not reproduce well in black and white. Also posted at this site are the slideshows from two workshops held for this project, which contain considerable additional information on biosensing R&D in the United States and abroad.

Roan Horning
WTEC, Inc.

EXECUTIVE SUMMARY

Jerome S. Schultz

The long-standing U.S. national interest in biosensing has encompassed broad requirements for reliable and sensitive sensing systems for medical diagnostics, environmental monitoring, and food safety assurance. National demands on biosensing systems have expanded and taken on a new urgency in the wake of the September 11, 2001, terrorist attacks and the anthrax attacks that followed.

In a broad sense, the study of biosensing includes any approach to detection of biological elements and the associated software or computer identification technologies (e.g., imaging) that identify biological characteristics. *Biosensing systems* incorporate a variety of means, including electrical, electronic, and photonic devices; biological materials such as tissue, enzymes, and nucleic acids; and chemical analysis, to produce detectable signals for the monitoring or identification of biological phenomena. This is distinct from *biosensors* that employ only biological materials or mechanisms for sensing. Biosensing is finding a growing number of applications in a wide variety of areas, including biomedicine; food production and processing; and detection of bacteria, viruses, and biological toxins for bio-warfare defense.

In late May 2002, the World Technology Evaluation Center (WTEC) embarked on a study to assess research and development activities related to biosensing in the United States and worldwide, under the sponsorship of the National Science Foundation (NSF), the National Institutes of Health (NIH), the United States Department of Agriculture (USDA), the National Aeronautics and Space Administration (NASA) and the Army Research Office (ARO). The goals of this study are to gather information and disseminate it to government decisionmakers, the research community, and the public on worldwide status and trends in biosensing R&D, and to critically analyze and compare the research in the United States with that being pursued in Japan, Europe, or other major industrialized countries. The information gathered through this study is intended to serve the purposes of identifying good ideas worth exploring in U.S. R&D programs; clarifying research opportunities and requirements for progress in the field; identifying opportunities for international collaboration; and evaluating the position of foreign research programs relative to those in the United States.

To achieve these goals, WTEC recruited a panel of seven experts in the field (see biographies in Appendix A) to carry out a series of three major tasks designed to deliver the maximum amount of quality information to the sponsors and the public within the constraints of time and resources:

1. Host a workshop of members of the U.S. biosensing R&D community to characterize the state of the art and current trends in biosensor technologies in the United States. [The WTEC Biosensing Study U.S. R&D Overview Workshop was held at NIH in Bethesda, MD, on 3–4 December 2002.]
2. Conduct site visits to gather first-hand information from many of the world's best university and industrial laboratories in biosensing research. [The WTEC panelists conducted site visits to laboratories in Europe, Australia, and Japan during January and March 2003.]
3. Report back findings in both a public forum and in writing to the U.S. sponsors, the scientific community, and the public at large. [The WTEC Workshop on Biosensing in Europe, Japan, and the United States was held on 13 May 2003 at the Bethesda, MD, Marriott Hotel.]

This report, the final phase of the study, details and analyzes the results of the WTEC biosensing panel's literature review, U.S. survey, and Europe and Japan site visits. It is available to the public on the Web at www.wtec.org/biosensing, as well as in print.

PRINCIPAL FINDINGS

Infrastructure

Biosensing research has exploded dramatically in recent years. Both NIH and NSF sponsored over 200 projects related to biosensing in 2002. Appendixes D and E lists these projects as examples of ongoing research, ranging from surface chemistry to intelligent agents, and Appendixes F-H give an insight into the depth and breadth of work funded by the Defense Advanced Research Projects Agency (DARPA), ARO, and the Department of Energy (DOE).

Expansion of research activity has been facilitated by major technological breakthroughs in the fields of microelectronics, microfabrication, surface science, photonics, and information sciences. In current terminology, "Bio-Nano-Info" has become a new paradigm for the convergence of research in the fields of biotechnology, nanotechnology, and information technology. In the United States, NSF has recognized this trend of connecting bio-nano-info in its report, *Converging Technologies for Improving Human Performance* (Roco and Bainbridge 2003). Further evidence for the overlap of fields are DARPA programs in BioComputational Systems, Bio-Molecular Microsystems (SIMBIOSYS), and Nanostructure in Biology.

Because of this technological convergence, it is difficult to separate out the human, technical, and financial resources that are being allocated to biosensing systems alone. Along with the multidisciplinary nature of the science advancing biosensing R&D, it is clear that Japan and Europe are increasingly building collaborative efforts to carry out biosensing projects; in some cases the teams are industrial/academic; in others, government/academic. It also appears there is an escalating interest in commercialization of biosensing technologies, and several large new biosensing-related R&D facilities are being built. A manifestation of these infrastructure trends is seen in various program initiatives in the United States, Europe, and Japan.

In Europe, an indicator of future goals and plans for research is provided by the EU's Sixth Framework Programme solicitations for 2002–2006 (see summary in Appendix I). Although, this framework document does not explicitly identify biosensing technologies as a program element, one can find references to biosensing systems under these program areas:

- Life Sciences, genomics, and biotechnology for health
- Information Society technologies
- Nanotechnologies and nanosciences, knowledge-based multifunctional materials, and new production processes and devices

The projected budget for these topics is about \$7 billion, and about \$1 billion of these funds will probably relate directly to biosensing systems.

Another feature of the European approach to building a research and commercial capability relating to biosensing products is the organization of collaborative partnerships between academic research centers and companies. For example, in the Berlin-Brandenburg region there are three Max Planck Institutes and two Fraunhofer Institutes located near the University of Potsdam that actively work on several collaborative projects. There are approximately 100 companies in this consortium with interests in diagnostics, biotechnology, and software that will accelerate the transfer of biosensing systems into the marketplace.

In Japan, the universities the WTEC panel visited all had programs relating biotechnology, nanotechnology, and computers. For example, the fields of interest stated by the President of the Tokyo University of Agriculture and Technology are (1) Biotechnology, (2) Information and Communications Technologies, (3) Environmental Science and Resource Science, and (4) Nanotechnology. This university has an extensive program of providing incubator facilities to promote technology transfer from the university to industry.

WTEC visits to various universities confirmed that a major change is underway in the ability of universities in Japan to interact with industry, as many state-owned institutions will be released from central government control in the next few years. This has resulted in a significant increase in patent application activity by Japanese faculty. Another example of the trend for the direct connection of university and corporate research is the new School of Bionics at the Tokyo University of Technology. A new US$250 million building with 15,000 m^2 of space opened in April 2003 to house industrial/academic research projects along with the traditional academic research and academic facilities. Four floors of the new facility were to be occupied by corporate research laboratories who will co-sponsor research in the institute. The university also plans to have a degree program in technology management.

There is extensive collaboration in Japan between government laboratories and academia. Visits to government laboratories indicate significant national spending, despite Japanese economic hardship. This suggests acceptance of the idea that technology is essential for future economic success.

To complement the WTEC panel's literature review, public forums, and first-hand observations of international biosensing research and development, this report includes in Appendix K a bibliometric study of international biosensors research in the period 1997–2002 that underscores the high activity in this field based on the number and quality of published biosensor studies in this period, particularly in the United States, Europe, and Japan.

Table ES.1 summarizes the key observations by the WTEC panel concerning the patterns of infrastructure development for biosensing in the United States, Europe, and Japan, to highlight the unique approaches and relative strengths of these regions.

TECHNOLOGY HIGHLIGHTS FROM SITE VISITS

In two separate, one-week rounds of visits in early 2003, the WTEC panel toured 40 premier research establishments in Europe, Australia, and Japan that have a focus or known activities in biosensing and related areas. These visits included universities, industry laboratories, and government research centers: 23 facilities in Europe and Australia, and 17 in Japan. The capabilities listed below reflect not a detailed analysis but rather highlights of first-hand interviews and observations of programs in the laboratories the panel visited. Site reports are included in Appendix B (Europe and Australia) and Appendix C (Japan) of this report.

Table ES.1
Comparative Patterns in Funding of Biosensing R&D and Commercialization, by Region†

	United States	Europe	Japan
Research Focus	**Fundamental science: Academia,** national labs **Applied science & engineering: small companies, national labs,** large companies	**Applied Science & engineering: academia**, national labs, small & large companies	Fundamental science, engineering
Teaming mechanisms	Individual investigators, interdisciplinary interactions	**Multidisciplinary teams**	Individual investigators, interdisciplinary teams
Types of collaboration	International collaborations	**Multinational teams**, major EU support	**National focus**
Support for new technology areas	**Federal support to open new technology areas: MEMS, microfluidics, nanotechnology**	Generally follow U.S. lead into new technology areas	Generally follow U.S. lead, but **industry has a longer time horizon** than the U.S.
Academic support for applications focus in R&D	Nonuniform support	**Strong support**	Widespread support
Support for technology transfer to industry, commercialization	Federal government, **individual entrepreneur, venture* support**	**Local government support**, national government support, **academic admin. support**	National government support and **increasing university emphasis**

† Bold indicates particular strength /emphasis
* In good economic times

Europe

- Highly automated 2D-gel ICAT (mass spectrometry) techniques are used to carry out high-throughput protein analysis at Oxford GlycoSciences (Dr. Christian Pohlff).
- A combination of lab-on-a-chip technologies and mass spectroscopy are used to tackle the challenging characterization of the proteome at the University of Twente, MESA+ Institute (Professor David Reinhoudt).
- Live cell analysis with the Biacore Procel fluorescence detection/microfluidic system is well established at Biacore in Uppsula, Sweden.
- Reflectometric interference spectroscopy is used for low-cost and highly miniaturized biosensing arrays at the Institute of Physical and Theoretical Chemistry, University of Tübingen (Professor Gunter Gauglitz).
- Low-energy electron point-source (LEEPS) microscopy appears to be leading towards resolutions of features below 1 nm at Ruprecht-Karls University Heidelberg (Professor Michael Grunze).
- Lipid bilayer vesicles and lipid nanotube-vesicle-networks are being investigated for encapsulation and support of reconstituted biological functions such as receptors, synaptic vesicles, and signal-transduction systems/pathways at Linköping University (Professor Ingemar Lundstrom).

Japan

- Uniform, nano-sized (50–100 nm), lipid-covered (containing fusion proteins) ferromagnetic particles produced by *magnetospirillium magneticum* are used as unique components of biosensors at Tokyo University of Agriculture and Technology, Department of Biotechnology and Life Science (Professor Tadashi Matsunaga).
- Ferrocenyl napthalene diimide (FND) is being used as a DNA hybridization indicator to enable charge transfer to microelectrodes producing an electrochemical signal proportional to the amount of target DNA at Kyushu University (Professor Shigeori Takanaka).
- Confocal microscopic imaging of molecular events in single living cells is being achieved by protein constructs of biorecognition molecules with fluorescent proteins at the University of Tokyo, Department of Chemistry (Professor Yoshi Umezawa).
- A thermal lens microscope technique has been perfected to measure concentrations in the zepto-mole range, or about 50–100 molecules,

on biochips at the School of Engineering, University of Tokyo (Professor T. Kitamori).

- Novel methods are being used to synthesize photo-induced electron transfer (PET) of organic species that are incorporated in the design of new sensing materials at the Graduate School of Pharmaceutical Sciences, University of Tokyo (Professor Kazuya Kikuchi).

COMPARATIVE REGIONAL STRENGTHS IN KEY BIOSENSING AREAS

The WTEC panel collected a vast amount of information from a preliminary literature review, the initial U.S. workshop, site visits to Europe and Japan, and the final workshop to report on and receive feedback from the research community about the study findings. Based on this information, the panel has made a comparative assessment of the status of biosensing research in Europe and Japan relative to that in the United States. Table ES.2 compares for each of the key areas of biosensing that are discussed in chapters 2 through 7 the panel's evaluation of the knowledge bases, work to date/in progress, and the relative approaches/strengths of the worldwide biosensing field generally, with a summary assessment of which region(s) lead the area.

Table ES.2.
Comparison of U.S., Japanese, and European R&D Activities in Biosensing

Area	Subarea	Topic	Knowledge Base	Work to Date	Leading Region
Optical Biosensing (see Chapter 2)		Arrays	Patterning Surface chemistry	Extensive, mature	U.S.
		Interfer-ometric, label-free	Surface plasmon resonance Interference	Old method, but ongoing efforts	Europe
		Cheap, distributed sensors	Screen printing Optical transduction	New and promising	Europe

Table ES.2.
Comparison of U.S., Japanese, and European R&D Activities in Biosensing

Area	Subarea	Topic	Knowledge Base	Work to Date	Leading Region
		Nano-technology	New signaling mechanisms New materials	Metal particles	U.S.
		Molecular biology	Genetic engineering Cell biology	Well-developed technology being applied to biosensing	U.S.> Europe> Japan
		Integration	Engineering, Chemistry, Computer science	Team science and engineering	Europe
Electro-Based Sensors (see Chapter 3)	Materials	Reagents	Advanced	Mature	U.S., Europe
		Electrodes	Advanced	Mature	All
	Surface Engi-neering	Surface chemistry	Advanced	Substantial	U.S., Europe
		Immob./ Pattern	Advanced	Substantial	U.S., Europe
	Trans-duction Strategies	Conventional	Advanced	Substantial	All
		Nanoscale	Early	Beginning	U.S.>Europe, Japan
	Systems	Arrays	Intermediate	Beginning	U.S.>Europe
		Integration	Intermediate	Substantial	Europe> Japan, U.S.
Cell- and Tissue-based Sensors (see Chapter 4)	Trans-duction Strategies	Electrical	Early	Beginning	All
		Optical	Intermediate	Progressing	U.S., Europe
	Interface Engi-neering	Surface chemistry	Advanced	Substantial	U.S., Europe
		Cell function	Early	Progressing	U.S., Europe
	Integration	Micro-technology	Intermediate	Substantial	Europe> Japan, U.S.
	Commerci alization	Drug discovery	Early	Progressing	U.S.
		Diagnostics	Little	Little	None

Table ES.2. Comparison of U.S., Japanese, and European R&D Activities in Biosensing

Area	Subarea	Topic	Knowledge Base	Work to Date	Leading Region
Mass Spectrometry (MS) (see Chapter 5)	Mass sensors, MEMS, and microfluidics	Mass sensors	Excellent	Europe, U.S., Japan	Equally advanced, commercialized
	Mass spectrometric methods	Compact instrument development	Excellent, but problems remain	Europe, U.S.	Europe, U.S.
		Portable MS development	Excellent, but problems remain	Europe, U.S.	Europe, U.S.
		Novel MS interface	Excellent, but problems remain	Europe, U.S.	Europe, U.S.
		Proteomics	Excellent; very active research area	Europe, U.S.	Europe, U.S.
Micro-fabricated Biosensing Devices: MEMS, Microfluidics, and Mass Sensors (see Chapter 6)	MEMS	Biosensing components	Advanced	Extensive	U.S.~ Japan ~ Europe
		Integrated systems	Incomplete	Significant	Europe
		Integration of biomaterials	Minimal	Isolated examples	Europe, U.S.
	Microfluidics	Discrete devices	Advanced	Extensive	U.S.
		Integrated systems	Incomplete	Minor	Europe ~U.S.> Japan
	Mass sensors	Piezo devices	Advanced	Extensive	None
		Si cantilevers	Incomplete (esp. liquid operation)	Significant (dry) Minor (wet)	U.S. ~ Europe
		Integrated biomaterials	Incomplete	Significant	Europe ~Japan

Table ES.2.
Comparison of U.S., Japanese, and European R&D Activities in Biosensing

Area	Subarea	Topic	Knowledge Base	Work to Date	Leading Region
	Nanotechnology	"Top-down" (nanofab) "Bottom-up" (molec. organized materials) Integration into complex (bio) systems	Incomplete Incomplete Incomplete	Significant Extensive Little	U.S. > Europe U.S. Japan, Europe Europe ~ U.S.
Information Systems for Biosensing (see Chapter 7)		Algorithms	Geometric segmentation	Theoretical/early experiments	U.S.
		Communications	Ad hoc and infrastructure	3G wireless and open spectra	U.S.– ad hoc; Japan– infra-structure; Europe– applications development
		System Integration	CMOS and MEMS/ Microfluidics	Lab on a chip demos	U.S.–MEMS, micro-fluidics; Europe– integrated electronics

The WTEC panel's conclusions regarding the relative strengths in Europe, Japan, and the United States of biosensing R&D may be summarized as follows:

- Europe leads in development and deployment of inexpensive distributed sensing systems.
- Europe also leads in integration of components and materials in microfabricated systems.
- Europe and Japan both have much R&D on DNA array technology, but the impact is likely to be only incremental.
- The United States leads in surface engineering applied to biosensing and in integration of analog-digital systems.

- Both Europe's and Japan's communication infrastructures are better suited for networked biosensing applications than those of the United States.
- Integrated biosensing research groups are more common in Europe and Japan.

CONCLUSIONS

Among the significant overall trends and emerging opportunities that the WTEC biosensing panel identified are the following:

- Increasing pervasiveness of systems on a chip and other integrated systems approaches
- Growth of microfluidic/micromechanical systems
- Emergence of molecular receptor engineering
- Development of sensor networks and advanced logistical strategies

There is also a general trend towards the development of biosensors as a low-cost, commodity-like technology that will find application in a wide variety of consumer products.

In addition to these trends, the U.S. research community has identified several broad requirements and goals for ongoing development of the field of biosensing systems:

- Rapid, inexpensive, and broad based tests for detection and identification of toxic materials and organisms
- Standards for validation and comparison of technologies
- Methods that can be fielded as sentinels in the environment to monitor food, water, soil, and air quality
- Improved sampling and preprocessing techniques
- System automation for unskilled operators

REFERENCES

Roco, M.C., and W.S. Bainbridge, eds. 2003. *Converging technologies for improving human performance: Nanotechnology, biotechnology, information technology and cognitive science*. Dordrecht; Boston, Mass.: Kluwer Academic Publishers.

CHAPTER 1

INFRASTRUCTURE OVERVIEW

Jerome Schultz

INTRODUCTION TO THE STUDY

Biosensing technologies comprise portable devices and systems used for identifying, monitoring, and controlling biological phenomena. Most of these technologies have come into use just in the last two decades; nevertheless, biosensing already garners major academic, government, and industry R&D funding; relies on highly sophisticated multidisciplinary technology; and enjoys well developed and growing markets. Prior to September 11, 2001, U.S. biosensing development was driven primarily by the requirements of medical diagnostics, environmental monitoring, and food safety assurance. Since September 11, worries about anthrax, smallpox, and or other biological "weapons" in the hands of terrorists have elevated the prominence of biosensing as a component of bio-warfare defense. Adding to the rapidly growing significance of biosensing is its place in the remarkable convergence of advanced bio-, nano-, and infotechnologies in a totally new scientific paradigm.

With these trends as background, in May 2002, five U.S. government agencies asked the World Technology Evaluation Center (WTEC) to investigate states of the art and trends in biosensing research and development worldwide in comparison to the United States. The National Science Foundation (NSF), National Institutes of Health (NIH), United States Department of Agriculture (USDA), National Aeronautics and Space Administration (NASA) and Army Research Office (ARO) intend for this biosensing study to

- aid government policymakers, the research community, and the public to identify good ideas worth exploring in U.S. R&D programs
- note technical, educational, and infrastructure requirements and prospects for better progress in the field

- ascertain opportunities for international and interdisciplinary collaboration
- identify ways to shorten the lead time for deploying new biosensing technologies emerging from the lab
- evaluate the status and funding of foreign research programs relative to those in the United States

The study's sponsors identified particular applications of interest to be *healthcare* (biomedicine), the *environment*, the *food industry*, and *defense* against the threats of chemical and biological agents. The sponsors further identified the following specific technical issues to be addressed:

- nucleic acid sensors and DNA chips and arrays
- organism- and cell-based biosensors
- bioelectronics and biometrics
- biointerfaces and biomaterials, biocompatibility, and biofouling
- integrated, multimodality sensors and sensor networks
- system issues, including signal transduction, data interpretation, and validation
- novel sensing algorithms, e.g., non-enzyme-based sensors for glucose, or mechanical sensors for prosthetics
- related issues in bio-MEMS and bio-NEMS (microelectromechanical and nanoelectromechanical systems), possibly including actuators

Approach and Methodology

To execute the biosensing study, WTEC recruited a panel of seven U.S. experts in the field, chaired by Professor Jerome Schultz, then of the University of Pittsburgh, now at University of California, Riverside. The panelists each represent distinct areas of specialization in the biosensing field. Table 1.1 lists the panelists and their areas of focus for the study, along with others who helped arrange, conduct, and evaluate the site visits. Panelists' biographies are provided in Appendix A.

With the goals and team established, the WTEC panelists carried out the study in four phases:

1. *Establish baseline information on U.S. activities* as a benchmark for the worldwide assessment by hosting a workshop of members of the U.S. biosensing R&D community. The WTEC Biosensing Study U.S. R&D Overview Workshop was held at NIH in Bethesda, MD, on 3–4 December 2002. Participants provided an overview of recent trends and advances in biosensing technology development in the

areas identified by the sponsors; addressed the barriers for translating these technologies into the marketplace; and identified several general needs and applications that should be addressed in future R&D plans and programs. Proceedings of the workshop are available online at wtec.org / biosensing / proceedings /.

2. *Conduct site visits to gather first-hand information* from a number of the world's best university and industrial laboratories in biosensing research. The WTEC panelists conducted two week-long series of site visits to 23 laboratories in Europe and Australia and 17 laboratories in Japan during January and March 2003, respectively. Site reports of those visits are included in this report as Appendixes B and C and are also listed by name in the Table of Contents of this report.

Table 1.1.
Key Members of the WTEC Team and Their Roles in the Biosensing Study

Name	Organization	Assignment	Technical Focus
Jerome Schultz	University of Pittsburgh	Panel chair	Infrastructure
Milan Mrksich	The University of Chicago	Panel vice chair	Electrochemical/surface treatment
David Walt	Tufts University	Panel member	Optical sensing
Sangeeta Bhatia	University of California, San Diego	Panel member	Biological/cellular sensing
Charles Wilkins	University of Arkansas	Panel member	Mass spectrometry
Antonio Ricco	ACLARA BioSciences	Panel member	Microfluidics
David Brady	Duke University	Panel member	Data fusion/system integration
Fred Heineken	NSF/Engineering	Sponsor/Observer	
Christine Kelley	NIH, Institute for Biomedical Imaging and Bioengineering (NIBIB)	Sponsor/Observer	
Hassan Ali	WTEC	Support staff	

3. *Report back findings in a public forum* to the U.S. sponsors, the biosensing scientific community, and the public at large. The WTEC Workshop on Biosensing in Europe, Japan, and the United States was held on 13 May 2003 at the Bethesda, MD, Marriott Hotel. This workshop served as an open forum for presentation, discussion, and critical review of the panel's findings among members of the panel and invited participants. Viewgraphs from this workshop are available online at wtec.org/biosensing/views/.

4. *Compile the results of the study findings* from the first three phases into a written report to be made available to the funding agencies, to policymakers, and to the public. Each panel member prepared a chapter describing and analyzing what has been found in a specific area of biosensing in Japan and Europe and compared that with the status of that R&D in the United States. Before publication of this report, sponsoring agencies and site visit hosts reviewed drafts of the chapters and site reports and made corrections of factual statements, as applicable. As well as being available in print, this report is available on the Web at www.wtec.org/biosensing.

The term "biosensing" has been used throughout the WTEC study and in this report to mean not just devices but *systems* that produce verifiable signals for detecting biological occurrences through a variety of means, for a variety of purposes. Biosensing systems can include electrical, electronic, photonic, or mechanical devices; biological materials such as tissue, enzymes, or nuceic acids; means to provide chemical analysis; and advanced imaging and information processing technologies. Biosensors, which are devices that employ biological mechanisms or materials to provide selectivity and amplification for sensing biochemical materials, often are components of biosensing systems.

Report Structure

This final report is organized by chapter along the lines of the discrete foci of the individual panelists, based on information obtained through their individual expertise, offline research (a literature review), Europe and Japan site visits, and the May 2003 U.S. workshop presentations. The core of this first chapter outlines the cross-cutting issues related to infrastructure, comparing the status and strategies for investment in research as well as in physical and human resources in the United States, Europe, and Japan. Chapter 2 by David Walt discusses activities in optical biosensing, highlighting the scientific findings and outlining the challenges ahead. In Chapter 3, Milan Mrksich provides an overview and regional comparison of the development and implementation of electro-based sensors and surface engineering. Sangeeta Bhatia discusses in Chapter 4 the power of cell-based sensors to push the frontiers of biosensing by leveraging the unique attributes of living systems; her chapter provides an overview and regional comparisons of the latest developments in cell- and tissue-based sensors for both clinical and non-clinical applications. Chapter 5 by Charles Wilkins reviews the major work and research centers on mass spectrometry and biosensing research in the three regions and reveals the emerging trends. Chapter 6 by Antonio Ricco reviews the R&D activities in biosensing that are based on microelectromechanical systems, or MEMS, including their relationship to the field now broadly known as nanotechnology. Finally, David Brady in Chapter 7 addresses how biosensing research integrates biochemistry, physical electronics, and information systems, highlighting how each of the three regions pursues research in biosensing information systems and pointing out the opportunities in system integration.

HISTORY OF BIOSENSING DEVELOPMENT

To put the current high level of interest and research activity into perspective and to set the context for the chapters that follow, it is useful to briefly review the history of chemical sensors and biosensors. The use of the term "sensor" usually refers to a device that is somewhat portable in nature and that can be placed into an environment of interest, often a liquid sample, to measure a specific chemical (an analyte) on-site. This is in contrast to the traditional procedure of sending samples to a chemical or clinical analytical laboratory, where a variety of instruments are employed. The earliest chemical sensor of this type is the glass pH electrode that was developed in 1922 and later implemented as a portable device. It took almost another third of a century before the next practical chemical sensor was developed, the oxygen electrode invented by Leland Clark in 1954. Dr. Clark later introduced the concept of a biosensor in 1962 through his invention of the glucose electrode. Since then, the introduction and development of many different kinds of sensor technologies have been increasingly rapid. Table 1.2 lists some of development highlights.

A brief description of the Clark glucose electrode is instructive, because the components of this sensor device recur in most biosensors that have been developed subsequently. Figure 1.1 shows the components of this device, as published by Dr. Clark in 1962. In brief, the operation of this sensor is based on the reduction of oxygen flux to the oxygen electrode due to the consumption of oxygen in the *biosensing layer* (labeled F in the figure, comprised of enzymes glucose oxidase and peroxidase), by the oxidation of glucose to gluconic acid. The greater the concentration of glucose in the external media (and also in layer F), the lower the flux of oxygen to the electrode.

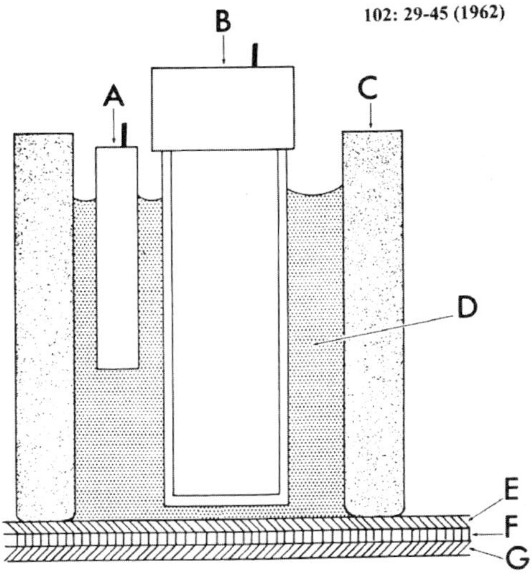

Fig. 1.1. First "enzyme" electrode—an electrode system for continuous monitoring in cardiovascular surgery. In this diagram, the elements marked A through E comprise the amperometric oxygen electrode. Addition of layer "F" containing the enzymes glucose oxidase and peroxidase converted this chemical sensor into a biosensor. Layer "G" is a semipermeable membrane that allows both glucose and oxygen to pass into the sensor. (Clark and Lyons 1962)

Table 1.2.
History of Chemical and Biological Sensors

Sensor Technology	Inventor	Date
Glass pH Electrode	Hughes	1922
Oxygen Electrode	Clark	1954
Carbon Dioxide Electrode	Stow and Randall	1954
Glucose Electrode	Clark	1962
Potentiometric Sensor	Guilbault	1969
Immunosensor	Janata	1975
Optodes	Lubbers	1975
Optical Affinity Sensors	Schultz	1979
Chip-Based Technologies	Fodor	1991

The essential components of a biosensor are a *detection capability* in this case, the oxygen electrode) and a *biological recognition capability* in this case, the enzyme layer). After Dr. Clark's invention, the research community realized that many detector systems can be used, and that many recognition materials can be found in nature. Figure 1.2 shows a simplified matrix that can lead to a variety of combinations of molecular recognition elements and transducers to produce biosensors, such as an antibody placed at the end of a fiberoptic system or a membrane receptor immobilized on a piezoelectric crystal. In the last few decades, the pace of biosensor research has increased dramatically, as described in thousands of journal articles, hundreds of patents, and dozens of books (several references are listed at the end of this chapter).

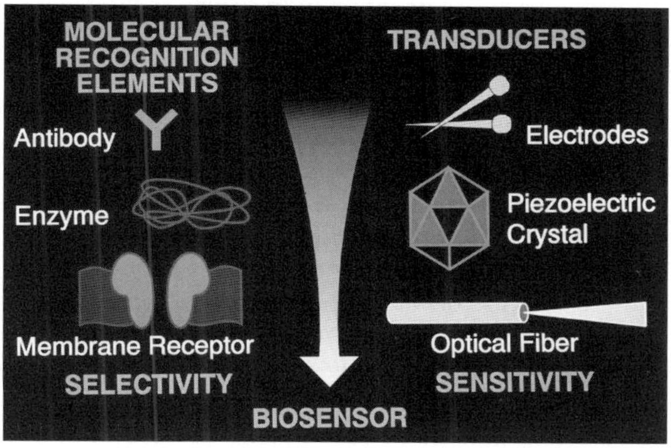

Fig. 1.2. A simplified matrix that can lead to a variety of combinations of molecular recognition elements and transducers to produce biosensors.

TECHNOLOGY DRIVERS

A number of factors have been driving interest and investment in biosensor research and product development. The primary driver has been the public's demand for healthcare aides, in particular ones to assist diabetics to cope with their disease. NIH conducted a major study during the period 1983–1993, called the Diabetes Control and Complications Trial, where approximately 1,500 individuals with Type 1 diabetes maintained close control of their blood sugar levels by self-testing about five times per day and administering insulin as needed to bring their blood sugar levels within the normal range. Those individuals who were able to maintain this regime of testing and control exhibited a remarkable reduction in the complications normally found in diabetics: risk of eye disease was reduced 76%, risk of kidney disease was reduced 50%, and risk of nerve disease was reduced 60%. Because of this fantastic outcome, development of more convenient blood glucose testing methods has become a major goal of the research and commercial communities. The current worldwide market for blood glucose testing equipment and test strips is estimated to be on the order of a billion dollars per year, and hundreds of millions of dollars have been spent on new sensing technologies for this purpose (see articles in the journal *Diabetes Technology and Therapeutics*).

This extended interest and investment in methods for blood glucose sensing has led to many new technologies, and researchers have been able to tap into this wealth of knowledge to apply sensing technologies for measuring biochemicals to other types of disease prevention and "wellness" maintenance. An example of this trend is the recent appearance of test devices for cholesterol self-testing for the general public. Corporations have recognized the desire of individuals to be able to monitor health indicators outside of the physician's office and have instituted research programs to fill this need. For example, Intel Corporation has a research group devoted to home healthcare that develops products for wellness, nutrition fitness, and mental health, as well as disease management. As will be described later in this report, health maintenance is an important issue for the Japanese.

Another driver for the development of biosensing systems is the need for new and expanded technologies for monitoring and controlling the environment. In addition to a long-standing concern to identify toxic materials in the environment, in recent years, recognition of the fragility of the environment and growth of the "green" movement worldwide have expanded so rapidly that robust and diverse environmental sensing technologies have become essential to achieving social goals. In addition to the need for selectivity and sensitivity in environmental sensors, two other requirements are for robustness to allow the systems to be fielded in remote locations and for methods for relaying information to monitoring centers (see, for example, the website of the Center for Embedded Network Systems at UCLA, cens.ucla.edu/).

Further, after the 9/11 tragedy, there has been a leap in interest in sensing for security and surveillance—sensing technologies capable of identifying chemical or biological materials that can result in diseases or death. All sorts of deployments are being considered to cover the immense range of threats, from immediate poisons such as sarin to biological agents such as smallpox that may take weeks or months to incubate.

ENABLERS OF BIOSENSING TECHNOLOGIES

The increasing sophistication of biosensing technologies has become possible because of national investments in other technologies, notably fabrication methods for integrated circuits; photonics and fiberoptics; and biotechnology, particularly genetic engineering. More and more, these technology fronts are coinciding, so that one sees programs called "Bio-Nano-Info" for biotechnology, nanotechnology, and information technology. An example of the coordinated thinking along these lines was highlighted in a 2002–2003 conference and report sponsored by the National Science Foundation and the Department of Commerce entitled "Converging Technologies for Improving Human Performance: Nanotechnology, Biotechnology, Information Technology and Cognitive Science" (Roco and Bainbridge 2003).

One can see the results of this confluence of technologies as related to biosensing in the field of clinical analytical chemistry (Figure 1.3). Several decades ago, the first breakthrough in analytical procedures was the development of the "Autoanalyser" by Technicon Corporation, shown on the left of the figure. This laboratory bench device allowed the robotic processing of many samples and could be "programmed" for different assays. Later, portable versions of the chemical laboratory were developed to bring the chemistry to the workplace (such as the surgery suite), rather than bringing the samples to the chemistry laboratory. In the past decade, "point-of-care" technologies have developed to the point where tests are accomplished at the patient's bedside, so that a physician can obtain critical information while examining the patient. In the example shown, the handheld i-STAT system on the right (www.istat.com/products) provides 6 different analyses from a drop of blood in about one minute.

Further miniaturization has occurred in the last five years, resulting in commercial products where the sensor elements have been made even smaller, on the order of millimeters in size, as shown in Figure 1.4. This is a glucose sensor 1 mm in diameter under development by Medtronic/ MiniMed Corp. (www.minimed.com/) for the continuous measurement of glucose subcutaneously.

As will be seen in the technology chapters of this report, the trends toward multianalyte and miniaturized sensors have produced array-type technologies where the active elements features are on the order of microns in size, allowing for thousands of target molecules (e.g., DNA sequences, RNA sequences, proteins) to be displayed simultaneously on chips only a few square centimeters in area. The identification of materials is obtained by binding patterns that are visualized by tags of fluorescent molecules (see Figure 1.5).

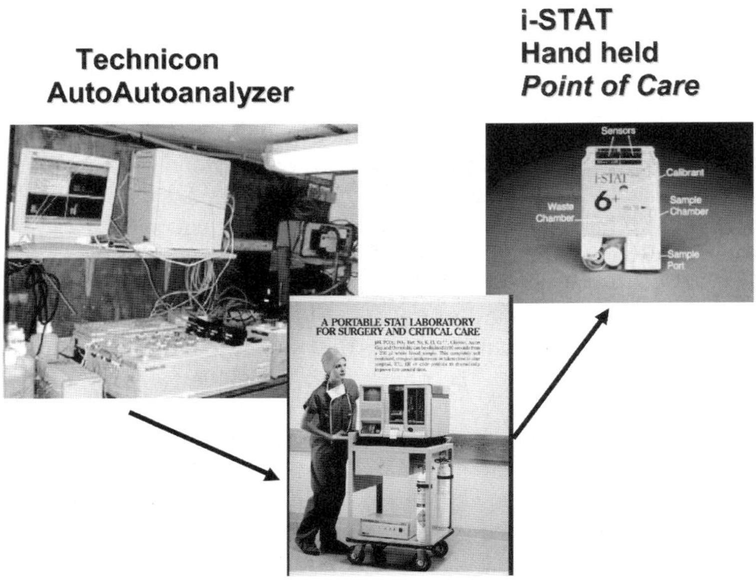

Fig. 1.3. Evolutions of the confluence of technologies as related to biosensing in the field of clinical analytical chemistry; examples from Technicon and Abbott Labs/i-STAT Corp.

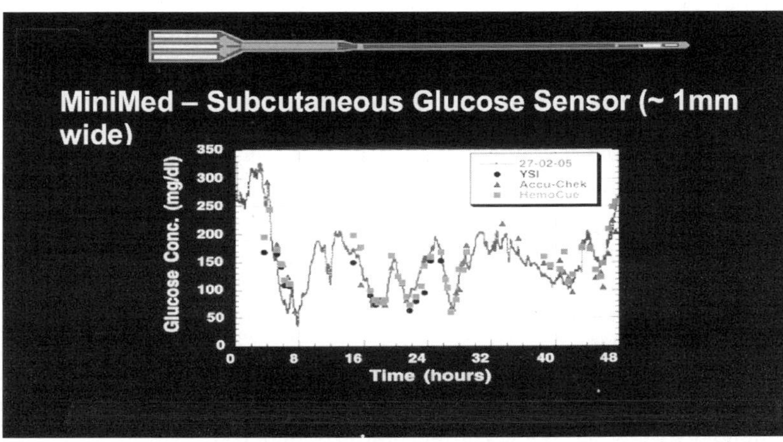

Fig. 1.4. Subcutaneous glucose sensor 1 mm wide under development by Medtronic/MiniMed Corp.

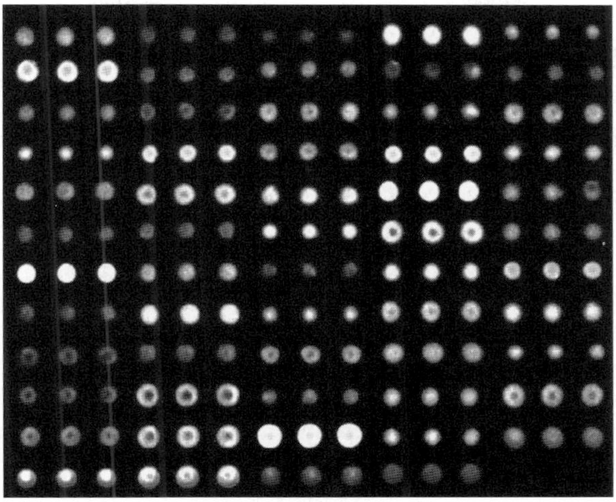

Fig. 1.5. Fluorescence pattern on an array chip for identifying DNA fragments.

BIOSENSING INFRASTRUCTURE / INVESTMENT TRENDS IN THE UNITED STATES

The WTEC biosensing panel observed that in the United States, Europe, and Japan, the strategies for investing in the creation of physical and human resources to support research—roughly defined here as "infrastructure"—played a critical role in the growth of the biosensing field in each of the regions. These strategies have guided each region in setting its government policies, seeking public funding, establishing research priorities, and putting the research results into commercial use. The panel has attempted to broadly identify and compare the infrastructure/investment circumstances in the United States, Europe, and Japan.

International marketing consulting and training firm Frost and Sullivan (2002) has made an analysis of biosensing trends in the United States and overseas, noting that the current analytical laboratory instrumentation market is about $10 billion per year, providing a great market opportunity for new biochip developments.

One only needs to note the U.S. companies that have significant development programs in genomics, proteomics, and other diagnostics to realize the significance of commercial investment in these technologies. Examples include Aclara, Affymetrix, Applied Biosystems, Beckman Instruments, Caliper, Cepheid, Curagan, Gene Logic, Genometrix, Hyseq, ID Biomedical, Incyte Pharmaceuticals, Molecular Tool, Mosaiz Technologies, Nanogen, Orchid Biocomputer, Synteni, and Vysis.

Despite the breadth of industry investment in sensing technologies and product development, much of the research funding that has enabled advanced concepts in biosensing has come from U.S. Government agencies. Although there is no comprehensive compendium of commercialization successes from this national research funding, a partial listing of some outcomes was provided at a workshop sponsored by the National Nanotechnology Initiative (NNI 2002), titled "Nanotechnology Innovation for Chemical, Biological, Radiological, and Explosive (CBRE) Detection and Protection." This listing is reproduced in Table 1.3.

The three primary U.S. government agencies that provide funding and guide directions for development of biosensing systems and technologies are the National Institutes of Health, the National Science Foundation, and the Department of Defense. The goals of these agencies are somewhat different; however, the basic research promoted by these institutions has a great deal in common.

Table 1.3.
Potential Near-Term Nanotechnology with CBRE* Impact (Source: NNI 2002)

Investigator	Institution	Technology	Company
Baker	University of Michigan	nanostructured bio decontamination	NanoBio Corp.
Doshi	—	polymer nanofibers	eSpin
Hellinga	Duke University	tailored biosensors	Johnson & Johnson
Klabunde	Kansas State University	nanocluster agent catalysis	Nanoscale Materials
Lieber	Harvard University	nanotube sensors	Nanosys
Martin	University of Florida	nanotube membranes	Broadley-James Co.
Mirkin	Northwestern University	nanoAu biological sensing	Nanosphere
Russell	University of Pittsburgh	sensing wipe	Agentase
Smalley	Rice University	carbon nanotube (CNT) for adsorbents	CNI
Snow	Naval Research Laboratory	nanoAu chemical sensing	MicroSensor Systems
Tatarchuk	Auburn University	CNT adsorbent media	IntraMicron Inc
Thundat	Oak Ridge National Lab. (ORNL)	cantilever sensing	Protiveris
Walt	Tufts University	nanoarray sensors	Illumina

*CBRE: Related to Chemical, Biological, Radiological, and Explosives R&D and commercialization

National Institutes of Health (NIH)

As an insight into NIH's priorities in biosensing, the proceedings from a Bioengineering Consortium (BECON) conference in 2002 gave an overview of biosensing research and provided guidance for future research in this field (BECON 2002). Suggestions from that report concerning opportunities and responsibilities for future NIH funding of research programs include the following:

- translation of [state-of-the-art] technology to the clinic or laboratory
- encouragement of awareness by researchers working in sensor development concerning the impact of their choice of biological models, which can impact sensor function dramatically
- encouragement of researchers to aim for utilization of complex mixtures, such as blood or saliva, in design of sensors that will permit the measurement of chemical, biological, and physical parameters
- emphasis on real-world field validation (for rescue work, third-world inaccessible populations, public health applications, etc.)
- application of computer science to sensor research needs in areas of data acquisition, data storage, analyses of dissimilar sets of data, algorithm development, performance modeling, telehealth, and medical information systems
- support of sensor materials research and development, specifically for materials with short response times; applicability to continuous or multiple measurement; and ability to deliver drugs, sense environments, detect therapeutic efficacy, and monitor physiology
- development of noninvasive sensors by the application of imaging technologies, with a focus on improvement of co-registration methods and development of high-performance optics to enhance the depth of measurement while maintaining molecular information
- support for the creation of a database/clearinghouse for building research teams with relevant skills and knowledge
- focus on gaps and deployment barriers that exist in sensor development, including the major problem of loss of sensor function in contact with complex mixtures such as blood, saliva, or interstitial fluid

In this BECON (2002) conference report, NIH also notes that standards and protocols are required, especially prior to the stage where the technology can be tested in animals or people:

- "functional standards" should correlate to the desired phenomenon (e.g., disease presence or analyte concentration)

- systems integration should combine the inputs from several sensors to yield useful integrated information from advances in miniaturization, materials, signal transduction, drug delivery, etc.
- micro/nano systems should integrate multiple functions to achieve performance and cost advantages
- research should define methods for the manufacture and transport of cell-based biosensors that are differentially sensitive to environmental stimuli (e.g., temperature, G-forces, culture medium, barometric pressure), and it should consider the condition of the cells attached to the sensor at the final place of use
- approaches to producing quantitative data from a large array of multiplexed data should overcome the major limitations in assays/sensors due to immobilized recognition and/or transduction events at interfaces

In order to characterize current research supported by NIH, the WTEC panel undertook a search on the NIH websites of all grants awarded in calendar year 2002, using the following keywords to select projects related to biosensors: biosens*, enzyme* and sens*, electro* and sens*, saw and sens*, antibody and sens*, optic and sens*, dna and sens*, gene and sens* (asterisks represent wildcards, to pick up various word endings in the search). The list of about 200 grants is given in Appendix D. Although the newly formed National Institute for Biomedical Imaging and Bioengineering (NIBIB) has a core interest in promoting sensor-related research, it is clear that many NIH Institutes have been supporting research in this field, attesting to the importance of these technologies across all of the health sciences.

National Science Foundation

With regard to biosensing, NSF traditionally has focused its funding activities on the fundamental sciences of materials, surface science, optics, and spectroscopy. In an open letter to the scientific community dated April 16, 2002 (www.nsf.gov/pubs/2002/nsf02112/nsf02112.pdf), NSF outlined its interest in sensing relative to its decision to provide added funding for R&D for next-generation sensors, particularly in multidisciplinary efforts:

> The goal of this effort is to speed advancements in the understanding, development, and applications of sensors. Specifically, improved and more reliable materials and protocols are sought which result in higher sensitivity, fewer false alarms, wireless operation, multifunctionality (e.g., simultaneous detection of both chemical and biological species), practicality, etc. Sensing principles

include but are not limited to optical, electrochemical, electrical, acoustic, and mass sensing phenomena. Multidisciplinary efforts are encouraged. Specific research areas might include but are not limited to:

1. Synthesis and testing of new low cost materials with high sensitivity, selectivity, robustness, and speed for defined sensor applications. Materials having predictable and tunable recognition properties, as well as robustness under anticipated manufacturing schemes, are desired. Work may include modeling of material/analyte interactions and design of specific binding sites. Also of interest are biologically sensitive materials and materials with biorecognition surfaces and membranes. Packaging materials and methodologies specific to sensing applications are also of interest.

2. New approaches for achieving sensitivity, selectivity, robustness, low cost and high speed for defined sensor applications. These might include but are not limited to: (a) development of biologically-motivated amplification schemes and sensing principles, (b) development of label-free assays for various pathogens (including recognition schemes for surface proteins, glycoproteins and other surface markers for rapid detection of pathogens), and development of functionally defined selectivity (e.g., neurotoxicity). Exploration of the dynamic behavior of sensors for various applications is another possible research area.

3. New approaches for the integration of diverse sensor data, including homogeneous arrays, higher order arrays, and superarrays. Development of new statistical algorithms and sampling theories tailored to specific sensor applications.

4. New approaches leading to miniaturization strategies, including lab-on-a-chip projects and power and vacuum pumping capabilities (for miniaturization of mass spectrometers or chromatographs, for example).

It should be noted that besides providing support specifically for sensing R&D, NSF also supports numerous programs in technologies that contribute both directly and indirectly to advancement of sensing technologies (the bio-nano-info connections). In addition to research grants, NSF also supports equipment facilities, workshops, educational programs, and small business grants.

The WTEC panel searched all grants awarded by NSF during 2002 for indications of programs with a focus on or application to sensing. The results were acquired from the Fielded Search (full text) on the NSF Awards website, using the keywords biosens*, enzyme* and sens*, electro* and sens*, saw and sens*, antibody and sens*, optic and sens*,dna and sens*, gene and sens*.The result was a compilation of about 400 awards; the results were then screened and approximately one-half discarded because they obviously were not related to biosensing. The ~200 NSF awards related to biosensing are listed in Appendix E.

Department of Defense

Although the Department of Defense (DOD) had been supporting programs in sensing technologies for a number of years through the Defense Advanced Research Projects Agency (DARPA), its efforts accelerated dramatically following 9/11/2001. In February of 2003, the Department of Defense released its *Fiscal Year (FY) 2004/FY 2005 Biennial Budget Estimates.* The following information is from that estimated budget document, specifically Volume 1 describing DARPA projects (DARPA 2003).

In the approximately $2.8 billion planned budget for DARPA's Research, Development, Test, and Evaluation Program in Fiscal Year 2004, about $291 million was allocated in programs that relate to biosensing. These programs are titled "Defense Research Sciences" (Program Element 0601101E) and Biological Warfare Defense (Program Element 0602383E). These programs consist of several sub-elements, including the programs BioComputational Systems; Simulation of Bio-Molecular Microsystems (SIMBIOSYS); Nanostructure in Biology; and Molecular Observation, Spectroscopy, and Imaging using Cantilevers (MOSIAC) program, all of which impact sensing research. There is a clear focus in several of these programs on multidisciplinary integration and exploration of phenomena at the nanoscale. Outlines of DARPA programs are given in Appendix F, taken from the published estimated budget information.

The focus of the DARPA-funded research is on DOD issues and products such as design of novel materials; sensing and computational devices or dynamic biological materials that utilize or mimic biological elements for force protection and medical intervention; new leads for the development of threat countermeasures; and improvement of human performance. Nevertheless, it is clear from DARPA statements in the estimated budget document that it planned to support a great deal of fundamental research at the interface between biology, materials, and information sciences, in order to "develop the basic research tools in biology that are unique to the application of biological based solutions to critical Defense problems" (DARPA 2003). The outcomes of these research projects will undoubtedly find applications in the public and private sectors, in keeping with the philosophy of "dual-use" now being promoted by many government agencies.

DARPA "Defense Research Sciences" and "Biological Warfare Defense" projects probably constitute the major sources of DOD support for biosensing research, but there are other agencies that actively support these kinds of activities as well. For example, the U.S. Army Corps of Engineers issued a solicitation for "Sensor Systems, Data Acquisition, Processing, and Transmission Systems" in support of military engineering, civil engineering, environmental engineering, and homeland defense. The Army Research Office has shared with WTEC a list compiled in March 2004 of about 50 active projects related to biosensing research for chemical and biological warfare defense; these are presented in Appendix G.

Programs at U.S. Government Laboratories

With the exception of NIH, the U.S. government agencies discussed above are primarily funding agencies that direct research by their funding priorities. In addition, there are several U.S. Government laboratories that perform extensive research and development programs in biosensing systems. The Naval Research Laboratory in Washington, D.C., has been particularly successful in taking biosensors from an initial concept to on-site application (www.nrlbio.nrl.navy.mil/, www.chemistry.nrl.navy.mil/). At least six biosensors invented at NRL are commercially available for uses including detection of drugs of abuse, explosives, pathogens in foods, bioterrorism agents, and research targets, with more biosensor technologies currently under commercialization. NASA (www.NASA.gov) also has a number of research programs related to biomedical and environmental sensing technologies at its various centers, including the Jet Propulsion Laboratory, Ames Research Center, and Johnson Space Center. The Army has also had a long-standing biosensor development and testing effort at Soldier Biological and Chemical Defense Command exploring military applications for biosensors and adapting them for field use.

One of the larger efforts is being undertaken by the Department of Energy, where the Office of Science devotes about $1.5 billion per year to programs in Basic Energy Sciences and Biological and Environmental Research. About one-third of these funds go to universities and the remainder to in-house projects. Although a breakdown relating to sensing research is not available, in 1999 DOE published an inventory of research conducted in its national laboratories that related to biomedical engineering research (www.osti.gov/sc73/doe-sc-1999-1.pdf). From that document, the WTEC panel identified and tabulated DOE projects that relate to biosensing systems; these are presented in Appendix H. They amounted to about 50 biosensing-related projects in 10 different DOE facilities. This represents a large amount of research sponsored by a single agency, however, the fact that the work is broadly distributed may limit its impact compared to some of the integrated research programs that the panel observed in Europe.

BIOSENSING INFRASTRUCTURE/INVESTMENT TRENDS IN EUROPE

A great deal of insight into biosensing research in Europe can be obtained from the research programs sponsored by the European Union (EU). During the WTEC panel's visits to various European research laboratories, we were informed that EU funding usually amounted to about 15% of a laboratory's total funding. However, the general scope of priorities as outlined in EU news articles and public documents provides a reasonable view into the interests and directions for future European research. Several of these are surveyed below.

Some general observations relating to biosensor R&D in Europe were reported in an article, "Biomedical Applications of Nanotechnology," by Ineke Malsch (2002). The article reported that the European Commission, which finances about one-quarter of the publicly funded research in the EU, was to spend about $300 million on nanotechnology projects in 2003, as compared to $700 million for the National Nanotechnology Initiative (NNI) budget in the United States. A portion of the European funds will go to biomedical applications that include diagnostics and biosensing technologies. Malsch noted that the focus of Europe's government nanotechnology R&D is on relatively short-term product development and is collaborative (Malsch 2002):

> In Europe, public research funding and networking for nanotechnology in industry tend to be more focused on applications with a time-to-market of 5 to 10 years. The international Network for Biomedical Applications of Micro & Nano Technologies (NANOMED), based in Newcastle upon Tyne (U.K.), has brought together 50 industrial and academic partners to develop bio medical applications of nanotechnology. In Germany, the Nanochem network, based at the University of Kaiserslautern, is organized in a similar public-private fashion and includes medical applications of nanotechnology. Germany has had by far the highest

> budget for nanotechnology research in Europe for several years; in 2000, funding was at a level of $56.7 million.

Malsch's article, which supports the WTEC team's observation of the emphasis in European biosensing R&D on public-private collaboration, gives this example:

> The Micro Electronics Material Engineering Sensors and Actuators (MESA+) research institute at the University of Twente (Enschede, The Netherlands) is engaged in high-throughput screening (HTS) research for Avantium in Amsterdam, an R&D company founded in 2000 by a consortium of chemical and pharmaceutical companies, venture capitalists, and three Dutch universities. Avantium aims to develop new strategies and equipment for screening active compounds for pharmaceutical and other products—specifically through development of highly sophisticated lab-on-a-chip systems.

For a more comprehensive view of public funding for science and technology research in the European Union, Table 1.4 gives an overview of the EU Sixth Framework Programme. The budget estimates are for the period 2002–2006. The actual implementation of this program is rather complex, and the reader is referred to the website www.cordis.lu/fp6/ for more detailed information.

An examination of the specific goals of the eight major research program elements reveals that there will be significant support for biosensing research in Elements 1, 2, 3 (and 5). General outlines for these programs follow; more details of the EU Sixth Framework Programme program objectives and research activities related to biosensing are given in Appendix I.

1. Life sciences, genomics, and biotechnology for health
   - Genomics and biotechnology for health
   - Advanced genomics and its application for health
   - Fundamental knowledge and basic tools for functional genomics in all organisms: gene expression and proteomics, structural genomics, bioinformatics, etc.
   - Application of knowledge and technologies in genomics and biotechnology for health: technological platforms, prevention, and therapeutic tools, etc.
   - Combating major diseases

Table 1.4.
EU Sixth Framework Programme, Research Budget

Thematic Priorities	€ million*
1. Life sciences, genomics and biotechnology for health	2,255
Advanced genomics and its applications for health	1,100
Combating major diseases	1,155
2. Information Society technologies	3,625
3. Nanotechnologies and nano-sciences, knowledge-based multifunctional materials, and new production processes and devices	1,300
4. Aeronautics and space	1,075
5. Food quality and safety	685
6. Sustainable development, global change, and ecosystems	2,120
Sustainable energy systems	810
Sustainable surface transport	610
Global change and ecosystems	700
7. Citizens and governance in a knowledge-based society	225
8. Specific activities covering a wider field of research	1, 300
Total	**†13,345**

* Conversion is approximately €1.00 = US$1.25; inverse, 0.80 (Dec. 2003).
† Including non-nuclear activities of the Joint Research Centre: €760 million.

- Application-oriented genomic approaches to medical knowledge and technologies: diabetes, cardiovascular diseases, resistance to antibiotics, brain, and ageing, etc.
- Cancer
- Major poverty-linked infectious diseases: aids, malaria, and tuberculosis

2. Information society technologies (IST)
    - Applied IST research addressing major societal and economic challenges: security, societal challenges, "ambient intelligence," electronic commerce, etc.
    - Communication, computing, and software technologies
    - Components and microsystems
    - Knowledge and interface technologies
3. Nanotechnologies and nanosciences, knowledge-based multifunctional materials, and new production processes and devices

- Nanotechnologies and nanosciences: long-term research, supramolecular architectures and macromolecules, nano-biotechnologies, applications in health, chemistry, etc.
- Knowledge-based multifunctional materials: fundamental knowledge; production, transformation and processing technologies, etc.
- New production processes and devices: flexible and intelligent manufacturing systems, systems research and hazard control, clean and safe production, optimisation of life cycles, etc.

Within the patterns of European Union R&D funding, there is a strong emphasis on building collaborative research centers that span country lines. As an example, during the Fifth EU program cycle, Cranfield University in the UK organized the research consortium SENSPOL (www.cranfield.ac.uk/biotech/senspol/). For the current Sixth Programme, Cranfield has expanded this effort and is in the process of developing a Network of Excellence in Sensing Technology (NEST), comprised of 120 biosensor labs selected from over 4,000 sensor labs in 24 countries. There are over 100 people at Cranfield working in this sensor network.

The WTEC panel observed firsthand a general pattern in Europe for the formation of integrated networks for enhancing research and technology, particularly with the goal of business generation. An excellent example of this trend is the growth of the biotechnology/biomedical capability in the Berlin-Brandenburg region of Germany. Three Max Planck Institutes and two Fraunhofer Institutes are located on the campus of the University of Potsdam, in addition to the University's own institutes. The focus of much of the science in these institutes is biotechnology and life sciences. A number of private companies are already emerging from this scientific synergy. The local political establishment is highly supportive of the region's focus on biotechnology, helping to fund infrastructure development, including several interdisciplinary technology parks. It also helps to coordinate biotechnology activities via a central office, BioTOP Berlin-Brandenberg, which among other functions hosts the BioTOP web site (www.biotop.de/index_e.asp?main=3) and produces the BioTOPics newsletter (e.g., see www.biotop.de/download/BraRep_eng.pdf, May 2002). The charts below, Figures 1.6 and 1.7, accessed from the *BioTOP* website, show the growth of the biotechnology industry in this region, and the product areas for these companies. At the writing of this report, over 100 companies in the Berlin-Brandenburg region have activities in diagnostics, instruments, or software that have some relationship to biosensing systems and technologies.

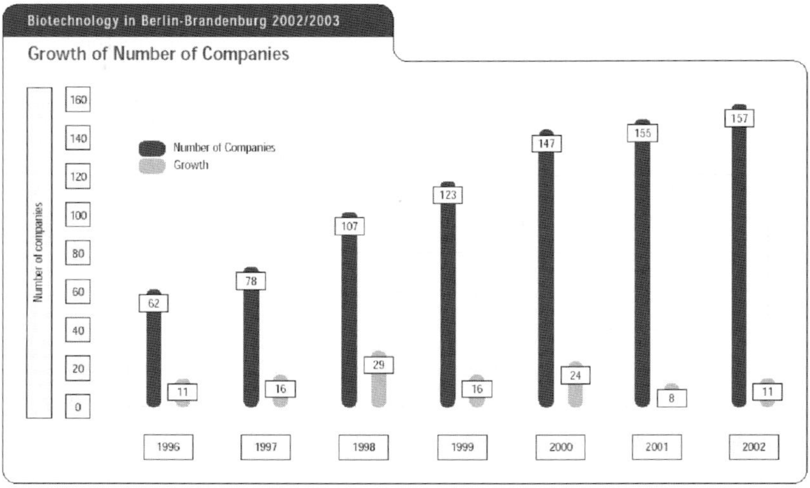

Fig. 1.6. Growth of the biotechnology industry in Berlin-Brandenburg region. (Source: BioTOP Biotech Report May 2003, available online www.biotop.de/download/Biotech_Report_02_03_en.pdf)

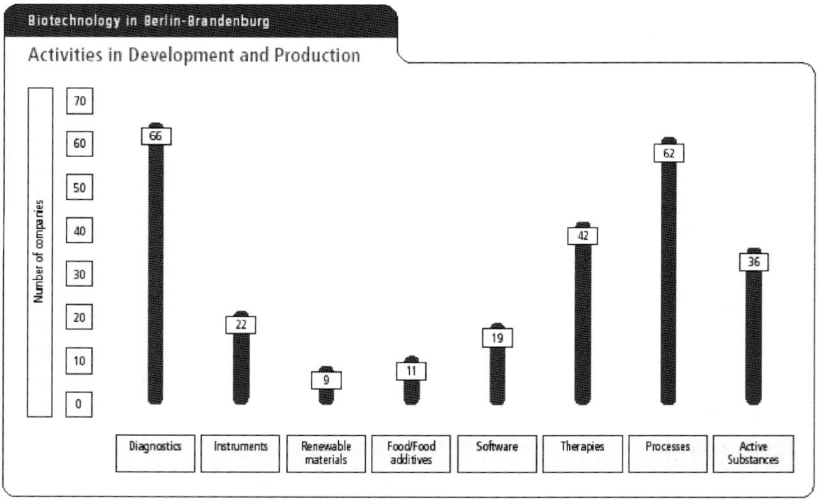

Fig. 1.7. Product areas for the biotechnology industry in Berlin-Brandenburg region. (Source: BioTOPics May 2002; www.biotop.de/download/BraRep_eng.pdf)

The WTEC panel conducted a survey of the patent literature for the sites that it visited in Europe in order to gain some appreciation for the range of commercialization activities for those centers that are involved in biosensing research. The results, tabulated in Table J.1 in Appendix J, indicate a significant effort in obtaining patents on the part of many university laboratories across Europe.

BIOSENSING INFRASTRUCTURE/INVESTMENT TRENDS IN JAPAN

Sites the WTEC panel visited in Japan included universities, government research laboratories, and companies. A common and important feature of these visits was the major change in attitude towards cooperative ventures between all these types of institutions for new product development. The panel's visit to the Tokyo University of Agriculture and Technology (TUAT) was indicative of this trend.

In TUAT's 2002 informational brochure (TUAT 2002), President Dr. Seizo Miyata is quoted as follows:

> For the sustainable development of the country, research in the following four fields will be of great significance in the 21st century:
>
> 1) Biotechnology, which will assist in the prevention and treatment of disease and help in solving future food problems;
>
> 2) Information and Communications Technologies, represented primarily by computers, cell phones and the Internet;
>
> 3) Environmental Science and Resource Science, which are essential to the survival of human kind;
>
> 4) Nano-technology (nanometer scale manufacturing technology) and research on new materials, which will have immense influences on our daily lives.

TUAT information also notes that patents jumped from 12 in 1999 to 136 in 2002 in the Graduate School of Bio-Applications and Systems Engineering, and that 116 cooperative research projects were carried out in 2002 by about 450 faculty and research associates. The School's major fields include Dynamics of Molecular Systems; Bio-modeled Sensory Systems; Molecular Mechanism of Bio-Interaction; and Biological and Environmental Sensing Systems.

With support from the Ministry of Education, TUAT has actively promoted cooperative ventures with private researchers since 1988, and it started providing advanced facilities for joint research in its Cooperative Research Center in 1989. The Center was expanded in 1996, and in 2001 it added a liaison office to better promote commercialization activities. The photograph in Figure 1.8 of the Cooperative Research Center building indicates the level of commitment to facilitating university-industry technology transference.

Fig. 1.8. Cooperative Research Center at TUAT.

Another example of the trend for the direct connection of university and corporate research is the new School of Bionics at Tokyo University of Technology . A new US$250 million building with 15,000 m^2 of space, pictured in Figure 1.9, opened in 2003 to house industrial/academic research projects, along with the traditional research and academic facilities. About twenty new faculty have been hired for a new bionics program led by Prof. Isao Karube, which is housed in this building, called Katayanagi Advanced Research Laboratories. Four floors of the new facility are occupied by corporate research laboratories that co-sponsor research in the institute. The university is building a degree program in technology management.

A similar emphasis on collaboration is also evident at RIKEN, one of Japan's premier national research institutes, as described by the Director of the Frontier Research System (FRS) (RIKEN 2002), Dr. Eiichi Maruyama. Dr. Maruyama noted that RIKEN's Frontier Research Program (FRP), in existence from 1986–1997, initiated an experimental research system consisting of fixed-term contract researchers that "introduced dynamism into Japanese research system and achieved remarkable research results." The FRP was succeeded in 1997 by the Brain Science Institute and then by the present Frontier Research System in 1999, with a more diverse project orientation towards novel, world-class basic scientific research, but maintaining FRP's organizational focus on bringing together "high caliber scientists from different disciplines to work together on cutting edge research projects...and to continue to develop and incubate novel interdisciplinary research areas" (RIKEN 2002).

Fig. 1.9. Tokyo University of Technology's Katayanagi Advanced Research Laboratories building, which houses the Bio-nanotechnology Center, Content Technology Center, Advanced IT Center, Creative Lab, Encoding Center, and Bionics Research Center, which is part of the academia-government-industry Collaborative Research Center.

FRS apparently is regarded as a unique approach by Japan's government to expand scientific knowledge via national and international scientific cooperation and project management that is flexible with regard to duration of projects; composition of research teams (including active recruitment of creative young researchers both within Japan and overseas); and involvement by international as well as national experts. Buoyed by successes in its predecessor programs, RIKEN is dedicated to fostering dynamic and flexible management in FRS, with the goals of creating new fields in science/technology, to benefit industry, the economy, and society at large.

These observations on changes in programs as related to new technologies were echoed in an article by Jean-Francois Tremblay, "Unleashing R&D in Japan," that appeared December 2002 in *Chemical and Engineering News* (pubs.acs.org/cen/topstory/8049/8049bus1.html). Tremblay describes several initiatives within Japanese funding agencies that are designed to increase funding to and economic benefit from R&D activities in universities and national research laboratories, especially in terms of increasing emphasis on patenting innovations and on transferring innovation from the laboratory to industry. These changes are reflected in the "new" National Institute of Advanced Industrial Science and Technology (AIST), which in 2001 grew out of a merger between 15 institutes run by the "old AIST" (the Agency of Industrial Science and Technology), and the Weights and Measures Training Institute, becoming the nation's largest public research organization (Tremblay 2002):

> The 3,200 scientists at AIST are now urged to conduct research that can be of use to industry, according to Takashi Goto, director of AIST's collaboration department. It's a 180- degree turn, he says, from the situation at the old AIST, which emphasized basic research.
>
> At the old AIST, Goto relates, researchers were primarily evaluated on the quality and quantity of their published research. Under the new system, "if a researcher does not publish particularly outstanding papers but comes up with useful patentable research, it will be looked upon very favorably." He adds that basic science is not dead at AIST. "Researchers can still go on simply publishing papers; it's just that there is another dimension to the way that they are evaluated," he says.
>
> Although the change occurred only one-and-a-half years ago, collaboration with private companies or universities is expanding rapidly. In 2000, the last year of the old AIST, 972 research projects were conducted with outside groups. In 2001, this had already grown to 1,131 projects. An additional dimension to the improvement, Goto says, is that several joint research projects now extend over several fiscal years, a type of arrangement that was prohibited at the old AIST.
>
> AIST also made a number of administrative changes to help technology transfer to private companies. The old AIST did not have a collaboration department, a technology licensing office, or even a patent policy office. Whereas before AIST researchers were not allowed to collect licensing fees exceeding $46,000, there is now no absolute limit on how much they can earn from their licenses—as long as AIST gets 75% of the proceeds.

Shin-ichi Kamei of Mitsubishi Research Institute in reviewing Japan's strategy for nanotechnology and its competitive position relative to the United States indicates that one of Japan's programs will be "nanotechnologies for observing the phenomena of biocompatible organisms and utilizing or controlling their mechanisms." Hideki Shirakawa, who won a Nobel Prize for chemistry in 2000, will head a nanotechnology effort. It appears that approximately $600 million of government funds will be allocated for this effort. An outline of Japan's research strategy is given in the government document, "The Science and

Technology Basic Plan, 2001–2005" (www8.cao.go.jp/cstp/english/basicplan01-05.pdf). Some Japanese companies are establishing partnerships with American universities, e.g., Fujitsu and the University of Maryland (pr.fujitsu.com/en/news/2001/02/26.html).

The increased emphasis on product development in Japan has resulted in a major increase in patent applications, especially from university faculty. Table J.2 in Appendix J lists the patents related to biosensing obtained from 1999 through February 2003 by the Japanese institutions visited in this WTEC study.

As a complement to this overview of relative organizational and funding patterns of the United States, Europe, and Japan in the fields of biosensing research and development, a bibliometric study of international biosensors research is included in Appendix K that underscores the growing activity in this area of all three regions, based on the number and quality of published biosensor studies in the period 1997–2002. (There is some insight, as well, into the interest in biosensing R&D in other countries not included in the WTEC study.)

SUMMARY

Several key factors may be used to provide a guide for assessing the relative approaches and strengths of infrastructure development in biosensing research: networking and consortia, product development, technology transfer, company development, and national priorities. Table 1.5 shows what drives each of these factors, how they are implemented, and the relative strengths of the three regions. Briefly, the WTEC panel finds that Europe is the trendsetter in developing networked consortia, both local and international, for interdisciplinary R&D. Europe and Japan are very active in university–industry collaboration for product development; Japan in particular is placing strong emphasis on technology transfer through newly enacted laws and funding policies. For company development, with its unique venture capital environment, the United States leads and will continue to lead in this area. Finally, the United States leads in setting national priorities and coupling them to biosensing research and related work.

Table 1.5.
Comparison of Infrastructure Development in Biosensing R&D: U.S., Europe, and Japan

Topic	Drivers	Implementation	Trend Leaders
Networking and Consortia	National/regional policy	Joint funding	Europe United States Japan
Product Development	National policy University policy Corporate	Faculty participation in companies	Europe Japan United States
Technology Transfer	National policy University policy	Accelerated patent procedures	Japan Europe United States
Company Development	Venture capital University policy	SBIR type programs	United States Europe Japan
National Priorities	Health Environment Security	Selected funding	United States Europe Japan

REFERENCES

BECON. 2002. *Sensors for biological research and medicine*. Report of the workshop held June 24–25. Washington, D.C: National Institutes of Health Bioengineering Consortium.

Bilitewski, U., and A. Turner, eds. 2000. *Biosensors in environmental monitoring*. London; New York: Taylor & Francis.

Buerk, D.G. 1995. *Biosensors: Theory and applications*. Lancaster, PA: Technomic Pub. Co.

Clark, L.C., Jr., and C. Lyons. 1962. *Annals, New York Academy of Sciences* 102:29–145.

Cunningham, A.J. 1998. *Introduction to bioanalytical sensors*. New York: Wiley-Interscience.

DARPA. 2003. *Fiscal Year (FY) 2004/FY 2005 biennial budget estimates: Research, development, test and evaluation, Defense-wide. Volume 1–Defense Advanced Research Projects Agency*. Available online at www.darpa.mil/ body/pdf/ FY04_FY05BiennialBudgetEstimatesFeb03.pdf.

Diamond, D., ed. 1998. *Principles of chemical and biological sensors*. New York: John Wiley & Sons.

DOE. 1999. *Biomedical engineering research at DOE national labs (DOE/SC—1999-1)*. Prepared for: U.S. Department of Energy, Office of Biological and Environmental Research, Office of Science, Washington, DC 20874–1290. Prepared by U.S. Department of Energy, Office of Scientific and Technical Information, Oak Ridge, TN 37830.

Eggins, B.R. 2002. *Chemical sensors and biosensors*. Chichester; Hoboken, NJ: John Wiley & Sons.

RIKEN. 2002. FRS: Frontier Research System. Tokyo: RIKEN. (Brochure.)

Frost & Sullivan (company). 2002. *Biosensors: Emerging technologies and growth opportunities* (Report D247). Houston: Technical Insights.

Ligler, F.S., and C.A. Taitt, eds. 2002. *Optical biosensors: Present and future. Amsterdam*; New York: Elsevier Science.

Malhotra, B.D., and A.P.F. Turner, eds. 2003. *Advances in biosensors: Perspectives in biosensors*. Greenwich, CT: JAI Press.

Malsch, I. 2002. Biomedical applications of nanotechnology. *The Industrial Phy*sicist June/July:15–17. Available online at www.aip.org/tip/INPHFA/vol-8/iss-3/p15.pdf.

National Nanotechnology Initiative (NNI). 2002. *Nanotechnology innovation for chemical, biological, radiological, and explosive (CBRE): Detection and Protection.* Final report of the workshop held May 2–3, 2002, in Monterey, CA. Available online at www.wtec.org/nanoreports/cbre/.

Ramsay, G., ed. 1998. *Commercial biosensors: Applications to clinical, bioprocess, and environmental samples*. New York: Wiley-Interscience.

Roco, M.C., and W.S. Bainbridge, eds. 2003. *Converging technologies for improving human performance: Nanotechnology, biotechnology, information technology and cognitive science*. Dordrecht; Boston, MA: Kluwer Academic Publishers. Also available online at www.wtec.org/ConvergingTechnologies/.

Taylor, R.F., and J.S. Schultz, eds. 1996. *Handbook of chemical and biological sensors*. Bristol; Philadelphia: Institute of Physics Pub.

Tokyo University of Agriculture and Technology (TUAT). 2002. *An introduction to Tokyo University of Agriculture and Technology*. Tokyo: TUAT. (Brochure.)

Tremblay, F. 2002. Unleashing R&D in Japan. *Chemical and Engineering News* 80 (49):13–15. Available online at pubs.acs.org/cen/topstory/8049/8049bus1.html.

CHAPTER 2

Optical Biosensing

David R. Walt

INTRODUCTION

Optical sensing offers a number of advantages relative to other transduction mechanisms. Optical methods are sensitive. For example, fluorescence is an intrinsically amplified method in that one fluorescent molecule can generate up to a million emitted photons. In addition, fluorescence is a black background technique in that the emission signal is at a wavelength separated from the excitation wavelength, thereby improving the detection sensitivity because one can measure a signal from a low background rather than try to detect a small signal difference from a high background (e.g., change in resistance). Attesting to these advantages, most of the single molecule detection research employs fluorescence, primarily due to its sensitivity. While methods such as fluorescence and absorbance have a long history, newer methods such as surface enhanced Raman spectroscopy (SERS) and surface plasmon resonance (SPR) have developed rapidly over the last two decades and are playing an increasing role in optical biosensing.

Optical methods are readily multiplexed—one can interrogate with many wavelengths simultaneously without the signals interfering with one another. Another advantage of optical methods is the ability to employ free path or remote interrogation without the need for wire connections. Finally, optical methods benefit from a developing infrastructure. Light sources, detectors, optical interconnects, and other optical technologies are being developed for the entertainment and telecommunications industries. The age of photonics is approaching, and optical methods will likely displace many of the electronic systems.

During the WTEC investigation of international research and development in biosensing, panelists identified a number of technology themes in the optical biosensing area:

- Surfaces
- Arrays
- Inexpensive sensors
- Distributed/networked systems
- Nanomaterials
- Molecular biology

Tremendous advances have been made over the last two decades in designing and preparing functional surfaces that can serve as attachment substrates for biosensing materials (Crooks 2003). In addition, surfaces, in conjunction with these new surface-binding methods, are being implemented in a variety of optical methods. The multiple methods employed for optical sensing correspond to the various optical transduction mechanisms:

- **Luminescence**. This category of methods encompasses fluorescence as well as chemiluminescence and bioluminescence. Luminescence methods are highly sensitive and probably the most prominent optical method employed today.
  - *Fluorescence: intensity, lifetime, polarization*. Fluorescence is the most commonly used optical technique. It involves the excitation of a fluorescent molecule at one wavelength followed by emission at a longer wavelength. The typical time between excitation and emission is the lifetime and is typically in the nanosecond timeframe. Fluorescence can be measured by its intensity, the lifetime (duration of the excited state), or its polarization (related to how rapidly the molecule is rotating during its excited state).
  - *Phosphorescence*. Phosphorescence is a phenomenon that results in a longer lived excited state leading to longer lived emission, typically in the micro- to millisecond timeframe.
  - *Fluorescence resonance energy transfer (FRET)*. FRET is a process whereby a donor molecule transfers its energy to an acceptor molecule. The process depends on distance and is a sensitive method for measuring interactions between two molecules.
- **Chemiluminescence and bioluminescence**. These processes arise when either chemical or biochemical energy is released in the form of

light. For example, fireflies and jellyfish glow from bioluminescent bacteria present in these organisms. These chemistries and biochemistries can be harnessed to prepare sensing materials.

- **Absorbance**. Absorbance is the simple phenomenon of a substance absorbing light at specific wavelengths and is proportional to the amount of absorbing material present. Absorbance can be used to measure the amount of a substance present; alternatively, indicators that bind to a substance and change their absorbance upon binding can be used to indirectly measure the substance of interest.
- **Scattering**. Scattering is similar to absorbance but measures the amount of light reflected. In a true scattering method, the scattering depends on particle size.
- **Surface methods**. Methods based on an optical phenomenon occurring at a surface include the following:
  - *Surface plasmon resonance* (SPR). SPR phenomena are those in which binding to a metal surface causes an optical change (due to refractive index change) at a metal-substrate (usually glass) interface.
  - *Surface-enhanced Raman scattering* (SERS). In SERS methods, an enhanced Raman effect occurs at certain metal surfaces.
  - *Interference*. Interference methods are those in which two optical signals are recombined to give an interference pattern due to a delay in one signal relative to the other caused by binding of an analyte.

SURFACE-BASED OPTICAL BIOSENSING

While much research and development related to fundamental surface chemistries is ongoing in both the United States and Europe, most of the present successes in surface-based optical biosensing are centered in Europe. The most prominent method is one commercialized by Biacore (Sweden, www.biacore.com) in 1990 and since refined that employs surface plasmon resonance (SPR). The Biacore instrument is a highly successful research apparatus that is able to measure both association and dissociation rate constants and therefore can determine binding constants. The success of the instrument is due to its full integration. The manufacturer has addressed the chemistry, fluidics, optical detection, and data processing and integrated them into a complete *system*. A variety of chemistries are available for immobilizing virtually any molecular entity to the sensing surface, making it a simple "plug and play" instrument for the end user. The Biacore SPR instrument is the industry standard and is used widely in both academic research laboratories and for performing binding studies in pharmaceutical laboratories.

Other surface-based methods include the promising work of Professors Günter Gauglitz (Institute for Physical Theoretical Chemistry, Eberhard Karls University, Tübingen, see site report Appendix B), and Michael Sailor (University of California, San Diego), both of whom are developing interferometric optical sensors in which surface binding shifts the absorption maximum.

BIOSENSING ARRAYS

Scientific advances in the last ten years have made it possible to display many different binding materials onto a single substrate and to simultaneously assay for binding to these materials. These abilities have revolutionized the fields of sensing/biosensing in particular and analysis in general. Optical biosensor array types include planar waveguide arrays, CMOS arrays, fiberoptic bundles, SPR arrays, and interferometry arrays. Such array types provide a comprehensive or "global" picture of the components in a complex mixture and enable subtle changes in composition to be monitored even in the presence of a constant background. Besides the ability to perform multianalyte sensing, other advantages of optical arrays include on-chip positive and negative controls, smaller size, lower cost, and higher speed.

Driving the development of optical biosensing arrays are the fields of genomics, integrated optics, microfluidics, and bioinformatics. Presently, the major research focus on arrays is in the area of proteomics. Optical methods, such as fluorescence, for observing binding to such arrays are the favored approach. The ability to capture global protein expression data by employing arrays will revolutionize our understanding of living systems. As the protein composition of cells changes rapidly, the ability to perform dynamic measurements using multiple arrays will be crucial; it is therefore important to address and solve the challenges associated with nonspecific binding to protein arrays, preparing arrays with a high degree of reproducibility, and attaching active materials to the array. While these challenges are all areas of active investigation, the problems are manifold. By solving these problems, however, there should be significant flow-through discoveries made that will be applicable to many other fields.

Commercially, the array field has been dominated by DNA arrays, with fluorescence detection dominating as the detection method. There is still a tremendous research effort concentrated on DNA arrays, and virtually any new transduction mechanism or biosensing system is applied to DNA detection. The WTEC panel's assessment is that while there is a major emphasis on DNA detection, it is a mature technology. Although work is this area is fashionable, any innovation is incremental, and additional developments will have low impact due to the established base of DNA array technology and users' desire to employ standard methods.

Other work on optical biosensing arrays is focusing on simplifying the supporting instrumental systems that enable optical sensors to be interrogated and read, and on wider applications, including detection of hitherto unknown hazards (e.g., toxins and biological agents), display of biological information, and monitoring of environmental changes.

INEXPENSIVE AND DISTRIBUTED SENSORS

A distinguishing feature of many European biosensor efforts is a focus on developing extremely inexpensive sensors for everyday applications. These sensors are primarily directed toward food and environmental applications and are intended to be widely incorporated in consumer products. For example, at Cambridge University in the UK (see site report in Appendix B), the laboratory of Professor Chris Lowe is developing holographic sensors that can measure a variety of parameters in food or can be emblazoned into consumer packaging. A visible hologram image functions as both the analyte-specific responsive media and the optical detection mechanism; further, it serves as the test result and therefore requires no additional electronic processing. Holograms can have presence/absence readout or can be designed with a built-in dial in which the dial moves as the concentration of the analyte increases. The holograms can even be written in the product material (e.g., food), providing a zero materials cost. Sample holographic sensor test results from Prof. Lowe's lab are shown in Figure 2.1 below.

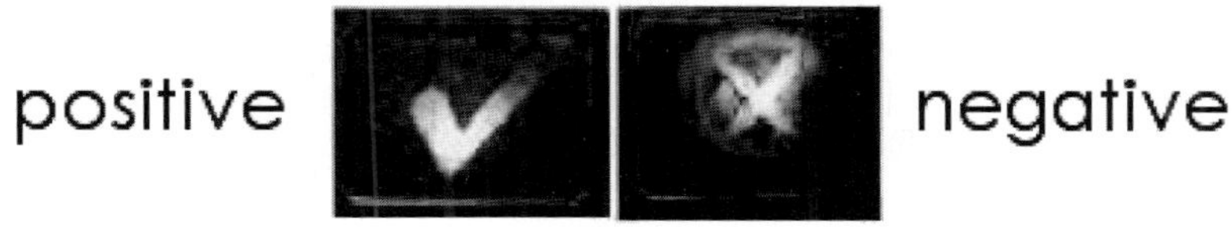

Fig. 2.1. Examples of holographic biosensing before and after a test. (University of Cambridge Institute of Biotechnology, www.biot.cam.ac.uk/ ~crl/crl6.html)

Another noteworthy example of development of inexpensive sensors for everyday applications is at Ireland's National Centre for Sensor Research at Dublin City University (site report in Appendix B; www.dcu.ie/~ncsr/ index_home.html), where researchers are printing CO_2 optical sensing films directly onto food packaging material (see Figure 2.2). Perishable foods, such as meats, are packaged under a CO_2 atmosphere. If there is a breach in the packaging material, the CO_2 sensing film changes color and signals to the consumer that the food is not fresh.

Fig. 2.2. Inexpensive optical sensor for testing integrity of meat packaging. (Dublin City University, www.dcu.ie/~ncsr/commercial/technologies.html)

Both of these techniques employ optical methods for readout, with the human eye as the detector. Such sensing systems should become increasingly popular and accepted as consumers become familiar with the capabilities of the technology and begin to demand this kind of quality assurance in their food as well as in some household products.

A related area of significant R&D effort, also centered in Europe, is that devoted to distributed sensors. These sensors are generally coupled to communications systems so that they can send data back to a central data repository for processing and possible action. For example, the group at Cranfield University in the UK has deployed sensors for monitoring environmental parameters (e.g., lead ion, pH, temperature) at hundreds of sites throughout the country. Some of the sensors are continuous while others require a discrete measurement by a technician at the site. The developing database will be a significant resource for the environmental community for both remediation and regulatory decisions.

Separate work at Dublin City University is directed at distributed temperature sensors for monitoring fish from catch to market. By using widely distributed sensors in fishing fleets, the people involved in the product chain have an incentive to maintain cold conditions, as the value of the catch will be reduced if the fish are exposed to temperatures outside the specified range. Neither the United States nor Japan has prominent efforts in either distributed or inexpensive biosensing.

Both of these areas—inexpensive and distributed sensors—underscore the potential for integration of academic research into the fabric of societal needs. While the United States is regarded as a bastion of entrepreneurship, both Europe and Japan have strong and transparent connections to industry and commercial applications. These direct links to industry help facilitate the commercial introduction of sensors developed in the research community. With pervasive and inexpensive wireless communications systems on the horizon as well as the technology to make inexpensive biosensors and sensors with additional capabilities, there needs to be an investment in developing such mundane and ubiquitous biosensing technologies.

NANOSTRUCTURED MATERIALS

At the WTEC Biosensing Study's U.S. R&D Overview Workshop held in Bethesda, MD, on 3–4 December 2002 (wtec.org/biosensing/proceedings/), a theme in U.S. biosensing R&D became apparent: nanostructured materials with built-in functionality and binding affinity are increasingly being used to perform optical sensing. Research in nanomaterials has led to the discovery of new optical (and other) transduction mechanisms. This area is particularly promising, as it leverages existing research investments. The Van Duyne group at Northwestern University (www.chem.northwestern.edu/~vanduyne/) is developing structured nanomaterials for surface-enhanced Raman-based biosensing, as illustrated in Figure 2.3.

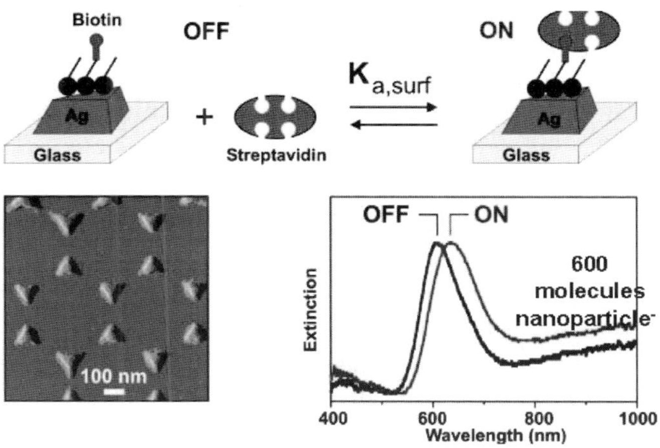

Fig. 2.3. *Top*: Nanoparticle array localized surface plasmon resonance spectroscopy (LSPR)—local refractive index change. *Bottom*: Nanostructured gold materials on a substrate provide local enhancement in the plasmon resonance. (Haes and Van Duyne 2002)

The Mirkin group at Northwestern (www.chem.nwu.edu/~mkngrp/) has been focusing on new optical methods for detecting DNA binding based on aggregation of gold nanoparticles. The Sailor group at the University of California, San Diego (UCSD: chem-faculty.ucsd.edu/sailor/), has been employing the unique material properties of porous silicon (called "smart dust") to detect binding to the silicon surface, resulting in an interferometric response manifested as a color change (see Figure 2.4).

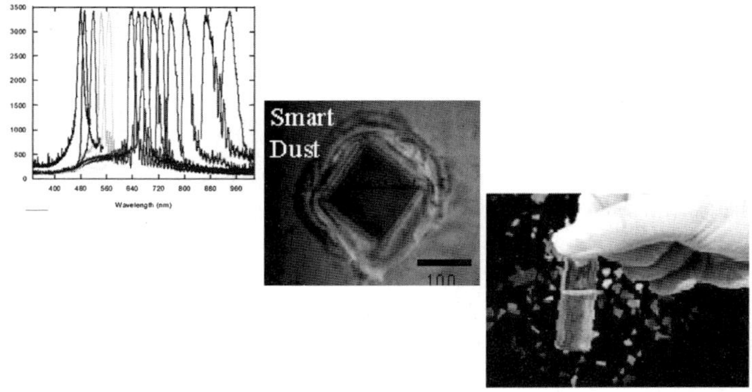

Fig. 2.4. Porous Si particles can be fabricated and used to sense analytes. *Left*: Spectral properties of different Si particles are due to different interference patterns. *Middle*: A single particle has nanometer dimensions. *Right*: A suspension of the Si material dispersed in solvent in a test tube. The background shows an enlargement of the suspension. (Cunin et al. 2002)

Multiple groups, such as those of Alivisatos at the University of California, Berkeley; Nie at Emory University; and Bawendi at Massachusetts Institute of Technology (MIT), are investigating the novel properties of quantum dots for potential use in optical sensing and biosensing applications. Complex optical sensing chemistries and biochemistries are being attached to nanoparticles called PEBBLES (Probes Encapsulated By Biologically Localized Embedding) by Kopelman at the University of Michigan (www.umich.edu/~koplab/research2/analytical/NanoScaleAnalysis.html) that can be introduced into living cells to report on intracellular concentrations of key metabolites.

Some of these methods are able to detect extremely low absolute numbers of molecules. The WTEC team found that many research efforts on optical biosensing were beginning to push toward single molecule detection limits. All of these methods are still at the research stage, and significant additional work will be required to bring them to the stage where they can be used for routine measurements.

Many of these investigators did not initially start out to develop biosensing materials. Serendipity has frequently played a role in the discovery of new transduction mechanisms and biosensing phenomena. With these new discoveries, the biosensing community is attracting new researchers to the field who are drawn to biosensing because they can see a direct application of their fundamental work to an application. Other materials researchers such as Swager at MIT (web.mit.edu/tswager/www/) are not working at the nanoscale but also are contributing new optical biosensing materials such as polymers that have amplified responses to analyte binding. This latter approach seems to be the model being pursued by investigators in Japan and Europe. They are either employing more traditional techniques of organic synthesis and polymer chemistry to generate new materials, or they are exploiting nanomaterials developed in the United States.

An often-overlooked advantage of moving to the nanoscale is the improved sensitivity exhibited by nanomaterial-based sensing. This sensitivity is a consequence of binding to very small structures, which localizes a small number of analyte molecules to an extremely small volume, causing a locally high concentration and/or significant perturbation in the vicinity of or on the surface of the nanomaterials. The United States leads efforts in this area relative to the rest of the world.

APPLICATION OF MOLECULAR BIOLOGY TO OPTICAL BIOSENSING

A major worldwide effort in optical sensing is aimed at using the tools of molecular biology to design new sensing schemes. There is exciting work being performed toward creating new protein constructs for optical sensing. Particularly exciting work is being conducted in Professor Umezawa's laboratory at the University of Tokyo (see the site report in Appendix C). In one approach, his laboratory is creating protein constructs of two types. The first construct contains cyan fluorescent protein (CFP) and yellow fluorescent protein (YFP) linked by a both a peptide substrate and recognition site (Figure 2.5).

In this construct, when the peptide substrate is modified, the recognition site binds to the modified peptide and changes the conformation of the construct. The conformational change causes a change in the efficiency of fluorescence resonance energy transfer from the CFP to the YFP. A second construct employs two proteins, each carrying a piece of a variant green fluorescent protein (EGFP). When the proteins are brought into proximity due to the presence of an analyte, the two proteins are spliced, and EGFP is reconstituted and green fluorescence begins to appear.

Molecular biological designs for optical sensing can be employed both for generating purified proteins that can be immobilized to various substrates and for whole cell biosensing in which the protein is expressed and functional in a living cell (see Chapter 4). Many optical transduction mechanisms are being exploited for cell-based biosensing. These methods open the door to performing functional assays, in contrast to biosensors based on affinity binding. The ability to construct and express functional molecules with optical tags in vivo presents a major opportunity and capability for observing localization and dynamics within single cells. This area presents a rich avenue for performing biosensing at the single cell level and should lead to both fundamental and practical outcomes.

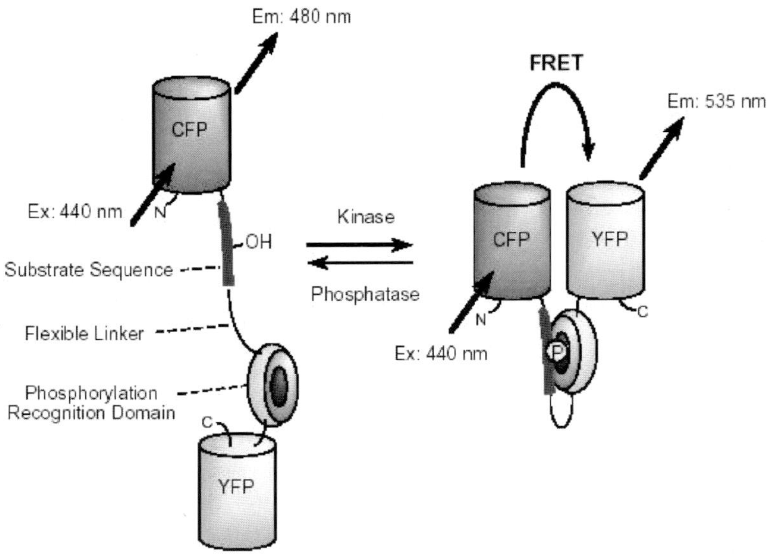

Fig. 2.5. A fluorescent indicator for protein phosphorylation in living cells, named "phocus." CFP and YFP are different colored mutants of green fluorescent protein (GFP) derived from *Aequorea victoria*. Upon phosphorylation of the substrate sequence within phocus by a protein kinase, the adjacent phosphorylation recognition domain binds with the phosphorylated substrate sequence, which increases the efficiency of fluorescent resonant energy transfer (FRET) between the GFP mutants within phocus. (Courtesy, Prof. Yoshio Umezawa, University of Tokyo Department of Chemistry)

GENERAL OBSERVATIONS

A salient feature of European research in biosensing is the presence of large integrated research groups and programs. Many groups in Europe have centers or laboratories with more than 100 researchers. Such large programs exist at the Universities of Dublin City, Potsdam, Twente, Cranfield, and Linköping; smaller but still substantial teams exist at Regensburg, Tübingen, and Cambridge. These groups combine expertise in all the areas necessary to research and develop a *sensing system*: molecular recognition, sensor fabrication, device assembly, data collection, and data processing. The WTEC group noted the presence of integrated biosensing programs in Japan, though generally not on the same order of complexity as those in Europe. There are no comparable programs for biosensing in the United States. Most U.S. biosensing efforts are single-investigator or several-investigator projects rather than biosensing team programs. Another observation—striking to the WTEC team—was the lack of interest (and funding) in Europe and Japan for chemical and biological threat agent detection. Every time the WTEC team broached this subject, the response was met with a complete lack of enthusiasm.

Overview of Scientific Findings Related to Optical Biosensing

- There is much work on DNA and DNA arrays with low commercial prospects.
- The United States leads in materials research resulting in new optical sensing phenomena. Materials researchers are being drawn into the optical biosensing field.
- Optical methods are pushing toward single-molecule detection levels.
- Molecular biological methods are being developed using optical (fluorescence) biosensing, particularly whole cells for functional assays in which the overall quality of the sample is analyzed (e.g., toxicity) rather than a specific analyte.

Challenges

Probably the most significant challenge for moving optical methods into the marketplace is integration. Unlike lithographic techniques employed for electronic devices, optical systems are most often fabricated by assembling many individual components into a functional device. These devices are generally hybrids in that they comprise both optical and electrical components. Full integration of optical components using lithography or other assembly methods remains a major challenge. Research in optical materials for biosensing and for optical signal transduction, microfabrication, and systems integration are all necessary to advance the field of optical sensing.

Comparison of Optical Sensing Expertise by Region

Table 2.1 summarizes the WTEC panel's findings with regard to optical sensing achievements in Europe, Japan, and the United States in the main areas of this field.

Table 2.1
Optical Based Sensing

Topic	Knowledge Base	Work to Date	Leading Region
Interferometric, Label-free	• Surface plasmon resonance • Interference		Europe
Arrays	• Patterning • Surface chemistry	• DNA arrays • Protein arrays	U.S.
Cheap, Distributed Sensors	• Screen printing • Optical transduction		Europe
Nanotechnology	• New signaling mechanisms • New materials	• Metal particles	U.S.
Molecular Biology	• Genetic engineering • Cell biology		U.S. Europe Japan
Integration	• Engineering, Chemistry, Computer Science		Europe

REFERENCES

Boisde, G., and A. Harmerand. 1996. *Chemical and biochemical sensing with optical fibers and waveguides*. Norwood, MA: Artech House.

Cooper, M.A. 2002. Optical biosensors in drug discovery. *Nature Reviews Drug Discovery* 1:515–528.

Crooks, R.M. 2003. Bio/chemical sensing using thin film recognition elements. In WTEC Biosensing Study U.S. R&D Overview Workshop Proceedings, Ch. 3, pp. 57–63. Baltimore, MD: World Technology Evaluation Center. Available online: wtec.org/biosensing/proceedings/03_session02.pdf.

Cunin, F., T.A. Schmedake, J.R. Link, Y.Y. Li, J. Koh, S.N. Bhatia, and M.J. Sailor. 2002. Biomolecular screening with encoded porous-silicon photonic crystals. *Nature Materials* 1:39–41.

Epstein, J.R., and D.R. Walt. 2003. Fluorescence-based fibre optic arrays: A universal platform for sensing. *Chemical Society Reviews* 32:203–214.

Gardner, J.W., and P.N. Bartlett. 1999. *Electronic noses: Principles and applications*. New York: Oxford Univ. Press.

Haes, A.J., and R.P. Van Duyne. 2002. A nanoscale optical biosensor: Sensitivity and selectivity of an approach based on the localized surface plasmon resonance spectroscopy of triangular silver nanoparticles. *J. Am. Chem. Soc.* 124:10596–10604.

Hurst, W.J. 1999. *Electronic noses and sensor array based systems: Design and applications*. Lancaster, PA: Technomic Publishing Company, Inc.

Kordal, R., ed. 2002. Microfabricated sensors: Application of optical technology for DNA analysis. *Journal of the American Chemical Society* 124:11224.

Ligler, F.S., and C.A.R. Taitt, eds. 2002. *Optical biosensors: Present and future*. Amsterdam: Elsevier Science.

Lopez-Higuera, J. 2002. *Handbook of optical fiber sensing technology*. West Sussex, UK: John Wiley and Sons, Ltd.

Murphy, C.J. 2002. Optical sensing with quantum dots. *Analytical Chemistry* 74:520A-526A.

Vercoutere, W., and M. Akeson. 2002. Biosensors for DNA sequence detection. *Current Opinion in Chemical Biology* 6: 816–822.

Wolfbeis, O.S. 2002. Fiber-optic chemical sensors and biosensors. *Analytical Chemistry* 74:2663–2677.

CHAPTER 3

Electro-Based Sensors and Surface Engineering

Milan Mrksich

INTRODUCTION

Electro-based sensing strategies played an important part in the early development of the sensing field, having predated methods based on fluorescence, mass spectrometry, and radioactivity; they retain a central position in the market today (Schuhmann and Bonsen 2003). The electro-based strategies are distinguished in that they are intrinsically interfacial, wherein biological recognition, or physical changes that follow from a recognition event, directly change the electrical properties of a contacting material. The simplicity of an interfacial assay and the sensitivity with which electrical currents and potentials can be measured are in large part responsible for the importance of these assays. This class of strategies also benefits from the localization of binding events to an interface, leading to an enhanced discrimination between specific analytes and background analytes. Additionally, these strategies are compatible with extension to array formats and integration with microfluidic structures.

The central importance of sensors in several industrial contexts, and the many successful products that have been developed and are now widely distributed, give this field a strong emphasis on product development and commercialization. Research and development activities in electro-based sensors maintain an even balance between basic research to develop novel transduction strategies, engineering activities to integrate established sensing mechanisms into prototype devices, and industrial activities to commercialize products. This maturation of the sensing field impacts basic research in academic and government laboratories, attaching an importance to market factors that will ultimately define cost structures, performance metrics, and reliability of sensors.

This chapter addresses the development and implementation of electro-based sensors in the United States, Europe and Australia, and Japan, including key underlying technologies critical to surface engineering, receptor design, and sample preparation. (These technologies are also important to sensors based on alternate transduction schemes, as indicated in other chapters of this report)

This chapter begins with a description of important current activities in electro-based sensors development, with an organization that reflects the various physical transduction strategies. The second section comments on critical underlying technologies that are important to the performance and development of these sensors: these factors are also important to sensing strategies based on non-electrical schemes and will be addressed in other chapters. The final section provides bullet-point summaries of the comparison of sensing programs in the United States, Europe and Australia, and Japan.

OVERVIEW OF R&D ACTIVITIES

Enzyme-Linked Assays

An historically important physical transduction sensing strategy that remains commercially important today relies on the use of enzymes that are immobilized to an electrode to recognize analytes and convert them to by-products that are electrically active and can be detected at the electrode (Wilson and Hu 2000). The many products now available for glucose sensing are based on this principle and have motivated the development of enzyme-linked assays for a host of other analytes. These strategies are best suited for the detection of low molecular weight analytes, which are more likely to give electroactive by-products than are enzymes that operate on peptide and protein analytes. Enzymes that effect oxidations and reductions of their target substrates are best suited to these strategies. Several options are available for immobilizing the enzymes to an electrode, such that the enzymatic activities remain intact and accessible to the diffusion of substrate analytes. Current work in this field, therefore, focuses on developing the enzymes that will give rapid and selective detection of new analytes. The use of protein engineering strategies and combinatorial/selection-based approaches are important in this regard.

Field-Effect Sensors

The binding of analytes to an electrode leads to an alteration of the field properties in the interfacial region and a corresponding field effect that can be measured (Kimura and Kuriyama 1990). This principle has been applied to a large class of field-effect sensors, ranging from chemical, to biomolecular, to cellular detection. This transduction strategy benefits from a simplicity in measurement and the elimination of labels; at the same time, these sensors respond to any molecule that can accumulate at the solid-liquid interface, leading to many sources of false-positive signals. The improvement of these methods relies on engineering interfaces to have much more stable field properties under a wide variety of solution conditions, yet still give clear changes in properties in response to analyte binding.

Electroactive Tags

The labeling of analytes with electroactive tags permits detection of analytes in designs that are analogous to common fluorescence-based assays (Mabayashi et al. 2003). Because it is not feasible to directly introduce labels onto the target analytes, these assays frequently employ a "sandwich" format, wherein the analyte mediates binding of a labeled moiety to the electrode by way of an immobilized recognition unit. For protein analytes, a pair of antibodies serves as the recognition motifs, whereas nucleic acid analogues serve this role for DNA and RNA analytes. One key consideration in these designs is that the electroactive label must approach the electrode sufficiently closely that direct electrochemical processes are feasible. For DNA-based assays, this approach is reliable and has led to a commercial technology for DNA detection (Drummond, Hill, and Barton 2003). For protein-based assays, however, the label is frequently too far from the electrode to be directly detected. In these strategies, small molecules that mediate redox processes must be added to the assay. The development of the redox tags and the mediator reagents is still an important activity in the sensing field (Hromadova et al. 2003). Research in all three regions visited by the WTEC team (Europe, Australia, Japan) as well as in the United States is applying synthetic approaches to address this need.

Nanoparticle-Based Sensors

The rapid emergence of nanoscience holds many new opportunities for creating biosensors with enhanced sensitivity (Reiss et al. 2002). The latter stems in part from the unique physical properties inherent to dimensionally confined materials and in part from the small number of molecules required to alter the properties of these nanoparticles. Research in the United States, Europe and Australia, and Japan is harnessing these attributes to develop and evaluate novel sensing designs. In the United States, the electrical conductivity of arrays of metallic nanoparticles is being exploited. In these schemes, biological interactions of the particles with an analyte are being used to bring soluble particles into distinct patterns on a surface, such that the particles complete a conduction path between two electrodes. In Europe, a program is using similar biomolecular interactions to bind nonconducting particles to nanoscale electrode patches, leading to an attenuation of current due to redox chemistry of a soluble probe. In Japan and in the United States, carbon nanotubes are being functionalized with bio-recognition groups to give single nanowire biosensors that respond to field effects (Cui et al. 2001).

Electrochemiluminescence-Based Sensors

A technology for coupling optical signals with electrical processes has been commercialized in the United States. In the electrochemiluminescence-based methods, tags are developed that undergo oxidation (or reduction) to give an excited state that subsequently relaxes with emission of a photon in the visible frequency (Armstrong, Wightman, and Gross 2001). Hence, a binding event can be recorded by measuring luminescence from the interface, but with an assay that retains the benefits of the immobilized formats. This design has been applied to assays of DNA/RNA, proteins, enzymes, and metabolites, and is now used in clinical diagnostic settings.

UNDERLYING TECHNICAL THEMES

Surface Engineering

The interfacial nature of electro-based sensing schemes places a high importance on methods that can tailor the electrode interfaces with biomolecular recognition units and at the same time prevent unwanted interactions of nonspecific analytes with the sensor (Mrksich 2000). These properties are determined entirely by the methods of surface engineering used to tailor the electrodes. The United States and Europe have made a substantial effort in developing monolayer strategies for this purpose. Work in the United States, for example, has lead in the development of inert surfaces and the development of immobilization chemistries to conjugate recognition motifs to these substrates. Work in Europe has led in developing a mechanistic understanding of the factors that are critical to the design of new inert surface chemistries (Figure 3.1, Feldman et al. 1999). Work in all three regions investigated by the WTEC team has invested in the development of polymeric surface coatings that provide these properties but that have the advantages that they are more versatile to apply to electrodes and that they provide higher loading densities of sensing interactions. Overall, the development of surface engineering methods has progressed substantially in the last decade and will see further important development in the next ten-year period.

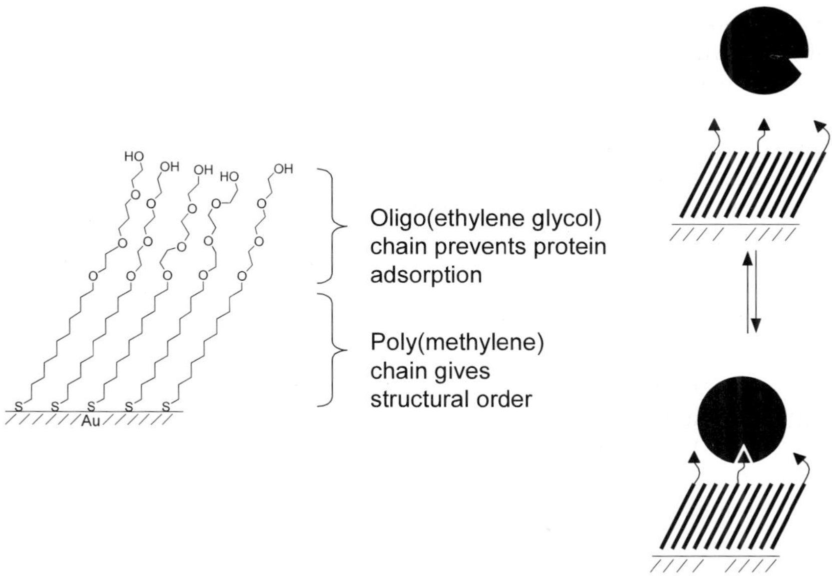

Fig. 3.1. Oligo(ethylene glycol)-terminated self-assembled monolayers. (G. Whitesides, Harvard; M. Grunze, Heidelberg)

Arrays

Electrical strategies for sensing are intrinsically suited for extension to array-based sensing formats, wherein a single substrate is patterned with multiple sensing chemistries and electrode arrays to conduct multiple assays on a single sample. The United States, Europe and Australia, and Japan all have active development programs in this area. These programs benefit from the mature technology available in the microelectronics industry, which offers far more functionality than is required for the sensing technologies. Indeed, multielectrode-based sensing devices have been commercialized and now play an important role in point-of-care diagnostics. The advent of systems based on electrochemiluminescence is one example.

Microscale Integration

Electro-based sensing technologies are intrinsically suited for integration with microfluidic technologies and other lab-on-a-chip technologies (Manz and Eijkel 2001). The match between these two areas stems again from the interfacial nature of electro-based sensing strategies, which leave the topside of an electrode accessible for direct integration with fluidic cassettes. Most work has used soft materials based on elastomers to construct microfluidic cassettes that can be joined to a substrate that is tailored with electrical components and sensing chemistries. The combination of microfluidics with electrical processes has also enabled new mechanisms for manipulating fluids and analytes present in a fluid. The use of dielectrophoresis, for example, provides a new dimension of control for manipulating analytes in a chip-based platform.

RELATIVE STRENGTHS OF REGIONAL PROGRAMS

Programs to engineer electro-based sensors must combine expertise from a wide range of technical capabilities in the areas of engineering, materials, and biological and chemical techniques. The relative strengths of the United States, Europe and Australia, and Japan in each area are indicated below. These comparisons do not address the state of development of a technical capability in a region, but rather they reflect the present importance of the subject in biosensing programs in the region. Industrial influence and funding sources are other considerations related to program strength and character.

Engineering

Programs in both Europe and Japan reflect a strong engineering base. Work in these regions is much more attuned to the development of prototype devices of the sort that are found in early-state industrial efforts. Work in the United States, by contrast, places a stronger emphasis on the development and evaluation of novel transduction schemes.

Materials

Programs in the United States place a greater emphasis on the development of new materials for sensing schemes. U.S. strengths include chemical and physical approaches to surface modification in order to install selective interactions. Europe places the greatest emphasis on molecularly imprinted polymers, even though the value of these materials remains unproven.

Biology

Programs in Europe and the United States make frequent use of molecular and cell biology techniques, particularly for the development of reagents for the selective recognition of analytes.

Chemistry

Work in all three regions reflects a strong investment in synthetic approaches to prepare novel tags and reagents for electrical detection.

Industrial Influence

Programs in Europe, and to a slightly lesser extent in Japan, reflect a strong commitment to industrial needs in the sensing field. Many university-based efforts take place with the active participation of an industrial partner or with an explicit focus on transitioning to a new venture. Programs in the United States emphasize the new venture model.

Funding Mechanisms

Research and development efforts in the United States, Europe and Australia, and Japan all take advantage of targeted federal investment in university laboratories. The development of technologies to combat biowarfare threats is a major driver in the United States, whereas environmental and health applications are drivers in Europe and Japan.

KEY FACTORS FOR FUTURE DEVELOPMENT

Several factors that are important to developing a broader program in biosensors are summarized below. In particular, it is important to promote the extensive level of collaboration that is required in these efforts and to understand the market needs for a particular sensor.

Multidisciplinary Teams

Programs in Europe and Japan benefit substantially from the organization of broad-scale research efforts that integrate multiple types of scientific and engineering expertise, both within the research organization and through collaborative efforts with outside groups.

Institutional Culture and Infrastructure

In all regions investigated in this WTEC report, the institutional character is conducive to biosensor research. Programs in Europe and Japan are team-oriented and focus on the development of prototype sensors, whereas programs in the United States tend more to be single-investigator-based and focus on fundamental work to move forward toward novel sensing strategies.

Fundamental and Applied Research

All three regions maintain a balance between fundamental and applied research.

OBSERVATIONS AND CONCLUSIONS

The Technology is Mature

The field of electro-based sensing is at a relatively mature stage of development. A number of technologies are now commercialized. Current work is focused on miniaturizing the sensors and extending them to multi-array formats.

Nanoscale Science Provides New Opportunities

The development of methods to synthesize materials with nanoscale resolution has lead to materials having properties that are different from those in the related bulk materials. Further, the development of methods that can pattern surfaces with nanometer-scale resolution has made it possible to harness these properties for sensing applications. This work will lead to sensors having unprecedented sensitivities and adaptability to multi-analyte formats.

Industrial-Academic Ties are Important

The modes of technology transfer and commercialization are different in the three regions. In the United States, new startup entities remain the most active vehicle. In Europe, both the startup and collaboration with large industry models are common. In Japan, most work is done in collaboration with large industrial entities.

Role for Targeted Investments

Each region maintains a baseline level of support for programs in biosensing to maintain evolutionary (as opposed to revolutionary) advances. Targeted investments will be necessary to realize sensors for markets that have only recently come to view. These markets include detection of biowarfare agents, consumer-level food safety, and household diagnostics.

REFERENCES

Armstrong, N.R., R.M. Wightman, and E.M. Gross. 2001. Light-emitting electrochemical processes. *Annual Review of Physical Chemistry* 52:391–422.

Cliffel, D.E., J.F. Hicks, A.C. Templeton, and R.W. Murray. 2002. The electrochemistry of monolayer protected Au clusters. *Metal Nanoparticles* 297–317.

Cui, Y., Q. Wei, H. Park, and C.M. Lieber. 2001. Nanowire nanosensors for highly sensitive and selective detection of biological and chemical species. *Science* 293 (5533):1289–1292.

Drummond, T.G., M.G. Hill, and J.K. Barton. 2003. Electrochemical DNA sensors. *Nature Biotechnology* 21 (10): 1192–1199.

Feldman, K., G. Haehner, N.D. Spencer, P. Harder, and M. Grunze. 1999. Probing resistance to protein adsorption of oligo(ethylene glycol)-terminated self-assembled monolayers by scanning force microscopy. *Journal of the American Chemical Society* 121 (43):10134–10141.

Hromadova, M., M. Salmain, R. Sokolova, L. Pospisil, and G. Jaouen. 2003. Novel redox label for proteins. Electron transfer properties of (η5-cyclopentadienyl) tricarbonyl manganese bound to bovine serum albumin. *Journal of Organometallic Chemistry* 668 (1–2):17–24.

Kimura, J., and T. Kuriyama. 1990. FET biosensors. *Journal of Biotechnology* 15 (3): 239 –54

Mabayashi, S.-I., K. Ban, T. Ueki, and M. Watanabe. 2003. Comparison of catalytic electrochemistry of glucose oxidase between covalently modified and freely diffusing phenothiazine-labeled poly(ethylene oxide) mediator systems. *Journal of Physical Chemistry B* 107 (34):8834–8839.

Manz, A., and J.C.T. Eijkel. 2001. Miniaturization and chip technology. What can we expect? *Pure and Applied Chemistry* 73 (10):1555–1561.

Mrksich, M. 2000. A surface chemistry approach to studying cell adhesion. *Chemical Society Reviews* 29 (4):267–273.

Reiss, B.D., R.G. Freeman, I.D. Walton, S.M. Norton, P.C. Smith, W.G. Stonas, C.D. Keating, and M.J. Natan. 2002. Electrochemical synthesis and optical readout of striped metal rods with submicron features. *Journal of Electroanalytical Chemistry* 522 (1): 95–103.

Schuhmann, W., and E.M. Bonsen. 2003. Biosensors. In *Encyclopedia of Electrochemistry*, vol. 3, *Instrumentation and electroanalytical chemistry*, ed. P.R. Unwin, 350–384. Weinheim: Wiley Europe.

Stenger, D.A., G.W. Gross, E.W. Keefer, K.M. Shaffer, J.D. Andreadis, W. Ma, and J.J. Pancrazio. 2001. Detection of physiologically active compounds using cell-based biosensors. *Trends in Biotechnology* 19 (8):304–309.

Wilson, G.S., and Y. Hu. 2000. Enzyme-based biosensors for in vivo measurements. *Chemical Reviews (Washington, D.C.)* 100 (7):2693–2704.

CHAPTER 4

CELL AND TISSUE-BASED SENSORS

Sangeeta N. Bhatia

INTRODUCTION

While the majority of existing biosensor technologies utilize biomolecules such as antibodies or nucleic acids as recognition elements, *live* cells and tissues offer potentially unique advantages over inanimate sensors. In particular, the development of hybrid (living/non-living) systems will leverage well-established microfabrication, microfluidic, and transduction technologies while exploiting the unique capabilities of living cells such as sensing, actuation, and computation. A major driver in this field has been the application of cell-based sensing to the drug discovery process; however, the field of cell-based sensors is not yet a recognized, well-defined area of research and development in any of the regions the WTEC panel studied. It follows that the observations put forth in this chapter provide only a snapshot of current research, and progress in the field is likely to evolve dramatically in the next several years.

SCOPE OF CELL-BASED SENSORS

What Are Cell-Based Sensors?

Cell-based sensors are sensors that combine living cells and tissues with conventional materials and microfabrication processes to form hybrid devices. Living cells have several common features that make them useful as sensor components (see also Figure 4.1): receptors with varying specificity for extracellular stimuli are embedded in the cell membrane and in the nucleus; signal transduction cascades amplify signaling events due to receptor/ligand binding events; production of a cellular response (e.g., release of Ca++ stores, transcription of genes, change in membrane potential, contraction, etc.).

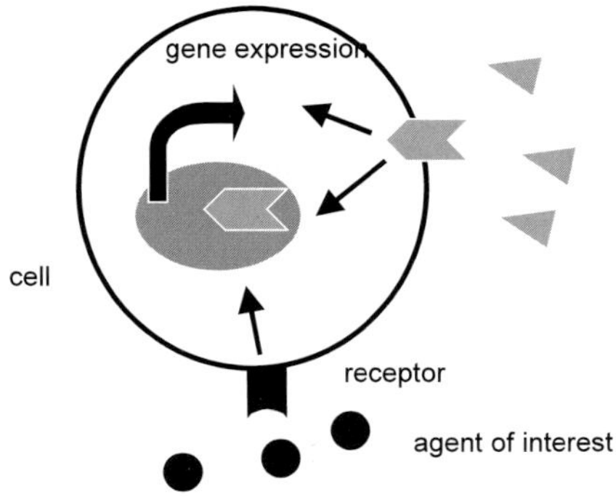

Fig. 4.1. Cell-based sensing. Cells sense extracellular species via membrane-bound or nuclear receptors. Multiple (triangle, circle) input stimuli can be integrated through complex signal transduction pathways into a limited number of detectable outputs (e.g., gene expression, second messengers). Thus, cells offer the ability to be both sensitive and specific sensors of both known and unknown agents.

As a result of these signaling pathways, cells have the potential to act as specific and sensitive sensors for other organisms (e.g., viruses, bacteria); mechanical forces (e.g., shear stress, tension); toxins; and biomolecules.

Advantages of Cell-Based Sensors

Cell- and tissue-based sensors offer numerous potential advantages over non-living sensors:

- *They are able to detect and/or classify unanticipated threats* (e.g., novel pathogens). As the ultimate site of action of chemical or biological agents, the mammalian cell acts as a robust functional reporter of toxicity, rather than relying on a surrogate marker such as nucleic acid or antibody-based detection. Therefore, mutated pathogens or novel chemical species will be characterized by their functional impact on cell physiology.
- *They can relate sensor data to human physiology/pathology* (e.g., toxicity). Living sensors can report on the integrated, physiologic response to an exogenous agent. These responses are often nonlinear, multifactorial, and exhibit hysteresis, and are therefore difficult to predict using biomolecular recognition alone.
- *They provide enhanced stability of enzymes, receptors, or antibodies in biological systems.* As self-renewing machines, cells continually replenish and repair their component biomolecules. This feature of

living cells may be critical to exploit enzymes or biomolecules that are particularly labile.

- *They can leverage genomic tools and assays such as DNA microarrays, polymerase chain reaction (PCR), reporter genes, and other high-throughput bioassay strategies.* The biotechnological and bioinformatics infrastructure developed in the wake of the Human Genome Project provides a large repertoire of tools that can be leveraged for biosensing applications. This complements and synergizes with the infrastructure developed in microtechnology for semiconductor (CMOS) and microelectromechanical (MEMS) applications.
- *Living cells are unique from a manufacturing perspective in that they are self-replicating micro- and nanoscale structures when provided with an appropriate energy source.*
- *Since the dynamic range of living systems is adaptive, they enable creation of sensors with sensitivity and specificity over a large range.*
- *They can leverage emergent phenomena being elucidated by microscale control of cells and tissues.* Microscale biological phenomena analogous to microscale physical phenomena exploited in microfluidics are now being uncovered. As this field of research develops, the findings will enable unprecedented engineering of cell fate and function in vitro.

Disadvantages of Cell-Based Sensors

Cell- and tissue-based sensors also introduce several potential challenges as compared to nonliving sensors:

- *They are environmentally sensitive.* Living systems introduce severe constraints on materials, processing, manufacturing, delivery, and operation. In particular, cells must be kept viable, sterile (free of bacterial/fungal growth), phenotypically stable, and maintained in a fluidic environment. Furthermore, the finite lifetime of living systems mandates strategies for preservation and storage.
- *Their variability calls for diverse computational strategies.* Even genotypically identical organisms differ functionally from each other. This variability arises from phenotypic differences resulting from the role of the cellular microenvironment, protein expression, receptor number, and numerous other variables. In order to effectively utilize cells as sensors, strategies for calibration, sampling rates, and data mining must therefore be developed in parallel with the solid-state components.

- *Requirements for a hybrid interface are complex.* Engineering the living/non-living interface is crucial to obtaining reproducible signals from living components. Non-fouling surface chemistries, biocompatible materials, strategies for localization of living components, and techniques to "communicate'" with the cells all must be considered.
- *Input/output interactions are poorly defined.* Despite the sensitivity of the cell to its environmental cues (soluble factors, extracellular matrix, cellular interaction, shear stress, etc.), the input/output relationship for mammalian cells is not well defined. Thus, systems that allow control over variability in the microenvironment are crucial. Further definition of cellular responses will emerge as a result of several multi-investigator teams, such as the U.S.-led, NIH-funded "Alliance for Cellular Signaling" (www.afcs.org/), which seeks to define the biochemical "state space" of two mammalian cell types in response to well-defined stimuli.
- *Transduction of cellular output to the solid-state signal is problematic.* Strategies to convert cellular responses to quantifiable signals typically require either engineering of cells (e.g., genetic modification to produce fluorescent reporter proteins) or interface engineering (e.g., growth of cells on electrodes) in order to automate data collection and analysis.

Applications

Applications of cell- and tissue-based sensors are potentially extensive:

- *Pharmaceutical drug development.* Automated cell-based assays are used to assess efficacy and toxicity of candidate drugs, primarily to eliminate candidates destined to fail later in development.
- *Neural networks.* Neurons and integrated circuits are fused with the goal of combining the advantages of the high speed and memory capability of chips with the advantages of pattern-based computation and adaptability of neural tissue.
- *Medical diagnostics.* Cell-based sensors are used on patient samples to predict clinical outcome. The mixed leukocyte reaction is an existing example of a cell-based assay that is used to predict immune rejection.
- *Cell-based therapies.* Cellular responses are used in schemes to replace tissue function. For example, beta-islets are used as glucose sensors to drive insulin release for diabetic patients.
- *Detection of chemical and biological agents.* Cell-based sensors are used to predict physiologic responses to both known and unknown

pathogens. This concept is an extension of the classic paradigm of the canary in the coalmine.

KEY SCIENCE/TECHNOLOGY ISSUES

Despite the enormous potential opportunities of using live cells and tissues as sensors, several critical hurdles remain before "live" sensors are available as "off-the-shelf" devices. Research activity to address several key challenges is in its early stages in the areas of interface engineering, transduction schemes, integration of microtechnology and biology, and commercialization.

Interface Engineering

The integration of cell and tissues with materials requires strategies for fusing biological and materials processes while preserving the biological responses of interest. First, one must consider the biological side of the interface—the cell. Control of cell function requires strategies to deal with the variability of individual cellular responses and methods to preserve the physiologic responses of the cell. Although research in this area is in its early stages, the United States is the leader in fundamental study of cellular responses at hybrid interfaces. In particular, alterations in cell fate (differentiation, division, apoptosis) that occur due to the cellular environment are active areas of research at Johns Hopkins University (Christopher Chen), Harvard Medical School (Mehmet Toner, Donald Ingber), and the University of California at San Diego (Sangeeta Bhatia). Figure 4.2 illustrates the work of Drs. Bhatia and Chen. These fundamental studies will provide insight into the "design criteria" for engineering of a hybrid interface that preserves biological responses of interest.

Another crucial aspect of development in cell-based sensors deals with the inherent variability in cells as sensor components. In order to improve specificity of the cellular response without comprising sensitivity, genetic approaches have been proposed where knockout cells (cells missing a key gene) are utilized as control populations for "background" estimation. In addition, DNA microarrays have been utilized to characterize the variability between different cells at the level of the transcriptome. Computational approaches for pattern recognition in cellular responses and data mining have also been proposed to improve specificity without reducing sensitivity. Methods to interpret, classify, and link data to physiologic responses of interest are also being explored. Finally, the need for uniform population of cells synergizes well with the current activity in stem cell biology and tissue engineering. It remains to be seen whether tumor-derived cell lines, adult stem cells, embryonic stem cells, tissue slices, primary cells, or immortalized cells are ideal for cell-based sensors.

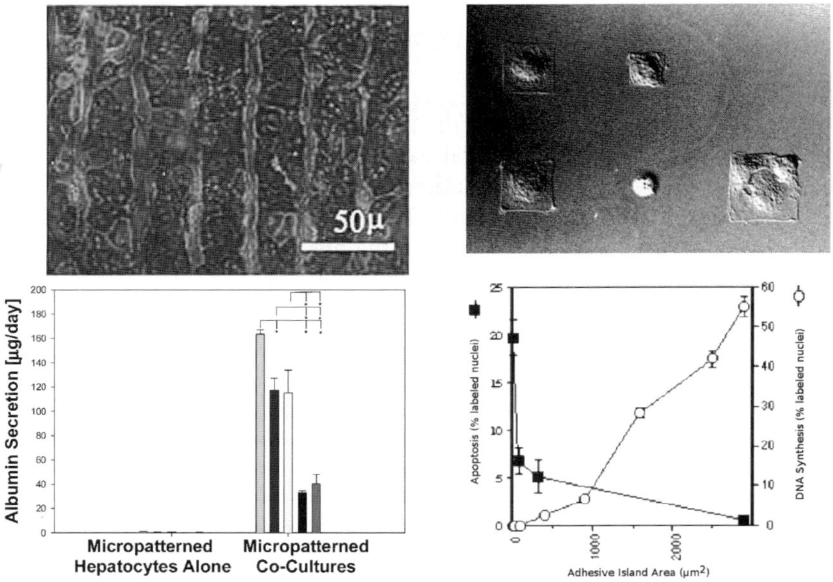

Fig. 4.2. Control of cell physiology using micropatterning. *Left*: Micropatterned co-culture of hepatocytes (liver cells, red) and stromal cells (green). Pattern configuration modulates level of liver functions (courtesy S. Bhatia, U.C. San Diego). *Right*: Micropatterned endothelial cells on laminin islands of different sizes modulate whether cells undergo programmed cell death or divide. (Courtesy C. Chen, Johns Hopkins University)

Fusing biological processes with materials also requires consideration of the inorganic side of the interface. Programs focused on this essentially center around techniques to modify and characterize surface chemistry. Europe and the United States have impressive programs in surface characterization, exemplified by the commercial surface plasmon resonance device sold by Biacore Sweden. Substantial expertise in modifying surfaces to mediate localized attachment of cells or "micropatterning" exists in the United States, Europe, and Japan; however, exciting progress in patterning lipid bilayers and vesicles has been reported at Stanford University (Gregory Kovacs, U.S.), Regensburg University (Germany), and École Polytechnique Fédérale de Lausanne (Switzerland). Finally, efforts to engineer surfaces that interact with cells dynamically are in their beginning stages. Responsive surfaces have been engineered using polymers that are temperature-sensitive (Tokyo Women's School of Medicine, Masayuki Yamato), electroactive (University of Chicago, Milan Mrksich), and sensitive to cellular enzymes (ETH Switzerland, Jeff

Hubbell). Taken together, these "dynamic" interfaces offer early examples of how interface engineering will progress in the years to come.

Transduction Schemes

Development of sensors from living elements also requires the development of strategies for transduction of a cellular response (e.g., Ca++ flux, membrane depolarization, gene expression) to a solid-state signal. The most well-developed strategy is based on interfacing excitable cells (neurons, hippocampal slices, cardiomyocytes) with microelectrode arrays. Both spontaneous and induced action potentials can be detected extracellularly by ion flux through membrane channel receptors. This strategy is well developed for detecting neurotoxins and other chemical agents that act against transmembrane targets. All regions the WTEC panel investigated have substantial programs in the development of cell-based devices that combine electrical and microfluidic engineering (e.g., in the United States, Stanford University, Greg Kovacs; Naval Research Laboratory, Joe Pancrazio). Moving forward, however, significant efforts are underway in Japan (e.g., at Matsushita) and Europe (e.g., at ETH) to build fully integrated drive circuitry and signal processing on customized chips that house excitable cells.

Biochemical secretion (e.g., insulin secretion in response to glucose in beta-islet cells) has also been utilized as a cellular output, though signals must then be converted by a secondary sensor technology to a solid-state signal. At Oregon State University (Phil McFadden), neurotransmitter release of a primary cell population was coupled to catecholamine-sensitive fish cells, producing pigment aggregation that was optically detected.

Cellular signals can also be detected fluorescently. Detection of drug metabolism by conversion of reporter compounds to fluorescent products is routine using commercially available biochemical probes in all three regions examined by the WTEC panel. Similarly, ion concentration (Ca++ concentration) can be detected remotely using fluorescent reporter dyes. Furthermore, transcriptional events have been detected using fluorescent reporter strategies such as those based on green fluorescent protein or beta-lactamase/ cephalosporin strategies (Vertex, U.S.). For example, engineered bacterial systems that are sensitive to TNT and organophosphates have been reported based on changes in expression of gene fluorescent protein (University of Wisconsin, Bob Burlage). Alternatively, translocation of proteins fused with fluorescent reporters can be monitored via fluorescent microscopy (Cellomics, U.S.). Increasingly sophisticated fluorescent assays are now emerging from combining expertise in optics (fluorescent resonance energy transfer) and molecular biology (University of Tokyo, Yoshio Umezawa; University of California

San Diego, Roger Tsien). For example, exploitation of protein splicing of "intein" (intervening protein sequence) domains allows recombination of green fluorescent protein as an indicator of colocalized species within cellular substructures. Advances in this area will provide new strategies for transducing cellular responses to optical signals for biosensing as well as providing new tools for fundamental cell biology research.

Finally, mechanical forces, local pH changes, alterations in dielectric permittivity, and thermal fluctuations are modes of detecting cellular responses that have been explored to varying degrees. As a rule, these responses are relatively nonspecific as compared to optical and electrical signals that are linked to unique biochemical events. However, classification of cellular responses in response to various stimuli is underway in several groups in all three regions of this WTEC study and may provide a useful approach to a subset of biosensing applications.

Integration of Microtechnology/Biological Species

Cell-based sensing requires detection of phenomena that occur at the micro- and nanometer length scales. Thus, innovation in micro- and nanotechnology (fabrication, MEMS, materials, microfluidics) will be required in order to incorporate biological species and processes into the next generation of "biochips" (Figure 4.3).

Europe is the leader in developing fully integrated "labs-on-a-chip" or "micro-total-analysis-systems." Both university programs (e.g., University of Twente, ETH Zurich, EPFL) and commercial ventures (e.g., DiagnoSwiss) are actively pursuing strategies to incorporate living components with microfluidics, integrated drive circuitry, controllers, signal processing, and biochemical detection. In the United States, integration of microtechnology and biological assays has occurred primarily in industry (Aclara, Caliper) and national labs, whereas in Japan, such research activity resides predominantly in universities (University of Tokyo). An innovative approach that is in its infancy is a reversal of the classic sensing paradigm where cellular responses are assessed by an external sensor (e.g., microelectrode)—that is, sensing of cellular responses by interrogating the contents of cells with invasive nanoscale probes.

Commercialization

In order to produce "off-the-shelf" cell-based biosensors, several engineering challenges must be addressed: miniaturization and portability; automation and parallel screening (see Figure 4.4); and preservation.

Overall, the United States is the leading region in commercialization. Specifically, efforts towards miniaturization and portability have been driven in the United States primarily by defense applications and funding

(Stanford, Naval Research Laboratory), whereas efforts towards automation and parallel screening have been driven by drug discovery applications (Cellomics, Aurora, Cyntellect, Surface Logix, Q3DM). Research related to storage of cell-based sensors (i.e., preservation) and device lifetime is in its early stages; however, a substantial scientific infrastructure exists in cryopreservation and aging research that should complement efforts towards successful commercialization of partially living devices.

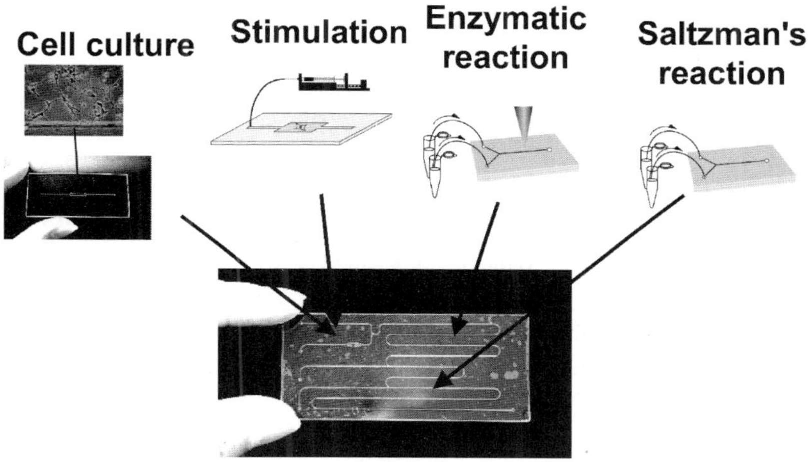

Fig. 4.3. Integration of microtechnology and biological species. Cultured cells are localized in microfluidic chips and stimulated biochemically. Cellular metabolites are detected enzymatically or by thermal lens detection. (Courtesy, T. Kitamori, University of Tokyo)

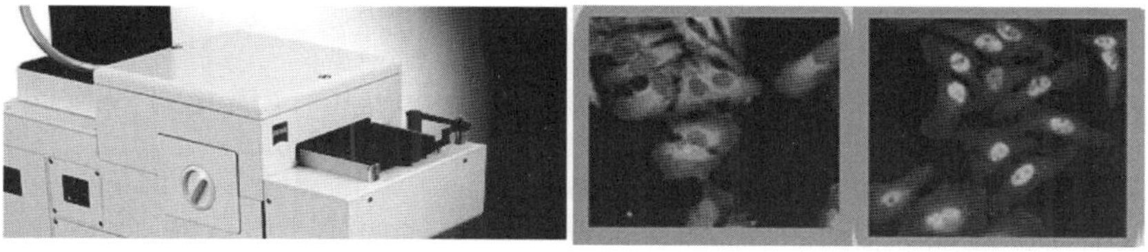

Fig. 4.4. Automation and parallel screening. *Left*: Commercial platforms for fluorescence-based cell screening. *Middle*: Cells are engineered with fluorescent reporters such as green fluorescent protein-tagged NF-KB. *Right*: Translocation of reporter to the nucleus upon stimulation can be visualized by digital image acquisition, and data is mined for drug discovery. (Cellomics, Inc., U.S.)

SUMMARY

The United States maintains the dominant position in cell-based sensors, in particular with regard to control of cell function through engineering of the hybrid interface. Motivation and resources have come primarily from the Department of Defense for sensors of pathogens and from the pharmaceutical sector for development of screening tools in the drug development pathway. In Europe and Japan, rapid gains are being made with regard to: (1) integration of microtechnology and biotechnology and (2) commercialization of integrated devices through broad-based programs to facilitate transfer of technology to industry.

Table 4.1 summarizes the WTEC panel's findings in terms of comparing the work being done in the United States, Europe, and Japan on the issues key to progress in cell-based sensors.

Table 4.1.
Comparison of International Research in Cell-Based Sensors

	Topic	Knowledge Base	Work To Date	Leading Region
Transduction Strategies	Electrical	Early	Beginning	ALL
	Optical	Intermediate	Progressing	U.S., EU
Interface Engineering	Surface Chemistry	Advanced	Substantial	U.S., EU
	Cell Function	Early	Progressing	U.S.
Integration	Microtechnology	Intermediate	Substantial	EU
Commercialization	Drug Discovery	Early	Progressing	U.S.
	Diagnostics	Little	Little	None

CONCLUSIONS

Cell-based sensors offer a powerful opportunity to push the frontiers of biosensing by leveraging the unique attributes of living systems. Early examples of functional sensing exist, primarily through use of excitable cells; however, the promise of cell-based sensing for drug discovery, diagnostics, tissue engineering, and pathogen detection is far from being realized. Moving forward, multidisciplinary teams and long-term funding mechanisms that target the broader development of cell- and tissue-based technologies will be critical to success. Given the progress to date and the level of international activity and expertise, the climate is ideal for a concerted effort to push the field forward. In return, a robust connection between the traditionally separate fields of microtechnology and biotechnology promises to yield new biosensing capabilities that are not yet available in either realm.

RECOMMENDED READING

(Multiple articles.) 2001. *Biosensors and Bioelectronics* (special issue focused on cell-based sensors) 16(7–8).

Bousse, L. 1996. Whole cell biosensors. *Sensors Actuat. B (Chem)* B34 (1–3):270–275.

Gross, G.W., S. Norton, K. Gopal, D. Schiffman, and A. Gramowski. 1997. Neuronal networks *in vitro:* Applications to neurotoxicology, drug development and biosensors. *Cellular Engineering* 2:138–147.

Kramer, K.J.M. and J. Botterweg. 1991. Aquatic biological early warning systems: An overview. In *Bioindicators and Environmental Management*. London: Academic Press, pp. 95–126.

McConnell, H.M., J.C. Owicki, J.W. Parce, D.L. Miller, G.T. Baxter, H.G. Wada, and S. Pitchford. 1992. The cytosensor microphysiometer: Biological applications of silicon technology. *Science* 257:1906–1912.

Pancrazio, J.J., J.P. Whelan, D.A. Borkholder, W. Ma, and D.A. Stenger. 1999. Development and application of cell-based biosensor. *Annals of Biomedical Engineering* 27:697.

Rudolph, A.S., and J. Reasor. 2001. Cell and tissue based technologies for environmental detection and medical diagnostics. *Biosensors & Bioelectronics* 16:429–431.

CHAPTER 5

Mass Spectrometry and Biosensing Research

Charles L. Wilkins

INTRODUCTION

At the outset of the WTEC study on biosensing research and development, it was recognized that mass spectrometry is playing an increasingly important role in the field of biosensing research. As noted during the WTEC December 2002 Workshop on Biosensing Research and Development in the United States (wtec.org/biosensing/proceedings/), historically a certain degree of ambiguity has existed with respect to the terms "biosensing" and "biosensor." For example, as Turner notes (1996), the term biosensor "...has been used to describe a thermometer, a mass spectrometer, daphnia in pond water, electrophysiology equipment, chemical labels for imaging, and ion-selective electrodes..." However, he concludes that, as defined in an earlier work, "...a biosensor [is] defined as a compact analytical device incorporating a biological or biologically-derived sensing element either integrated within or intimately associated with a physicochemical transducer. The usual aim of a biosensor is to produce either discrete or continuous digital electronic signals which are proportional to a single analyte or a related group of analytes" (Turner, Karube, and Wilson 1987). Within this context, a mass spectrometer clearly qualifies as a biosensor.

On the other hand, when one considers the desirable characteristics of a biosensor (specificity, sensitivity, stability, wide applicability, low cost, and portability), there are a number of respects where generally available mass spectrometry technology falls short, most notably in the areas of low cost and portability. For biosensor applications, present laboratory-based mass spectrometry provides superior performance. There are also a number of options available for so-called "field-portable" applications. Finally, and perhaps of most interest, there is considerable research directed toward the long-range goal of achieving truly portable (or perhaps personal) mass spectrometers.

The WTEC panel's charge with respect to reviewing relevant recent advances in mass spectrometry are focused on the degree to which they broaden the potential for biosensor applications of the field. In addition to addressing this issue, this chapter identifies some of the factors that currently limit biosensor applications of mass spectrometry and considers the prospects for addressing these limitations. Finally, as charged by the panel's sponsors, the chapter summarizes and compares the current status of mass spectrometry research and development in Europe, Japan, and the United States.

MASS SPECTROMETRY BACKGROUND

Technical Advances

There have been a number of important developments in mass spectrometer design that are highly relevant to their possible importance as biosensors. Key developments in mass spectrometer sources have been the development of matrix-assisted laser desorption/ionization (MALDI) mass spectrometry (Tanaka et al. 1988; Karas, Hillenkamp, and Chem 1988) and of electrospray ionization (Yamashita and Fenn 1984); the impact of these developments on bioanalytical chemistry has been so important that it led to the Nobel Prize in Chemistry being awarded to Fenn and Tanaka in 2002 (www.nobel.se/chemistry/laureates/2002/)(Smith and Felton 2002). From a practical standpoint, these techniques have allowed extension of mass spectrometry to biomolecules with masses extending well above 100,000 Daltons.

Equally important, advances in developments of mass analyzers have continued, with notable examples being *Fourier transform mass spectrometry* (Dienes et al. 1996); *quadrupole ion trap* (Patterson et al. 2002; Riter et al. 2002); and new *time-of-flight mass spectrometer* designs (Cornish and Cotter 1994). As mentioned above, there has been a good deal of recent attention devoted to development of smaller mass analyzers (Henry 1999 and Figure 5.1). These efforts have been driven by the recognized need for capable, in situ mass analysis systems that are compact and easily portable. Obviously, successful development of such mass spectrometric equipment would be of great interest for a very wide variety of applications, particularly those in the biosensor area. Badman and Cooks (2000) have provided an excellent perspective on this research.

Fig. 5.1. A miniaturized cylindrical ion trap (right) with a commercial ion trap (left) for comparison. (Henry 1999, reprinted with permission)

Field Portable Mass Spectrometers

The 12th Sanibel Conference on Mass Spectrometry in early 2000 was devoted to the topics of field-portable and miniature mass spectrometry. This meeting was reviewed in some detail by Sparkman (2000). An interesting aspect of his report is the perspective of the opening speaker, J. Franzen (Bruker Daltonics, Bremen, Germany), who was reported to have said that there is little or no market for field portable mass spectrometry instrumentation. He attributed this to the fact that the limitations of the instruments were more involved with the skills required for the interpretation of the data rather than the actual performance of the analysis. This perspective, if accurate, suggests the need for continued research into data interpretation algorithms, with the possible objective of making computer interpretation much more effective than it is at present.

Critical Parameters

Virtually all types of mass analyzers have been or are under investigation for miniaturization and potential field-portable applications. Thus, small magnetic sector analyzers, linear quadrupole and quadrupole ion trap (QIT), Fourier transform, and time-of-flight (TOF) mass spectrometers are being developed and evaluated, many for biosensor applications. As evident in the publications that have resulted so far, most of the miniaturization efforts have necessarily resulted in mass spectral performance compromises, to a greater or lesser degree, depending on the type of mass analyzer involved. In a great many possible applications, these compromises may well be analytically acceptable. However, the factor currently limiting further success in miniaturization and enhanced field portability is the lack of correspondingly miniaturized vacuum (pumping) and electronic systems. As Henry (2002) reported in her article discussing a half-day symposium on miniature mass spectrometers at the 53rd annual Pittsburgh Conference and Exposition on Analytical Chemistry and Applied Spectroscopy in 2002, "One after another, the speakers reiterated that, until the ancillary parts of the system are also reduced in size, it won't do much good to continue to shrink the mass analyzer."

As documented in a report appearing in *Analytical Chemistry* (Harris 2003), there is continued progress in miniaturization of mass spectrometry, and quadrupole ion trap development recently has resulted in a new mass spectrometer design with a mass of 17 kg and a mass range of 500 m/z (see Figure 5.1). Table 5.1 presents parameters for miniature mass analyzers typical at the time of the writing of this report.

Table 5.1.
Typical Parameters for Miniature Mass Analyzers*

Analyzer Type	Dimension	Mass Range	Resolution
Cooks QIT	2.5 mm radius	250 m/z	100 m/Δm
Ramsey QIT	0.5 mm radius		
Quadrupole	0.5 mm radius 10 mm long 4 x 4 array	300 m/z	600 m/Δm
Quadrupole	0.5 mm diam. 10–30 mm long	150 m/z	14 m/Δm
Cotter TOF	7.5 cm long	66 K m/z	300–1200m/Δm
Double-focusing EB	17 x 37x 57 cm instrument	39–255 m/z	131 m/Δm

*Examples only; parameters change often as designs are improved

Mass Spectral Instrumentation Research Needed

Progress toward combining capable mass spectrometry sources with new miniaturized mass analyzer designs has been reasonable over the past few years. Research with a number of promising innovative approaches, including quadrupole arrays, cylindrical ion trap arrays, curved field reflectron time-of-flight (Figure 5.2), and small permanent magnet Fourier transform mass analyzers, has continued during the last few years.

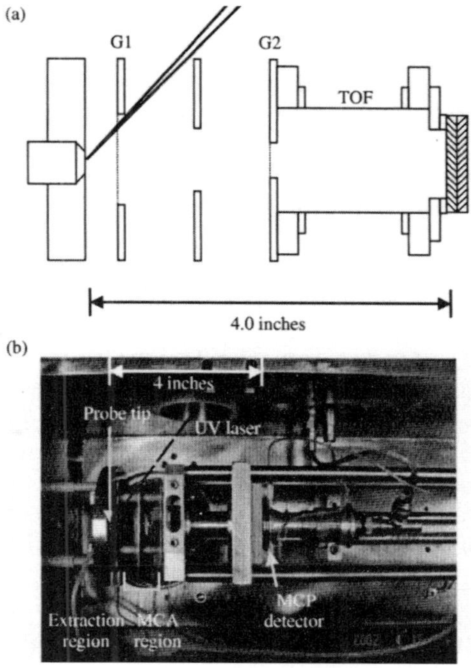

Fig. 5.2. A miniaturized time-of-flight mass spectrometer showing the sample probe, the end cap, and the coaxial detector. (English and Cotter 2003, reprinted with permission).

It is quite clear that development of miniaturized low-power vacuum systems should be an area of the highest priority, if the promise of compact and analytically effective mass spectrometers is to be fully realized. It does not seem that there are any fundamental reasons that this would not be possible. One approach, suggested by Cotter (Henry 1999), might be to develop combined mechanical and turbomolecular pump systems, which could be potentially smaller and more compact than the present alternatives.

Potential Applications

Among the most intriguing possible applications for mass spectrometry as a biosensor tool is the identification of biomarker signals that are expressed by viruses, bacteria, and spores, with the interpretation aided by comparison with genomic information for each organism. The successful approach will most likely involve identification of the biomarkers with high-performance laboratory instruments, with subsequent routine analysis and detection by lower-performance inexpensive instruments.

Another possible application is the use of mass spectrometry as a tool for clinical diagnostics, such as protein biomarkers for cancer (Phillips et al. 1999). Environmental analysis and emergency response applications involving chemical and biological warfare or terrorism are obvious applications of mass spectrometry, both with high-performance laboratory instruments and with lower-performance mass spectrometers. For example, there is contemporary evidence that rapid chemical taxonomy of bacteria is possible by MALDI mass spectrometry, using detected lipid patterns and analysis of proteins detected. Such applications, in common with the previously mentioned need for field-portable instruments, will require improved and sophisticated data analysis algorithms and software. Therefore, research in that area also should be of high priority.

Summarizing, the primary new thrusts in mass spectrometry instrumentation appear to be in technologies that will facilitate development of compact new mass spectrometers. Additional important efforts will involve the ongoing exploitation of electrospray and matrix take assisted laser desorption/ionization mass spectrometry to allow analysts to full advantage of mass spectrometry's speed, sensitivity, and selectivity advantages.

MASS SPECTROMETRY RESEARCH IN EUROPE

Recognizing that the WTEC panel's survey could not be comprehensive, the team visited four leading European laboratories in mass spectrometry research as applied to biosensing. It is understood that these visits only provide a snapshot of the current status of such research in Europe but can convey some impressions of the current issues. Chosen for participation in the study were Professor Renato Zenobi's laboratory at the Eidgenössische Technische Hochschule (ETH) in Zurich; Dr. Peter Derrick's laboratory at the University of Warwick; Professor Simon Gaskell's laboratory at the University of Manchester Institute of Technology; and Oxford Glycosciences. The first three are university laboratories; the last is a commercial biomedical research laboratory. The full site reports for these institutions may be found in Appendix B.

Eidgenössische Technische Hochschule (ETH)

Professor Zenobi's group in ETH's Department of Chemistry is focused on the development of analytical tools—laser-assisted analytical chemistry near-field scanning microscopy and laser mass spectrometry—that are of great importance to biosensing research and development. Articles from his group detail their work in the applications of soft ionization mass spectrometry analysis to the study of noncovalent interactions, the acquisition of topological information about biomolecules, and analysis of water and aerosol samples (Zenobi 2001; Daniel et al. 2002; Friess and Zenobi 2001; Friess et al. 2002; Bucheli et al. 2000; Morrical and Zenobi 2002).

Zenobi advocates a method he calls "two-step laser mass spectrometry," which employs an infrared laser in the first step for ablation of the sample and a tunable ultraviolet laser in the second step for ionization. There has been a debate in the scientific literature between Zenobi's group and those who advocate the single-step laser desorption approach to aerosol particle analysis (e.g., Haefliger, Bucheli, and Zenobi 1999; Reilly et al. 1999). This debate exemplifies the laboratory's interest in the development of MS instrumentation and sample preparation methods that have evolved from MALDI theory (Zhang et al. 2002), to development of a MALDI sample preparation method applicable to insoluble polymers (Skelton, Dubois, and Zenobi 2000), to construction of an atmospheric pressure nanosampling interface for mass spectrometry based on near-field laser ablation (Stöckle et al. 2001). The latter represents a linkage of mass spectrometry with the combination of scanning near-field optical microscopy (Zenobi and Deckert 2000) and optical spectroscopy (see Figure 5.3). Furthermore, Zenobi and his students have realized important progress in the area of near-field Raman spectroscopy measurements, further establishing the feasibility of this new technique (Stöckle et al. 2000).

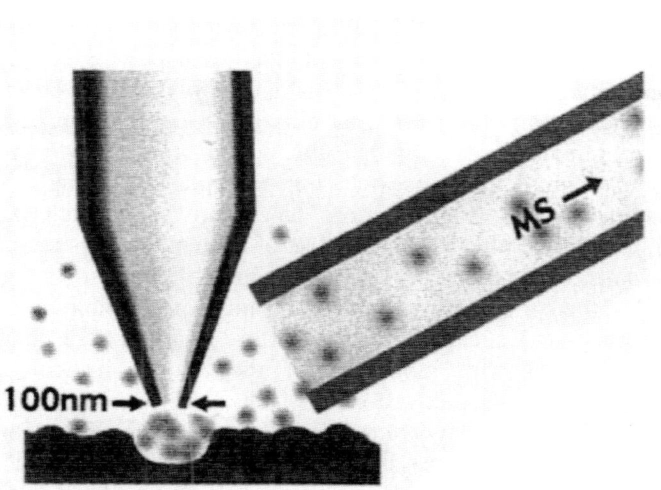

Fig. 5.3. Laser ablation MS through scanning near-field optical spectroscopy (SNOM) tips (200 nm spatial resolution). (Zenobi and Deckert 2000, reprinted with permission)

University of Warwick

At the University of Warwick, Professor Peter Derrick, Chair of the Department of Chemistry and Director of the Institute of Mass Spectrometry, reviewed the state-of-the-art mass spectrometry research underway at Warwick. This work is focused on atmospheric pressure mass spectrometry, ion funnel technology, and protein-protein interactions (see sample publications Lafitte et al. 1999 and Heck et al. 1998). He noted that one of the issues potentially having a great impact on pursuit of biologically oriented mass spectrometry research is the increasing difficulty in attracting properly qualified individuals to participate in the research. He perceives this as a general problem in the United Kingdom.

University of Manchester Institute of Science and Technology

University of Manchester Institute of Science and Technology is a small university undergoing plans to merge administratively with the University of Manchester. Due to UMIST's unique policy that any intellectual property developed as a result of research is vested in the faculty, to date it has been a particularly productive source of "startup" companies that have facilitated technology transfer.

The WTEC team's hosts were Professor Douglas Kell of the Department of Chemistry and Professor Simon J. Gaskell, Director of the Michael Barber Centre for Mass Spectrometry. Dr. Kell directs research on metabolomics and machine learning applicable to a broad range of data and chemical analysis problems. With regard to mass spectrometry, his group reported in 2000 on detection of Bacillus spores based on Fourier transform infrared spectroscopy (FT-IR) and pyrolysis mass spectrometry data (Goodacre et al. 2000). Dr. Gaskell's group covers a broad range of projects centered on developing and applying state-of-the-art mass spectrometry to biological research. For example, the group's work on understanding peptide fragmentations has improved characterization of structurally modified proteins through use of tandem mass spectrometry.

In the context of the characterization of biomolecules such as proteins and peptides, the compelling advantages of mass spectrometry are those of high sensitivity and a capability for mixture analysis. The analysis of peptides associated with molecules of the Major Histocompatibility Complex provides an extraordinary challenge in both respects. Dr. Gaskell's research in this arena has included collaboration with several immunology research groups both within the UK and outside. Other collaborations, with UMIST's Department of Biomolecular Sciences and other Manchester biological science departments, are centered on proteomics research "Conventional" biochemical techniques and mass spectrometry are frequently of complementary value; thus, for example, Gaskell has developed (with Dr. J. Brookman of the University of

Manchester) the combination of immunoaffinity adsorption and mass spectrometry for the characterization of minor components of complex cell lysates.

Oxford Glycosciences (UK), Ltd.

Oxford Glycosciences (OGS) is a company of about 200 employees that specializes in applying glycobiology and glycoproteins to proteome research. The primary tool used for this work is 2D gel-based mass spectrometry, employing highly automated 2D-gel isotope-coded affinity tag (ICAT) techniques to do high-throughput protein analysis. The Institute for Systems Biology (Seattle, WA) licenses its patented ICAT technology to OGS, and the two companies seek joint patents for intellectual property developed as a result of their collaborative research. OGS has developed methods for automated processing and archival storage of 2D separated gel samples of clinical test groups. Its primary analytical mass spectrometry approaches involve matrix-assisted desorption/ionization time-of-flight mass spectrometry (MALDI-TOF) and quadrupole time-of-flight mass spectrometry sequencing. Several of OGS's principal research and commercial activities are centered on identifying biomarkers for clinical studies and, longer range, on the development of a human protein atlas.

MASS SPECTROMETRY RESEARCH IN JAPAN

The WTEC panel was unable to obtain invitations to any laboratory in Japan specifically known for its work in mass spectrometry, but the panelists did inquire about MS as applied to biosensing R&D in the premier laboratories, like AIST and Matsushita, that panelists visited. In the laboratories of the Biosensing Technology Research Group of AIST's Division of Biological Resources and Functions, as well as one or two others that were visited, effective use of quartz crystal microbalance (QCM) technology was evident. This analytical tool appears to be widely accepted and well-developed in Japan (as it is also in Europe and the United States). WTEC panelists did not see widespread application of mass spectrometry to biosensing R&D in the laboratories visited in Japan.

CONCLUSIONS

With regard to mass spectrometry as it relates to biosensing, several trends emerge from this WTEC study. The primary centers of mass spectrometry instrument development are in the United States and Europe, with significant effort being devoted to development of miniaturized and portable mass spectrometers. There is also considerable effort being devoted to development of novel interfaces and new instrument combinations. One good example is the innovative combination of near-field optical scanning microscopy with mass spectrometry as exemplified by Zenobi's recent work (Zenobi and Dekert 2000). Major efforts, both in the United States and Europe, are being concentrated on proteomics, both experimentally and in the data analysis aspects. Although there are mass spectrometer manufacturers in Japan (notably, Shimadzu Corporation), there does not seem to be an especially active interest in mass spectrometry research in Japan. There is relatively widespread use of quartz crystal microbalance technology there, and research that involves applications of the methodology is active.

Table 5.1 summarizes the WTEC panel's assessment comparing mass spectrometry efforts applied to biosensing in the United States, Europe, and Japan.

Table 5.2
Comparison of Research in Mass Spectrometry Applied to Biosensing

	Topic	Knowledge Base	Work to Date	Leading Region
Mass sensors, MEMS, and microfluidics	Mass Sensors	Excellent	Europe, U.S., Japan	Equally advanced; commercial
Mass spectrometric methods	Compact instrument development	Excellent; problems remain	Europe, U.S.	Europe, U.S.
	Portable MS development	Excellent; problems remain	Europe, U.S.	Europe, U.S.
	Novel MS interfaces	Excellent; problems remain	Europe, U.S.	Europe, U.S.
	Proteomics	Excellent; very active research area	Europe, U.S.	Europe, U.S.

Regarding possible future international collaborative efforts, in view of the widespread mass spectrometry activity in both Europe and the United States, collaborative mass spectrometry research in all the fields of study mentioned seems to be a promising possibility. There appear to be more limited prospects for this in Japan, where the research interests are more heavily focused on mass sensors than on mass analysis as a way to tackle biosensor problems.

REFERENCES

Badman, E.R., and R.G. Cooks. 2000. Special feature. Perspective: Miniature mass analyzers. *J. Mass Spectrometry* 35:659–671.

Bucheli, T.D., O.P. Haefliger, R. Dietiker, and R. Zenobi. 2000. Analysis of water contaminants and natural water samples using two-step laser mass spectrometry. *Anal. Chem.* 72:3671–3677.

Cornish, T.J. and R.J. Cotter. 1994. A curved field reflectron time-of-flight mass spectrometer for the simultaneous focusing of metastable product ions. *Rapid Communications in Mass Spectrometry* 8:781–785.

Daniel, J.M., S.D. Friess, S. Rajagopalan, S. Wendt, and R. Zenobi. 2002. Quantitative determination of noncovalent binding interactions using soft ionization mass spectrometry. *Inter. J. Mass Spectrom.* 216:1–27.

Dienes, T., S.J. Pastor, S. Schurch, J.R. Scott, J. Yao, S.L. Cui, and C.L. Wilkins. 1996. Fourier transform mass spectrometry - Advancing years (1992–mid-1996). *Mass Spectrometry Rev.* 15:163–211.

English, R.D., and R.J. Cotter. 2003. A miniaturized matrix-assisted laser desorption/ionization time of flight mass spectrometer with mass-correlated acceleration focusing. *J. Mass Spectrom.* 38:296–304.

Friess, S.D., J.M. Daniel, R. Hartmann, and R. Zenobi. 2002. Mass spectrometric noncovalent probing of amino acids in peptides and proteins. *Inter. J. Mass Spectrom.* 219:269–281.

Friess, S.D., and R. Zenobi. 2001. Protein structure information from mass spectrometry? Selective titration of arginine residues by sulfonates. *J. Am. Soc. Mass Spectrom.* 12:810–818.

Goodacre, R., B. Shann, R.J. Gilbert, E.M. Timmins, A.C. McGovern, B.K. Alsberg, D.B. Kell, and N.A. Logan. 2000. Detection of the dipicolinic acid biomarker in bacillus spores using Curie-point pyrolysis mass spectrometry and Fourier transform infrared spectroscopy. *Anal. Chem.* 72:119–127.

Haefliger, O.P., T.D. Bucheli, and R. Zenobi. 1999. Comment on "Real-time characterization of the organic composition and size of individual diesel engine smoke particles. *Environ. Sci. Technol.* 33:3932.

Harris, C.M. 2003. Mini MS shows big results. *Anal. Chem.* – A: 75, 250A.

Henry, C.M. 1999. The incredible shrinking mass spectrometers. *Anal. Chem.* 71:264A-268A.

———. 2002. Itsy-bitsy mass spectrometers. *Chemical and Engineering News* 80 (14): 34–35.

Heck, A.J., J.T.D. Jorgensen, M. O'Sullivan, M. von Raumer, and P.J. Derrick. 1998. Gas-phase non-covalent interactions between vancomycin-group antibiotics and bacterial cell-wall precursor peptides probed by hydrogen/ deuterium exchange. *J. Am. Soc. Mass Spectrom.* 9:1255–1266.

Karas, M., F. Hillenkamp, and A. Chem. 1988. Laser desorption ionization of proteins with molecular masses exceeding 10,000 Daltons. *Anal. Chem.* 60:2299–2301.

Lafitte, D., A.J.R. Heck, T.J. Hill, K. Jumel, S.E. Harding, and P.J. Derrick. 1999. Evidence of noncovalent dimerisation of calmodulin. *Eur J Biochem.* 261:337–344.

Morrical, B.D., and R. Zenobi. 2002. Detection of polycyclic aromatic compounds at Jungfraujoch High-Alpine Research Station using two-step laser mass spectrometry. *Inter. J. Environ. Anal. Chem.* 82:377–385.

Patterson, G.E., A.J. Guymon, L.S. Riter, M. Everly, J. Griep-Raming, B.C. Laughlin, Z. Ouyang, and R.G. Cooks. 2002. Miniature cyclindrical ion trap mass spectrometer. *Anal. Chem.* 74:6145–6153.

Phillips, M., K. Gleeson, J.M.B. Hughs, J. Greenerg, R.N. Cataneo, L. Baker, W.P. McVay. 1999. Volatile organic compounds in breath as markers of lung cancer: A cross-sectional study. *Lancet* 353 (9168):1930–3.

Reilly, P.T.A., R.A. Gieray, W.B. Whitten, and J.M. Ramsey. 1999. Response to comment on "Real-time characterization of the organic composition and size of individual diesel engine smoke particles." *Environ. Sci. Technol.* 33:3933–3934.

Riter, L.S., Y. Peng, R.J. Noll, G.E. Patterson, T. Aggerholm, and R.G. Cooks. 2002. Analytical performance of a miniature cyclindrical ion trap mass spectrometer. *Anal. Chem.* 74:6145–6152.

Skelton, R., F. Dubois, and R. Zenobi. 2000. A MALDI sample preparation method suitable for insoluble polymers. *Anal. Chem.* 72:1707–1710.

Smith, M.D., and M. J. Felton. 2002. Analytical chemists win Nobel prize. *Anal. Chem.* 74:567A.

Sparkman, O.D. 2000. The 12th Sanibel conference on mass spectrometry: Field-portable and miniature mass spectrometry. J. *Amer. Soc. Mass Spectrom.* 11:468–471.

Stöckle, R., P. Setz, V. Dekert, T. Lippert, A. Wokaun, and R. Zenobi. 2001. Nanoscale atmospheric pressure laser ablation-mass spectrometry. *Anal. Chem.* 73:1399–1402.

Stöckle, R., V. Dekert, C. Fokas, D. Zeisel, and R. Zenobi. 2000. Sub-wavelength Raman spectroscopy on isolated silver islands. *Vibr. Spec.* 22:39–48.

Tanaka, K., W. Hiroaki, Y. Ido, S. Akita, Y. Yoshida, and T. Yoshida. 1988. Protein and polymer analyses up to m/z 100000 by laser ionization time-of-flight mass spectrometry *Rapid Commun. Mass Spectrom.* 8:151.

Turner, A.P.F. 1996. Biosensors: Past, present and future. Paper published by Cranfield University, Institute of BioScience and Technology. Available online: www.cranfield.ac.uk/biotech/chinap.htm.

Turner, A.P.F., I. Karube, and G.S. Wilson. 1987. Biosensors: Fundamentals and applications. Oxford: Oxford University Press.

Yamashita, M. and J.B. Fenn. 1984. Electrospray ion source. Another variation on the free jet theme. *J. Phys. Chem.* 88:4451–4459.

Zenobi, R. 2001. Laser-assisted analytical chemistry and mass spectrometry. *Chimia* 55:773–777.

Zenobi, R., and V. Dekert. 2000. Scanning near-field optical microscopy and spectroscopy as a tool for chemical analysis. *Angew. Chem.* (Int. Ed.) 39:1746–1756.

Zhang, J., T.-K. Ha, R. Knochenmuss, and R. Zenobi. 2002. Theoretical calculation of gas-phase sodium binding energies of common MALDI matrices. *J. Phys. Chem. A.* 106:6610–6617.

CHAPTER 6

MICROFABRICATED BIOSENSING DEVICES: MEMS, MICROFLUIDICS, AND MASS SENSORS

Antonio J. Ricco

INTRODUCTION

The ability to microfabricate sensors, actuators, and the components of microsystems has become commonplace in the past decade. The term microelectromechanical systems, or MEMS, commonly describes devices and integrated microsystems in the micrometer to millimeter size range, fabricated using technologies akin to the lithographic patterning and physical/chemical feature definition processes developed for electronic semiconductor chips. As the acronym implies, MEMS devices differ from traditional electronic components in the inclusion of mechanical features: moving parts, or simply structures for which physical parameters such as pressure, stress, or acceleration perturb the device mechanically to produce a signal, or where mechanical effects are used to implement device function, e.g., in an actuator. In the past ten years, MEMS devices have found increasing commercial success in applications ranging from intravenous blood pressure transducers to automobile airbag accelerometers and digital light projectors.

MEMS is expanding from its roots in electromechanical devices in multiple directions. Explosive growth in optical telecommunications applications—followed in the commercial world, unfortunately, by a smaller but nearly as impressive implosion—has stretched the capabilities of, and demands upon, this technology. The emergence and integration of nanotechnology, manifested in both nanostructured materials and in submicron fabrication approaches, has increasingly pushed device feature size into the nanometer range.

The context of "mechanical" in the acronym MEMS expanded in the early 1990s from the solid phase to include the liquid: microfluidic systems use similar fabrication approaches to traditional MEMS, but manipulate and interrogate liquid streams and droplets rather than solid structures. MEMS has cautiously pushed the range of materials of construction beyond those of the semiconductor industry, with increasing use of polymers that offer the promise of lowering device costs and integrating diverse materials that enhance functionality. Discrete devices are giving way to integrated subsystems that include input/output capabilities, data processing, closed-loop sensing and actuation, and multiparameter measurements from a single microsystem.

The WTEC study's investigation of international R&D activities to develop biosensing systems based upon MEMS included specific focus on microfluidic systems and mass-sensitive devices and some examples from the field now broadly known as nanotechnology. The study found significant emphasis in this field on the challenges of selectively, sensitively, and robustly coupling biochemical analytes to MEMS in general, and to micro- and nanodevices that respond to mass or mechanical perturbations in particular. The complexity of biological samples is addressed by the implementation of a range of laboratory processes in integrated chip format to both reduce the complexity of the sample and to make it more readily detectable. The role of interfacial chemistry is central to biosensing with such systems, and there is a key enabling role and opportunity for structured as well as molecularly defined materials. The sorts of biosensing applications where an effective combination of MEMS and interfacial materials can have major international impact include diagnostic devices that rapidly measure cellular, genetic, and proteomic signatures and patterns, as opposed to single analytes; new approaches for the massively parallel, high-information-content drug discovery process; and the high-sensitivity, low-false-positive multiplexed detection of biological and biochemical pathogens.

DEFINITIONS AND SCOPE

Included in this WTEC survey are approaches to biosensing based upon microfabricated devices or systems that incorporate a mechanical component or measurement, for example, a micropump, a pressure sensor, or a mass-sensitive nanocantilever. Relevant related areas and subdisciplines include the following:

- *BioMEMS*, implying biological or biochemical functions or components

- *Optical* integrated devices*, known as MOMS (micro-optical mechanical systems) or MOEMS (micro-optical electromechanical systems)
- *Nanoelectromechanical systems* (NEMS)
- *Microfluidics*, drawing upon many of the same fabrication methods as classic MEMS and sharing many of the most active researchers with the MEMS community
- *Mass sensors*, including acoustic wave/piezoelectric devices, along with micro- or nanofabricated oscillating or deflecting cantilevers and beams
- *Electrochemical* devices*, including bioFETs (field-effect transistors), chemFETs, and systems incorporating amperometric or potentiometric sensors
- *Electronic devices*, including chemiresistors, gas-sensing diodes, integrated Kelvin probes, and scanning tunneling probes
- *Mechanical sensors* measuring force, pressure, stress, displacement, velocity, or acceleration on the atomic, molecular, thin-film, or bulk scale

Of potential interest to the reader of this report is a 2002–3 comprehensive study of the status of MEMS R&D in Japan, chaired by Professor Roger Howe of the University of California at Berkeley (Howe et al. 2003; available online at www.wtec.org/mems1/).

R&D: DRIVERS, TRENDS, AND CHALLENGES

Drivers: Why Take the MEMS Path?

Incorporating a "simple" biosensing device into a complex integrated system can be a costly, time-consuming process, so one must ask, "Why go to such trouble?" For a mass-produced commercial product, the low costs associated with batch fabrication are attractive, and most MEMS technologies, having been adapted from integrated circuit production, are relatively low in cost when utilized at high volumes. This approach carries a low cost for device parallelism: although area on a silicon wafer is not free, making 100 copies of a device on a single substrate costs much less than 100 times the cost of making one. Such parallelism enables device duplication (redundancy) and easy incorporation of reference or control devices, two important contributors to high-reliability systems.

---

* Biosensing using optical and electrochemical methods are discussed in chapters 2 and 3 of this report, respectively.

While manufacturing considerations are important for adapting MEMS to biosensing, the WTEC evaluation found that system performance advantages provide some of the most compelling arguments. Device multiplicity, for example, is well suited to the creation of bio/chemically diverse arrays based on a common transducer. Increasingly, scientists and engineers are showing that everything from odor recognition (Zubritsky 2000) to the diagnosis of some early-stage cancers (e.g., prostate: Petricoin et al. 2002) are most effectively accomplished using the output from not just one sensitive material or receptor, but from an array of chemical or biochemical interactions (Ricco, Crooks, and Osbourne 1998) in concert with pattern-recognition methodologies (a key component of the collection of methods known as bioinformatics). MEMS and related batch-manufacturing methods often utilize the increasingly complex fabrication and integration processes developed by the massive commercial infrastructures of the microelectronics, optical telecommunications, and plastics industries, the last of these playing an increasingly important role.
A number of additional performance drivers result from the capability to integrate diverse functions on one substrate, including device-to-device and batch-to-batch reliability, as well as fewer device-to-world interfaces, manifested in both enhanced physical robustness and the elimination of error-producing manual steps such as sample transfers. Integration usually diminishes required quantities of expensive reagents or precious samples. On-chip data processing simplifies the task of communicating and interfacing with the outside world while providing the amplification, digitization, or noise reduction without which many weak signals would be unusable; this improves the limits of detection and dynamic range. For example, Figure 6.1 shows a highly integrated "multisensor" that includes several transducer types as well as all necessary data acquisition and control electronics; the capacitive sensor in particular benefits from the integration of the measurement and control circuitry with the transducer. Furthermore, integration of data processing softens the impact of the biotechnology data explosion by enhancing the [information-content] to [output-bytes] ratio. Finally, with wireless communication becoming routine, integrated electronics facilitate distributed, multiplexed, and networked sensor systems that provide more comprehensive, useful responses in everything from process control to the detection of acts of biological terrorism.

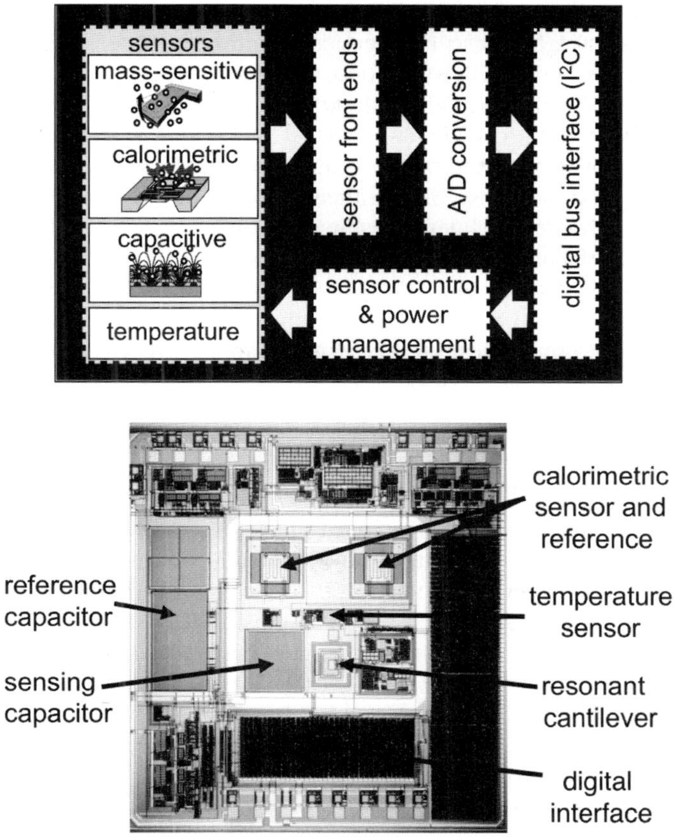

Fig. 6.1. System architecture (*top*) and chip photograph (*bottom*) of an integrated MEMS multisensor combining calorimetric, mass-sensitive, and capacitive sensors to provide chemical discrimination (Hagleitner et al. 2001). In addition to the transducers, this single chip integrates analog-to-digital converters, a digital interface bus, and power management. (Courtesy Dr. A. Hierlemann, Physical Electronics Laboratory, ETH, Zürich)

Along with noting manufacturing and performance drivers, the WTEC panel observed a number of biosensing applications providing a palpable "market pull" to complement the "technology push" of bioMEMS solutions developers. Detection of chemical or biological hazards, including chemical agents, pathogens, and biologically derived toxins, is an enunciated need of military and civil defense organizations. This is particularly true in the United States, where defense organizations define much of the publicly funded sensor and microsystem R&D; however, all the geographical regions the WTEC panel evaluated were advancing their ability to accurately sense biochemical hazards in order to address the public's need for confidence that food, air, and water supplies are safe.

Another application providing market pull is medical diagnostic devices, including both point-of-care and central laboratory systems. These are taking on increasing importance as the population of the industrialized world ages and the cost of healthcare soars. A market also may be emerging for home "wellness" testing, although much will depend on a regionally variable combination of supportive social structures and reimbursement parameters for such tests. For much of the bio/chemical/medical process and manufacturing industry, including the pharmaceutical companies, the needs are to more effectively monitor and control a range of processes, speed the discovery of new drugs, help improve the efficiency of clinical trials, and multiply the present output of fundamental biological research.

Trends: To Integrate or Not to Integrate

Despite the performance advantages of system integration, the extent to which research and early-stage developmental biosensing devices are integrated with supporting electronic, optical, and fluidic components varies widely. At the simplest level, examples of *discrete devices* presently under study in many laboratories include mass-sensitive cantilevers, as exemplified by Figure 6.2 (Yang et al. 2003; Tamayo, Alvarez, and Lechuga 2003; Ming, Li, and Dravid 2003; Liu et al. 2003; Arntz et al. 2003; Subramanian et al. 2002; Cleland and Roukes 2002; Wu et al. 2001) and capillary electrophoresis chips (Manz et al. 1992; Gottschlich et al. 2001; Woolley and Mathies 1994).

At the next level of sophistication are *multichip integrated systems*, built using separate, interconnected sets of chips having different functions. Examples of the popular "lab on a chip" concept (Oosterbroek and van den Berg 2003; Northrup, Jensen, and Harrison 2003; Baba, Shoji, and van den Berg 2002) include a space bioreactor system for culturing yeast (Walther et al. 1994, and Figure 6.3) and an integrated polymerase chain reaction (PCR) system, known as the "GeneXpert," to process, amplify, and detect particular fragments of DNA (www.cepheid.com). Both of these systems include, on separate chips or miniature circuit boards, various pumps, valves, transducers, biosensors, optical components, and microelectronics.

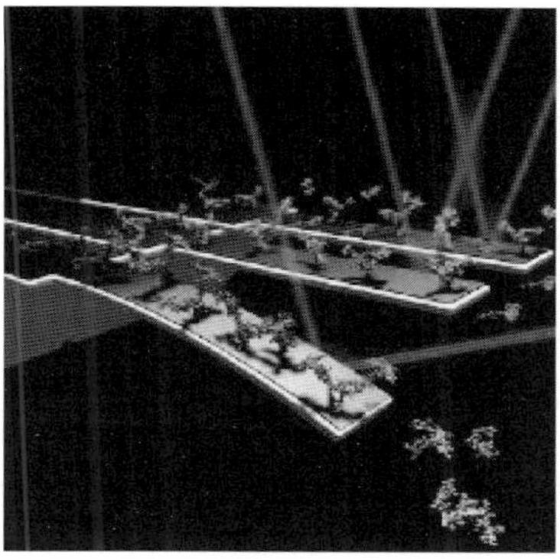

Fig. 6.2. Artist's concept of a "diving board" microcantilever biosensor developed at the University of California, Berkeley, and Oak Ridge National Laboratory (Wu et al. 2001). Cantilevers are coated with antibodies to PSA, a marker for prostate cancer. When PSA binds to the antibodies, the cantilever is deflected, which is detected with a laser beam. (Courtesy Kenneth Hsu, U.C. Berkeley, and the Protein Data Bank)

Fig. 6.3. MEMS space bioreactor system developed by the Institute of Microtechnology at the University of Neuchâtel. This unit, which supports the growth of yeast in the micro-gravity environment of outer space, was flown by the European Space Agency in 1994 and 1996. (Courtesy Professor N. F. de Rooij, University of Neuchâtel)

At the pinnacle of complexity are *monolithic integrated microsystems*, which are only now beginning, in just a few laboratories, to make the difficult transition to aqueous-phase biosensing. Examples are

- one-chip sample-preparation microfluidic systems that include capture, wash, PCR, preconcentration, and capillary electrophoretic separation and detection (Burns, Johnson, and Brahmasandra 1998; Lagally, Emrich, and Mathies 2001; Koh et al. 2003)
- a multisensor chip developed by ETH in Switzerland (www.ethz.ch) that includes sensing materials, three different transducer types, drive electronics, signal conditioning, microprocessor, and electrical interface on one chip (Hagleitner et al. 2001 and Figure 6.1 above)

While examples of the entire range of integration can be found, trends and patterns became apparent in the course of this WTEC study. First, the development of a complex multichip system or a monolithic integrated system requires significant financial resources, a high level of technical sophistication, and a coupling of diverse technical disciplines that can be realized only by a handful of the top universities, research institutes, and the more stable and well-funded of the small startup companies.

Second, as the underlying transducer technology matures, there is a trend towards greater integration and less emphasis on fundamental studies of a particular biosensing platform. For example, quartz microbalances are the heart of several commercial instruments (www.initium2000.com/index_e.html; www.maxtekinc.com/; intel.ucc.ie/sensors/universal/; www.par-online.com; see Table 1 in Handley 2001), while micromachined micro- and nanocantilevers are being studied as discrete devices, in some cases with the integration of drive electronics. Many issues have yet to be resolved to confer optimal biochemical sensitivity and selectivity.

Finally, taking as cues the more mature MEMS technologies such as pressure sensors, accelerometers, and micromirror arrays, the WTEC panel found it to be typical that application details dictate the optimal degree of integration. BioMEMS devices increasingly emphasize single-use capability, the threat of cross-contamination being a major consideration in biochemical measurements. Another important consideration, particularly in medical diagnostics, is extraordinary reliability. These two factors play against one another, with low-cost consumability favoring disposal of the smallest discrete component that can be separated from the rest of the system, while robustness and reliability favor a fully integrated device with the fewest external connections and interfaces.

The Biosensing Interface and Device Packaging: BioMEMS' Grand Challenges

While it is an area with enormous promise, biosensing is arguably the most challenging area into which MEMS and allied technologies are expanding. The interface between wet, salty biological samples and materials and devices adapted from the dry, sterile microelectronics industry is not an easy one. Electrical devices and connections must be well encapsulated while leaving biochemical or biomechanical interfaces exposed to the sample. Added to the handful of physical parameters measured or controlled by traditional MEMS devices are many thousands of biochemical measurands, often requiring a unique, tailored interfacial sensing material for each and every analyte: a different antibody for each protein, a different strand of nucleic acid for each gene. This study found the range of materials that must be used in device manufacture to be far more vast than for physical or even chemical sensors, and many pose unique challenges for deposition, characterization, and maintenance of long-term viability; these challenges are opportunities for high-impact technological advances.

In biosensing, the manipulation of materials properties is critical, for the interface between the physical device and the biological measurand requires simultaneously satisfying conditions for selective, predictable biological interactions and for providing reproducible perturbation of a magnitude sufficient for reliable detection. Further complicating the design of the biochemical interface are the ultralow limits of detection demanded by some applications, which therefore require exceptional stability and high sensitivity. A single molecule, a single surface receptor, or a single copy of a gene from one cell may be the ultimate analytical target for the early detection of cancer or the analysis of a virulent pathogen.

While there are several identifiable engineering and manufacturing challenges, the WTEC team found that packaging plays a role in nearly all of them. The input (and for a few concepts, the output as well) of bio/chemical samples and reagents presents a challenge in terms of attaining leak-free, low-volume, automation-compatible connections (Fu et al. 2002); conformity to extant sampling techniques; and freedom from sample-to-sample cross-contamination. Clever manufacturing strategies are needed if the package and its interface connections are to cost less than the biosensing system within—at present, not a likely prospect for most biosensing systems. Developing a new manufacturing process or subprocess when current industrial methods are lacking can be extraordinarily costly, and projected sales volumes must justify the expenditure.

Typically, the best packaging approach is not readily borrowed from established methods, needing instead to be tailored to a convolution of the manufacturing process and the end application. The conjunction of biology, chemistry, electronics, optics, and mechanics invariably leads to difficult materials compatibility questions and, once again, the system packaging approach can play a critical role in facilitating integration, or it can or exacerbate the difficulties associated with materials incompatibilities. Maintaining the extended viability of integrated biochemical reagents depends primarily on creating a stable, hospitable environment for them; to realize the full potential of the "lab-on-a-chip" concept, the chip must integrate some of the chemical reagents, not only the apparatus. Integration of biological reagents is, of course, one of the most difficult packaging challenges.

MEMS Fabrication Methods and Materials

MEMS biosensing systems, including microfluidic systems, have adopted many of the microfabrication methods common to integrated circuit (IC) manufacture, including lithography, dry and wet chemical processing, surface and thin-film coating technologies, bulk and deep etching, and ion implantation and etching, to name a few. Unlike state-of-the-art IC manufacture, MEMS does *not* generally push the limits of technology in terms of

- number of process steps (the IBM/Apple G5 processor requires some 500 steps; sophisticated MEMS devices use an order of magnitude fewer)
- size of silicon wafers (the G5 and Intel's Pentium 4 are built on 12" [300 mm] wafers, while this study showed that 3", 4", and occasionally 6" wafers are the rule for MEMS)
- density of features (a G5 or Pentium 4 packs in 55–58 million transistors, several orders of magnitude more than those in the most highly integrated MEMS biosensing systems)

Consequently, it is often possible for academic and national laboratory researchers to use second-hand, sometimes donated equipment that has grown obsolete for commercial IC manufacture, but nonetheless remains fully functional.

There are some areas where the WTEC team found that (bio)MEMS *does* push current microfabrication technologies beyond the present comfort limits of IC technology. For example, the 130 nm feature sizes used in state-of-the-art microprocessors are sometimes bettered by the small gaps and widths of nano-mechanical sensors and features (Möller et al. 1999), for which costly direct-write processes are feasible—at least for research quantities of devices. Also, costly high-aspect-ratio fabrication approaches, such as deep reactive ion etching and LIGA, are used to provide the deep trenches and tall structures needed for chip-based chromatography and unique mechanical structures. Further, the rapidly expanding suite of "bottom-up" approaches to fabrication on the nanometer scale, wherein interfacial processes and energetics are manipulated to guide atomic and molecular assembly, is inherently free from many of the size constraints of traditional microfabrication. Figure 6.4 is an example of a collection of nanometer-scale features not manufacturable with current microfabrication technologies (Ng et al. 2003).

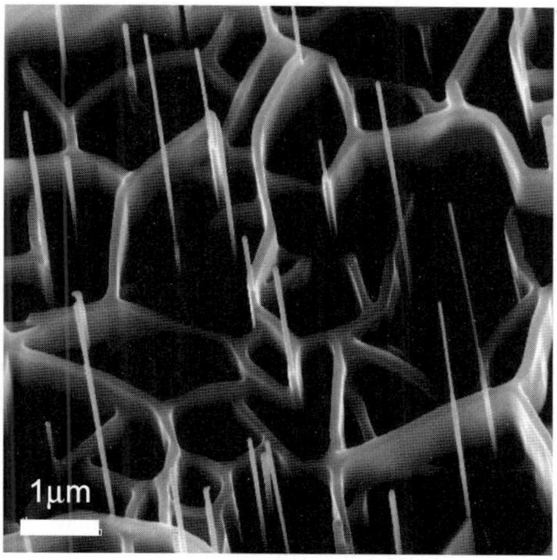

Fig. 6.4. Metal nanowires and nanowalls grown at 925°C in a process utilizing gold surface diffusion and aggregation at nodes. (Courtesy Dr. M. Meyyappan, NASA/Ames Research Center)

Finally, there is one area where this study found that biosensing consistently, aggressively pushes beyond the comfort limits of IC technology: integrating new and unconventional materials with ICs. The materials forays are well beyond the IC mainstays of silicon and its oxides, nitrides, and silicides; aluminum, copper, and occasionally gold; and the small handful of polymer dielectric layers (various polyimides and the photopatternable epoxy, SU-8, being the most common). Biosensing interfaces, in contrast to conventional IC fabrication, require unusual polymers, metals, ceramic dielectrics, "smart" (responsive) films, and biological materials, including antibodies, enzymes, nucleic acid oligomers, aptamers, whole cells, or tissue slices. This WTEC evaluation found that such materials present challenges beyond any previously encountered in the IC world in such realms as compatibility with other device materials, patterning, packaging, and long-term maintenance of viability.

In fact, some researchers interviewed by WTEC panelists noted increasing divergence between the mainstream IC industry and the industries using IC technologies for chemical and biological sensing. For sensing there is no upward pressure on feature density or wafer size, but capabilities for deep etching, through-wafer vias, creative packaging approaches, and integration of diverse materials are key. The sensing-specific need to combine well-established with leading-edge technologies renders the typical biosensor fabrication facility dated by some measures and state-of-the-art by others. The use of replication-based manufacturing methods (stamping, molding, embossing) and the incorporation of plastics as substrates that provide very low-cost "real estate" for single-use disposability is also outside the realm of current IC process innovation.

Nevertheless, the WTEC team found some trends in the IC industry to be mirrored and utilized to advantage in the nascent biosensing device fabrication industry. In both industries, dry processing technologies such as reactive ion etching and plasma-enhanced chemical vapor deposition are used increasingly, though bioMEMS does cling more frequently to wet chemical steps for greater process and materials flexibility. Both industries are pushing aggressively into the nanometer feature size regime as well.

In a move to utilize the standardized infrastructure of IC manufacture to maximum advantage for sensing, groups at ETH in Zurich, Switzerland (www.ethz.ch/; see also the ETH site report in Appendix B), the U.S. National Institute of Standards and Technology in Gaithersburg, MD (www.nist.gov), and elsewhere are devising process sequences that rely on a constrained combination of IC-standard and MEMS-unique fabrication steps. In particular, their sensing systems are designed such that a series of "front-end" conventional microfabrication steps can be carried out by any one of several commercial integrated circuit foundries (particularly those that follow the MOSIS IC Fabrication Service set of conventions; see www.mosis.org) to yield signal processing, analog-to-digital conversion, data processing, digital interfacing, and overall control and power management functionalities. "Back-end" processing—deep or through etching, addition of selective sensing films, and unique encapsulation processes—is then executed at the university or national laboratory facility on the same wafers, resulting in a complex integrated sensing system (Hagleitner et al. 2001; Cavicchi et al. 1995). This approach requires that the back-end MEMS steps do not destroy or degrade the conventional electronics already on the chip.

In a separate processing approach, Sandia National Laboratories pioneered the integration of MEMS and microelectronics in a process sequence christened SUMMiT™ Technology (mems.sandia.gov/scripts/

index.asp) that is the converse of that just described. A core element of many MEMS processes is the deposition, annealing, patterning, and release (suspension of structures above a gap) of thin layers—a fraction of one to several micrometers—of polycrystalline silicon, from which complex interconnected moving mechanical structures are made. This powerful film-based approach to MEMS is often known as surface micromachining. The deposition and/or obligatory annealing of the mechanical polysilicon layers requires, however, temperatures well above those tolerated by conventional microelectronic devices. Sandia therefore devised a method to fabricate the polysilicon mechanical structures in a depression in the surface of the Si wafer, cover them temporarily with a protective dielectric layer, polish the entire wafer to flatness, fabricate the conventional electronic circuitry, and finally strip away the protective dielectric layer to expose the mechanical subsystems.

MICROFLUIDIC SYSTEMS

The Promise of Sample-to-Answer Devices

Because direct measurement of scarce targets in a dilute and complex biological milieu is so challenging, the drive to miniaturize, integrate, and automate the techniques of the biochemical laboratory in a "MEMS-like" fashion has spawned major activity in microfluidics. Foremost among the goals is the so-called "sample-to-answer" device that accepts a raw biological sample, performs a complex series of biochemical manipulations—everything from filtration to "amplification" (replication), purification, and separation—and then detects multiple target analytes with high sensitivity, high biochemical selectivity, and wide dynamic range.

The advantages touted for microfluidics by many researchers interviewed in the course of this study include

- miniaturization to conserve costly reagents and limited samples
- parallelization to handle many samples at once
- multiplexing to analyze multiple targets for each sample
- automation and integration to save time, labor, and manual sample transfers, decreasing the chance of human error and thereby improving reliability and accuracy

These benefits (a number of them being inherent advantages of integrated MEMS in general) can be realized using a toolbox of microcomponents that includes channels, reservoirs, fluid interconnects, valves, pumps, filters, electrodes, electrical interconnects, sensors, and detectors. The building blocks in turn implement such functions as dispensing, distributing, mixing, filtering, preconcentrating, diluting, binding, releasing, washing, heating, separating, and detecting. Combining several diverse building blocks to accomplish multiple preparative tasks is now far from routine but will have major impact when it becomes commonplace. Figure 6.5 shows a plastic microfluidic system that integrates target amplification using PCR in a volume of about 50 nL with subsequent on-chip separation and analysis of PCR products by capillary electrophoresis (Koh et al. 2003). Components integrated on the plastic chip include fluid reservoirs, channels, gel-based valves, and printed-ink electrodes, heaters, and temperature sensors. The limit of detection for this system, developed to detect DNA signatures of pathogens such as salmonella and *E. coli,* is about 5 copies of a given DNA fragment of interest.

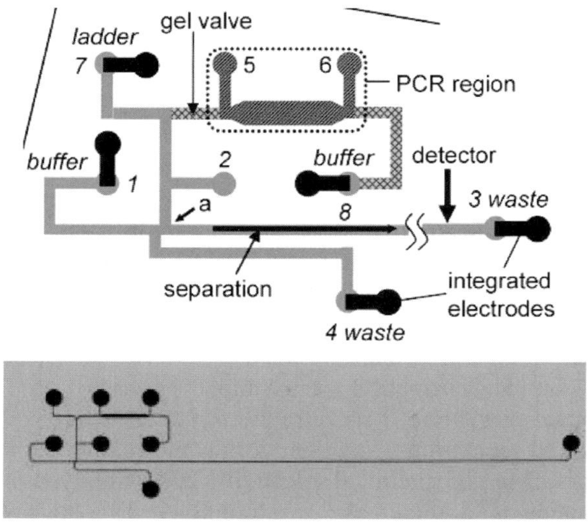

Fig. 6.5. *Top*: Schematic of a microfluidic system developed by ACLARA BioSciences (Mountain View, USA) that amplifies (replicates) DNA characteristics of biological pathogens, mixes in a set of sizing standards ("ladder"), adds fluorescent dye for detection, concentrates and separates the amplified products electrokinetically, then sends each DNA band past the detector for quantitation. *Bottom*: Photograph of a dye-filled plastic microfluidic device, prior to printing of electrodes and heaters, that implements the above functionalities, including printed conductive ink electrodes and heaters (Koh et al. 2003). (Courtesy ACLARA BioSciences)

The greatest academic activity and hundreds of millions of dollars worth of commercial attention have focused recently on systems that use electrokinetic means to motivate and separate species in fluidic channels (Andersson and van den Berg 2003; Boone et al. 2002; Bousse et al. 2001; Figeys and Pinto 2000; Hadd, Jacobson, and Ramsey 1999; Lagally, Emrich, and Mathies 2001; Li et al. 2002; Locascio, Hong, and Gaitan 2002; Sanders and Manz 2000; Sato et al. 2002; Soper et al. 2000; Tang et al. 2002). Figure 6.6 is a photograph of a protein separation chip, using microchannels etched in glass to separate complex protein mixtures by first subjecting them to micellar electrokinetic chromatography (MEKC), then further separating each MEKC-derived band of proteins using capillary electrophoresis. The collection of thousands of proteins found in even a single cell type is sufficiently overwhelming that such "two-dimensional" approaches to protein separations are a mainstay of the field of protein study known as proteomics.

In general, attractive features of electrokinetic-microfluidic systems are their

- lack of moving parts—only the application of high voltage to on-chip electrodes is necessary
- ability both to pump liquids and to effect high-resolution separations according to molecular size-to-charge ratios

Nonetheless, the clever utilization of capillary-flow and pressure-driven fluid movement have shown significant promise and versatility as well as major commercial success (e.g., the Triage® panel tests from Biosite, Inc., www.biosite.com), with pressure being supplied by external pumps, pressurized gas, chip-mounted micromachined pumps, on-chip electrolysis of water, or by spinning the chip to achieve centrifugal pumping (Madou et al. 2001; Gustafsson et al. 2004; Duffy et al. 1999).

As with other MEMS approaches to biosensing, the critical parameters for fluidic systems are materials compatibility, manufacturability, and longevity. In contrast to most MEMS, microfluidic devices have been developed most often on glass (silica) or organic polymer substrates, owing to bio/chemical materials compatibility, use of very high electric fields, and/or the use of high sensitivity optical detection at relatively short wavelengths.

MASS SENSING: MATURE QUARTZ AND EVOLVING SILICON TECHNOLOGIES

A biosensing approach in which MEMS technologies are now playing an increasingly important role is mass sensing. A key strength of mass-based biosensing is its "label-free" character, i.e., the inertial mass of the analyte molecules provides the detector response; hence no fluorophore or electroactive tag need be attached. Mass detection does not, however, obviate specific interfacial biochemical recognition; analyte molecules must be selectively recognized and bound in preference to all other species. Herein lies a key limitation of label-free detection: nonspecific adsorption. Solving this problem presents an opportunity to advance the state of the art, and it was noted by several host researchers during the course of the WTEC site visits (e.g., see Linköping University site report in Appendix B) that suitable reference devices and clever, well-controlled surface chemistries are beginning to demonstrate their potential to prevent false positive signals that arise from unintended physical adsorption of one or more components of the sample matrix.

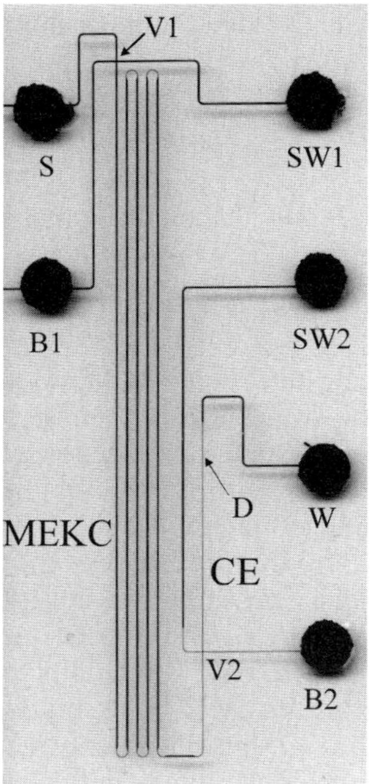

Fig. 6.6. Glass microchip with arrangement of microchannels to accomplish "two-dimensional" protein separations. Developed at Oak Ridge National Laboratory, the chip first separates a complex protein mixture into a series of multicomponent bands using MEKC, then further separates each band via capillary electrophoresis (Gottschlich et al. 2001). (Courtesy Dr. J. M. Ramsey, Oak Ridge National Laboratory)

Mass-sensitive micro- and nanodevices can be divided into two broad categories:

1. *Piezoelectric crystal-based devices*. These utilize a small "slab" or film of piezoelectric material (quartz, zinc oxide, lithium tantalate,

lithium niobate, gallium arsenide) to generate, by application of the appropriate time- and spatially varying electrical signal, traveling or standing acoustic waves whose propagation characteristics are perturbed by changes in the mass or mechanical properties of matter on the moving device surface (Ballantine et al. 1997).

2. *Silicon MEMS-based devices*. These rely on thermal, electromagnetic, or direct mechanical means to periodically or statically deflect a micro(nano)fabricated beam, cantilever, or membrane from some nonpiezoelectric material, most often silicon (Yang, Ji, and Thundat 2003; Tamayo, Alvarez, and Lechuga 2003; Ming, Li, and Dravid 2003; Liu et al. 2003; Arntz et al. 2003; Subramanian et al. 2002; Cleland and Roukes 2002), with the oscillation characteristics or extent of bending being a measure of the mass of sorbed analytes.

Piezoelectric Crystal-Based Devices

The best known of the piezoelectric devices are those that utilize surface acoustic waves (SAWs) or thickness-shear modes (TSMs); resonators based on the latter mode are popularly known as "quartz (crystal) microbalances" (QCMs or QMBs). A principal limitation of both types of oscillating mechanical device when used in biosensing is the potential for intolerable levels of damping of the acoustic wave by the liquid.

A "classic" SAW (a Rayleigh wave), while an excellent basis for a gas sensor, is ill suited to liquid-phase detection applications, as the surface-normal component of its motion leads to excessive damping by the contacting liquid. Close relatives of the SAW, including the shear-horizontal acoustic plate mode (SH-APM), the Love wave, the surface transverse wave (STW) and the leaky SAW (LSAW) carry much or all of their energy (as do TSM resonators) in modes that cause in-plane motion of the device surface, leading to manageable attenuation of the wave. The flexural plate wave (FPW) has significant surface-normal displacement, but unlike the other modes described above, its velocity is slower than that of sound in water, so energy transfer from wave to liquid is relatively inefficient, and damping is therefore quite manageable.

One important trend noted by the WTEC team in TSM resonators and other acoustic wave biosensing devices is operation at ever-higher frequencies, leading to enhanced sensitivity and, in some cases, lower limits of detection—provided the associated circuitry is carefully designed so as to not introduce additional noise, which can offset the gains in sensitivity as frequencies go higher. Where TSM resonators running at 5 and 9 MHz were the rule a number of years ago, devices over the 5–30 MHz range are now commercially available (for example, from International Crystal Manufacturing Co., Inc., http://www.icmfg.com/quartzmicrobalance.html#frequency), and devices up to 100 MHz are being evaluated in research labs. Note, however, that the thickness of the crystal is inversely proportional to the fundamental frequency and, in practice, quartz TSM devices above about 30 MHz are quite fragile. Improving the stability of the oscillator circuitry and sample temperature control system to provide, for example, 0.1 Hz short-term stability rather than the more typical 1 Hz is often a more effective means to improve the limit of detection. MEMS methods have also been used to provide a localized thin, "energy trapping" region within a quartz substrate that is thick elsewhere to maintain mechanical robustness (Smith and Senturia 1995). Notably, many of the other acoustic modes, though less widely used than TSMs, are either independent of substrate thickness or dependent in a way that allows realization of higher sensitivities without unreasonably thin substrates. In a twist on the traditional measurement of surface-bound mass by tracking changes in TSM resonant frequency, Figure 6.7 details how a group at Cambridge University and Akubio is using the acoustic signature produced when particles as small as viruses are dislodged from the moving surface by high-amplitude motion of the crystal surface (Cooper et al. 2001).

Despite this example of a new transduction approach, a second general finding of the WTEC team with regard to piezoelectric crystal-based devices is the relatively mature state of the technology. For these transducers, the fundamental biosensing advances are predominantly in the interfacial chemistry, while the basic platform is static, save for a gradual increase in operating frequencies. A growing number of commercial operations supply complete TSM resonator-based systems, in some cases including oscillator circuitry, temperature control apparatus (critical for high-sensitivity measurements), and integral flow cells (Handley 2001; Gizeli and Lowe 2002, 296).

Silicon MEMS-Based Devices

Among academic and national laboratory researchers around the world, silicon MEMS-based micro/nano cantilevers and beams are receiving an increasing share of the visibility formerly focused on piezoelectric devices. Being silicon themselves, the new MEMS mass-sensitive devices are simpler to integrate with control and measurement electronics. These devices operate in two principal modes:

1. by vibrating, where the drive can be electrothermal (e.g., using resistors incorporated in the silicon chip as heaters), electromagnetic (using the force of an external magnetic field acting upon a current passing along the structure) (Hagleitner et al. 2001), or even piezoelectric, using an added-on transduction material

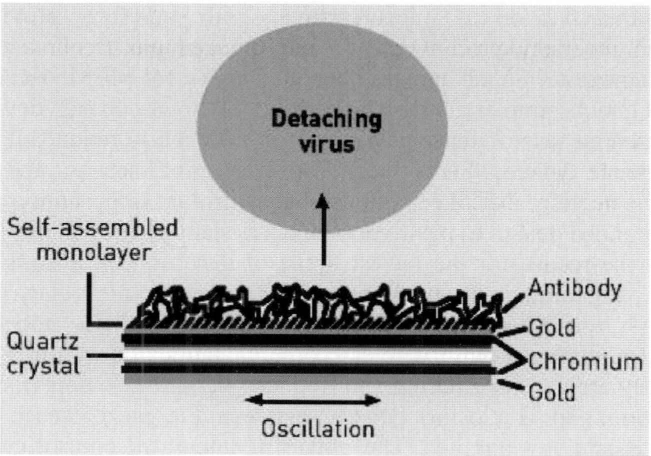

Fig. 6.7. A particle-type-specific piezoelectric biosensor developed at Cambridge University and being commercialized by Akubio, Inc., of Cambridge, UK. A quartz crystal resonator, coated with a receptor, binds an analytical target particle such as a virus. Transverse oscillation of the quartz, induced by applying an alternating voltage across the disc, is deliberately increased in amplitude to the point of bond breakage, releasing acoustic energy that is detected by an external circuit connected to the crystal electrodes. (Adapted from Cooper et al. 2001)

2. by bending, where a biomechanical transduction layer deposited on one side of the cantilever creates a mechanical bimorph that bends in response to binding of the target species

In the case of the vibrating structures, changes in the mass on the cantilever tip lead to changes in resonant frequency that are readily measured with integrated circuitry, typically resistive or capacitive. Smaller cantilever effective mass leads to better sensitivity and, noise and background issues being appropriately addressed, to improved limits of detection. The fact that GHz frequencies are now routinely achieved in microprocessors and other silicon microelectronics means that high frequencies no longer preclude complete integration of drive and measurement electronics.

For bimorph measurements, readout can be optical, using angular or interferometric changes of light reflected from the cantilever tip; or capacitive, if a plate-to-plate gap varies with bio-target-induced mechanical stress; or based upon integrated resistive measurement of variation in the strain over some portion of the cantilever. Though it is not piezoelectric, silicon is piezoresistive, and therefore provides the opportunity to include a convenient means of direct electrical readout.

The manufacturing advantages described above for MEMS device types makes it much simpler to fabricate arrays of mass sensing devices that include diverse sets of sensing materials in addition to redundant, control, and reference devices. Such integration of multiple sensors and controls has yet to be fully exploited, offering an important opportunity for chip level integrated design to positively impact system performance.

While the record for the lowest limits of detection on a mass-per-area basis is arguably still held by high-frequency piezoelectric SAW resonators, which are pushing from the hundreds of MHz into the GHz regime (in synchrony with wireless communications of various types, for which they are used as filtering and frequency-control elements), the limit-of-detection gap is closing quickly between the piezoelectric and the MEMS technology families. The power of integrated control electronics combined with sophisticated temperature-control strategies is beginning to be developed for integrated silicon mass-sensing systems; this area is ripe for further advances.

Progress has been made on another front that offers unique challenges to micro/nano mechanical biosensing devices; this front comprises the tasks of reproducibly depositing selective, fully viable biointerface materials onto one surface or onto the tip of a cantilever whose dimensions are measured in micrometers or nanometers. Advances in ink-jet, pin-based, and similar dispensing technologies are being driven by the needs of the burgeoning DNA microarray industry, as well as of the nascent field of protein microarrays. The needs of such spot-based biomaterial arrays have provided impetus for improved hardware as well as solution matrices specifically designed to place a micrometer or smaller "spot" of material in a precise location. Funding is significant for such technologies; in many cases, they should be directly applicable to the needs of micro/nanomechanical biosensors.

A major challenge facing micro/nanomechanical biosensors is their operation in liquids. In the case of bimorph-based transducers, to the usual consideration of protecting the electronics must be added the following:

- concerns related to the mechanical fragility of such devices (and associated consequences for surface tension effects in tiny gaps)
- difficulties of optical deflection measurements when multiple air/water interfaces must be traversed
- added thermal instabilities that follow from immersion in a medium with much greater thermal conductivity than air or vacuum.

In addition to these issues, vibrating structures face the additional, greater challenge that, like the SAW, their motion is an effective generator of compressional waves in liquids: this is an overly effective damping mechanism. Thus, while the so-called Q or quality factor (a measure of how "well-defined" and therefore how stable the mechanical resonance is), easily reaches into the thousands and beyond for vibrating microstructures in the gas phase or vacuum, it is only recently that reports have appeared of Q factors exceeding ten for cleverly designed combinations of mechanical structure and driving circuitry. Such Q values are sufficient for reasonably stable biosensor systems; nonetheless, this is clearly an opportunity for continuing engineering innovation.

SUMMARY FINDINGS: GENERAL TRENDS AND SPECIFIC OPPORTUNITIES

Key to Success: Multidisciplinary Teams

An important result of this WTEC study is the finding that strong multidisciplinary science-plus-engineering teams hold many of the keys to rapid advancement of biosensing systems. This is likely true to a greater extent for the MEMS approaches (including microfluidics and mass sensors) than for many of the discrete devices analyzed by this report, because of the critical interplay between mechanical, electrical, and chemical engineering, and the close involvement of the fundamental chemical, biological, materials, and information sciences. Researchers in many or, in some cases, all of these disciplines, together with industrial specialists having skills in areas ranging from device packaging to clinical assay development, must work together smoothly to implement each new bioMEMS approach to solving an important problem.

The WTEC panel observed that merging the methods and materials of molecular biology with MEMS and fluidics is

- immensely powerful
- in its infancy, with the exception of discrete sensors and simple fluidic devices
- an area where Europe currently leads in overall level of academic and national laboratory effort
- an area where Japanese industry is now focusing

Packaging Opportunities

The broad area represented by packaging and sample I/O (input/output), particularly fluidic interconnects, probably represents the largest ratio of payoff opportunity to current effort. Awareness of the criticality of this topic, along with availability of funding to address it, has improved substantially in the past ten years, but is still suboptimal. Three general challenges seem to hamper progress in this area:

1. It has been perceived to be a rather pedestrian area for research—or perhaps not even perceived as research at all, though it certainly is—without the glamour of a new class of materials or a novel device platform. Thankfully, this perception is changing, albeit slowly.
2. There is a degree of the "chicken-and-egg" problem: the best packaging and I/O approaches are likely to be dictated by the specifics of the application, which are often best dealt with by the commercial developer of a finished product, but the toolbox of fundamental approaches is not filled for general use unless university and national laboratory researchers work and share results in this area.
3. There have been attempts to minimize the size of the toolbox, e.g., to devise and codify a "universal standard" fluidic interconnect structure; what is needed, rather, are highly flexible strategies and approaches and, above all, successful demonstrations of clever solutions that will stimulate the thinking of those who must solve similar problems.

In Japan and Europe, the WTEC team observed several operations that seem to be on the cusp of a number of powerful solutions to complex integrated systems. For examples, see the site report for Dr. Kitamori's Lab at the University of Tokyo in Appendix C, and the site reports for ETH Zürich and the University of Neuchâtel in Appendix B. The United States has demonstrated success in the packaging of complex devices like the digital micromirror array (Digital Micromirror Device, DMD) developed by Texas Instruments and a range of medical diagnostic devices such as Biosite's Triage® panels and meters (www.biosite.com/products) or the I-STAT® portable clinical analyzers (www.i-stat.com/products).

The Promise of Nanotechnology

Nanotechnology clearly represents a broad area of intense interest and high-visibility effort. Generally, it can be divided into three general categories: (1) clever innovation and advances that have been made possible by the availability of new tools or new understanding of how to manipulate and exploit materials properties on the nanometer scale; (2) the renaming of advances, typically in the field of structured materials, that first occurred as little as five or as much as 50 years ago—though in some cases with updates and variations rendering them newly effective; and (3) building devices or structures that are smaller simply because it is now possible, without a convincing reason or rationale other than the availability of funds to do "nanostuff." Because this WTEC study sought out the most innovative researchers around the world in the various disciplines related to biosensing, no examples in the last category were found, but examples are numerous on the World Wide Web, and they have given the field a somewhat unsavory flavor for potential private investors.

From a technical perspective, nanotechnology can be viewed either as "top-down" or "bottom-up"; the former is the use of advanced fabrication tools to move from the domain of micrometer structures into the tens-to-hundreds of nanometers (Cleland and Roukes 2002); the latter is the use of molecular self-assembly and related organizational strategies to create structured materials from atoms and molecules (Scott et al. 2003). Notably, CMOS (the technology of modern integrated microelectronics) has already forged well into the top-down nanometer domain, with 130 nm feature size being a standard for mass-produced microprocessors, and thus it is accessible around the world to well-funded institutions having access to the expensive tools of microelectronics fabrication: electron-beam lithography, ion milling, synchrotron light sources, and so forth.

In the bottom-up arena, the United States seems to have a slim lead over the European Union as measured by the number times the impact of nanostructured materials innovations. New funding sources are now coming on line in Europe following the lead of the U.S. Nanotechnology Initiative, which will likely close this gap. Japan, too, has recognized the importance of innovations in nanostructured materials, and activities are supported in private industry as well as in academic and government laboratories. In all three regions, "mesoscale" thin-film materials, many having been rechristened "nanomaterials," are being exploited for the advantageous characteristics of molecular monolayers, such as well-defined binding sites; however, since these materials are tens to thousands of molecular diameters in thickness, they can simultaneously include controlled porosity to provide selective access to the internal binding sites.

The very young field of nanofluidics appears promising in selected applications, particularly in the manipulation and analysis of individual cells, as is being demonstrated at the University of Twente in the Netherlands (see site report in Appendix B and Figure 6.8). Given the size of most any biological organism larger than a virus, however, the characteristic size often creeps back into the micrometer range. For solution-phase analysis, a critical issue pointed out time and again in the comparatively mature field of microfluidics presents a much greater concern for nanoscaled devices: finding at least one molecule to analyze. Protein biomarkers of early-stage cancer, for example, must be detected at such low concentrations (1 femtomolar or less) that a full cubic micrometer of solution (10^{-15} L) has less than a one-in-one-million chance of containing even a single molecule. Thus, the requirement and the real opportunity for using nano strategies in fluidics is not so much to build nanoscale structures, but to devise powerful molecular capture strategies that enable one to find, selectively, a single molecule or two in many, many cubic micrometers of solution matrix. A synergistic approach to such capture, which also represents a significant opportunity, is to devise for proteins and other biological targets the sort of "amplification" (replication) that the polymerase chain reaction offers for nucleic acid (DNA, RNA) analysis: from a single molecule, 10^8 identical copies can be generated in less than half an hour in a portable, self-contained system (e.g., Cepheid's Smart Cycler® II System, www.cepheid.com/pages/products.html). In this direction, enzymes are being used in conjunction with electrochemical biosensors not to replicate the analyte molecule itself, but to provide a molecularly amplified signal that is 10^4–10^6 times easier to detect (Wang et al. 2001; Rossier and Girault 2001).

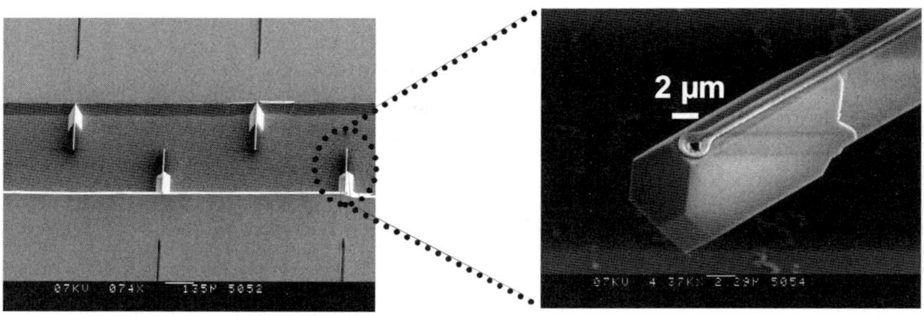

Fig. 6.8. *Left*: Scanning electron micrograph of a microfluidic channel containing a series of micromachined pipets (*close-up at right*). This is part of a system under development at the University of Twente for the manipulation and analysis of the contents of single living cells. (Courtesy Professor A. van den Berg, University of Twente, The Netherlands)

A final nanotechnology area of both challenge and promise, which demands careful, case-by-case justification, is the combination of top-down nanodevices with bottom-up nanostructured materials that are biochemically specific (i.e., "bionano-nano"). The potential impact here is in the development of appropriately selective devices that can detect very small numbers of molecules, specifically for those platforms, like nanocantilevers, where the limit of detection can be made to improve as the area or volume of the device shrinks. However, the caution raised above about uniting the transducer with the molecule to be detected is more critical than ever.

CONCLUSION: IMPORTANT TARGETS FOR BIOMEMS

A clear international trend in the selection of analytical targets for bioMEMS is the manipulation and measurement of single cells. A major driver for this is the discovery of new drugs: many of the most potent new pharmaceuticals are based on the selective manipulation of cellular ion channels, and it is clear that parallelized MEMS-enabled "patch-clamp" technology, wherein the conductivity through the ion channels of many isolated single cells are probed directly, is an important recent advance where commercial products are in the offing (see the Matsushita site report in Appendix C; Kiss et al. 2003; Xu et al. 2001). In addition, the concept of a "canary on a chip"—a microdevice-supported living organism that responds to the presence of a biologically hazardous substance in a manner detected by the same chip—is an approach valued particularly for national security applications in the United States and is a recognized pathway to selective biosensors of many sorts in Europe and Japan. See site reports for the University of Neuchâtel (Switzerland) Institute of Microtechnology in Appendix B; and the University of Tokyo (Japan) in Appendix C.

Going a step beyond single cells in complexity, the experiments that have been conducted for many years on making electrical interfaces to living tissue (Bell et al. 1998) are now being extended to supporting living tissue in thin-slice form on a host chip that allows monitoring of the response of the tissue to agonists or antagonists, and/or the use of such constructs for biochemical analysis (Kristensen Bjarne et al. 2001).

The surface has barely been scratched, however, on what is arguably the most important biosensing application with regard to understanding disease and developing new therapies: the analysis of biochemical pathways in their entirety. This will require simultaneous, multiplexed, *dynamic* analyses of gene expression; multiple proteins, including excreted, membrane, and intracellular species; cell functional characteristics such as the state and function of ion channels; and the products of metabolism. Putting together this "cellular big picture" will challenge every advance that researchers in biosensing and microfluidics can muster for years to come, but the potential payoff is enormous.

REFERENCES

Andersson, S.M.H., and A. van den Berg. 2003. Microfluidic devices for cellomics: A review. *Sensors and Actuators B (Chemical)* 92:315–325.

Arntz, Y., J.D. Seelig, H.P. Lang, J. Zhang, P. Hunziker, J.P. Ramseyer, E. Meyer, M. Hegner, and C. Gerber. 2003. Label-free protein assay based on a nanomechanical cantilever array. *Nanotechnology* (UK) 14 (1):86–90.

Baba, Y., S. Shoji, and A. van den Berg, eds. 2003. *Micro Total Analysis Systems, 2002: Proceedings of the μTAS 2002 symposium.* Boston, MA: Kluwer Academic Publishers.

Ballantine, D.S., R.M. White, S.J. Martin, A.J. Ricco, G.C. Frye, E.T. Zellers, and H. Wohltjen. 1997. *Acoustic wave sensors: Theory, design, and physico-chemical applications*. San Diego: Academic Press.

Bell, T.E., K.D. Wise, and D.J. Anderson. 1998. A flexible micromachined electrode array for a cochlear prosthesis. *Sensors and Actuators A* 66:63–69.

Boone, T.D., Z.H. Fan, I. Gibbons, A.J. Ricco, H. Tan, and S.J. Williams. 2002. Plastic advances microfluidic devices. *Anal. Chem.* 74:78A–86A.

Bousse, L., S. Mouradian, A. Minalla, H. Yee, K. Williams, and R. Dubrow. 2001. Protein sizing on a microchip. *Anal. Chem.* 73:1207–1212.

Burns, M.A., B.N. Johnson, and S.H. Brahmasandra. 1998. An integrated nanoliter DNA analysis device. Science 282: 484–487.

Cavicchi, R.E., J.S. Suehle, K.G. Kreider, M. Gaitan, and P. Chaparala. 1995. Fast temperature programmed sensing for micro-hotplate gas sensors. *IEEE Electron Device Letters* 16:286–288.

Cleland, A.N., and M.L. Roukes. 2002. Noise processes in nanomechanical resonators. J. *Appl. Phys.* 92:2758–2769.

Cooper, M.A., F.N. Dultsev, A. Minson, C. Abell, P. Ostanin, and D. Klenerman. 2001. Direct and sensitive detection of a human virus by rupture event scanning. *Nature Biotech.* 19:833–837.

Duffy, D.C., H. L. Gillis, J. Lin, N.F. Sheppard, Jr., and G.J. Kellogg. 1999. Microfabricated centrifugal microfluidic systems: characterization and multiple enzymatic assays. *Anal. Chem.* 71:4669–4678.

Figeys, D., and D. Pinto. 2000. Lab-on-a-chip: A revolution in biological and medical sciences. *Anal. Chem.* 72:330A-335A.

Fu, A.Y., H.P. Chou, C. Spence, F.H. Arnold, and S.R. Quake. 2002. An integrated microfabricated cell sorter. Anal. Chem. 74:2451–2457.

Gizeli E., and C. R. Lowe, eds. 2002. *Biomolecular Sensors*. London: Taylor and Frances.

Gottschlich, N., S.C. Jacobson, C.T. Culbertson, and J.M. Ramsey. 2001. Two-dimensional electrochromatography/capillary electrophoresis on a microchip. *Anal. Chem.* 73:2669–2674.

Gustafsson, M., D. Hirschberg, C. Palmberg, H. Jörnvall, and T. Bergman. 2004. Integrated sample preparation and MALDI mass spectrometry on a microfluidic compact disk. *Analytical Chemistry* 76:253–502.

Hadd, A.G., S.C. Jacobson, and J.M. Ramsey. 1999. Microfluidic assays of acetylcholinesterase inhibitors. *Anal. Chem.* 71:5206–5212.

Hagleitner, C., A. Hierlemann, D. Lange, A. Kummer, N. Kerness, O. Brand, and H. Baltes. 2001. Smart single-chip gas sensor microsystem. *Nature* 414:293–296.

Handley, J. 2001. Quartz crystal microbalances. *Anal. Chem.* 73:225A-229A.

Howe, R.T., M. Allen, A. Berlin, E. Hui, D. Monk, K. Najafi, and M. Yamakawa. 2003. *WTEC Panel Report on Microsystems Research in Japan.* Baltimore, MD: World Technology Evaluation Center, Inc. Available from National Technical Information Service (Springfield, VA) and online at www.wtec.org/mems1/.

Khandurina, J., T.E. McKnight, S.C. Jacobson, L.C. Waters, R.S. Foote, and J. M. Ramsey. 2000. Integrated system for rapid PCR-based DNA analysis in microfluidic devices. *Anal. Chem.* 72:2995–3000.

Kiss, L., P.B. Bennett, V/N. Uebele, K.S. Koblan, S.A. Kane, B. Neagle, and K. Schroeder. 2003. High throughput ion-channel pharmacology: planar-array-based voltage clamp. *ASSAY and Drug Development Technologies* 1:127–135.

Koh, C.G., W. Tan, M. Zhao, A.J. Ricco, and Z.H. Fan. 2003. Integrating PCR, valving, and electrophoresis in a plastic device for bacterial detection. *Anal. Chem.* 75:4591–4598.

Kristensen Bjarne, B.W., J. Noraberg, P. Thiébaud, M. Koudelka-Hep, and J. Zimmer. 2001. Biocompatibility of Si-based arrays of electrodes coupled to organotypic hippocampal brain slice cultures. *Brain Research* 896 (1):1–17.

Lagally, E.T., C.A. Emrich, and R.A. Mathies. 2001. Fully integrated PCR-capillary electrophoresis microsystem for DNA analysis. *Lab-on-a-Chip* 1:102–107.

Li, J., T. LeRiche, T.L. Tremblay, C. Wang, E. Bonneil, D.J. Harrison, and P. Thibault. 2002. Application of microfluidic devices to proteomics research. *Molecular & Cellular Proteomics* 1.2:157–168.

Liu, T., J. Tang, M. Han, and L. Jiang, 2003. A novel microgravimetric DNA sensor with high sensitivity. *Biochem Biophys Res Commun.* 304:98–100.

Locascio, L.E., J. Hong, and M. Gaitan. 2002. Liposomes as signal amplification reagents for assays in microfluid channels. *Electrophoresis* 23 (5):799–804.

Madou, M.J., L.J. Lee, S. Daunert, S. Lai, and C.H. Shih, 2001. Design and fabrication of CD-like microfluidic platforms for diagnostics: Microfluidic functions. *Biomedical Microdevices* 3 (3):245–254.

Manz, A., D.J. Harrison, E.M.J. Verpoorte, J.C. Fettinger, A. Paulus, H. Lüdi, and H.M. Widmer. 1992. Planar chips technology for miniaturization and integration of separation techniques into monitoring systems: Capillary electrophoresis on a chip. *J. Chromatogr.* 593:253–258.

Ming, S., S. Li, and V.P. Dravid. 2003. Microcantilever resonance-based DNA detection with nanoparticle probes. *Appl. Phys. Lett.* (USA) 82 (20):3562–4.

Möller, M., J.P. Spatz, M. Moessmer, P. Eibeck, P. Ziemann, and B. Kabius. 1999. Formation of chemical nanopattern by means of block copolymers. *Polymeric Materials Science and Engineering* 80:3.

Ng, H.T., J. Li, M.K. Smith, P. Nguyen, A. Cassell, J. Han, and M. Meyyappan. 2003. Growth of epitaxial nanowires at the junctions of nanowalls. *Science* 300:1249.

Northrup, M.A., K.V. Jensen, and D.J. Harrison, eds. 2003. *Proceedings of μTAS 2003 (Seventh International Conference on Micro Total Analysis Systems)*. Cleveland, OH: Transducers Research Foundation.

Oosterbroek, R.E., and A. van den Berg, eds. 2003. *Lab-on-a-chip: Miniaturized systems for (bio)chemical analysis and synthesis* and references therein. Elsevier: Amsterdam.

Petricoin, E.F., III, D.K. Ornstein, C.P. Paweletz, A. Ardekani, P.S. Hackett, B.A. Hitt, A. Velassco, C. Trucco, L. Wiegand, K. Wood, C.B. Simone, P.J. Levine, W. M. Linehan, M.R. Emmert-Buck, S.M. Steinberg, E.C. Kohn, and L.A. Liotta. 2002. Serum proteomic patterns for detection of prostate cancer. *J. Natl. Cancer. Inst.* 94:1576–1578.

Ricco, A.J., R.M. Crooks, and G.C. Osbourn. 1998. SAW chemical sensor arrays: New chemically sensitive interfaces combined with novel cluster analysis to detect volatile organic compounds and mixtures. *Accts. Chem. Res.* 31:289.

Rossier, J.S., and H.H. Girault. 2001. Enzyme linked immunosorbent assay on a microchip with electrochemical detection. *Lab on a Chip* 1:153–157.

Sanders, G.H.W., and A. Manz. 2000. Chip-based microsystems for genomic and proteomic analysis. *Trends Anal. Chem.* 19:364–378.

Sato, K., H. Kawanishi, M. Tokeshi, T. Kitamori, and T. Sawada. 1999. Sub-zeptmole molecule detection in a microfabricated glass channel by thermal lens microscope. *Anal. Sci.* 15:525–529.

Sato, K., M. Yamanaka, H. Takahashi, M. Tokeshi, H. Kimura, and T. Kitamori. 2002. *Electrophoresis* 23:734–739.

Scott R.W.J., A.K. Datye, and R.M. Crooks. 2003. Bimetallic palladium-platinum dendrimer-encapsulated catalysts. *J. Am. Chem. Soc.* 125:3708–3709.

Smith, J., and S. Senturia. 1995. Self-consistent temperature compensation for resonant sensors with application to quartz bulk acoustic wave chemical sensors. *Proc. Transducers* 95 (2): 724–727, Stockholm: Foundation for Sensor and Actuator Technology.

Soper, S.A., S.M. Ford, S. Qi, R.L. McCarley, K. Kelly, and M.C. Murphy. 2000. *Anal. Chem.* 72:642A-651A.

Subramanian, A., P.I. Oden, S.J. Kennel, K.B. Jacobson, R.J. Warmack, T. Thundat, and M.J. Doktycz. 2002. Glucose biosensing using an enzyme-coated microcantilever. *Appl. Phys. Lett.* (USA) 81(2):385–7.

Tamayo, J., M. Alvarez, and L.M. Lechuga. 2003. Digital tuning of the quality factor of micromechanical resonant biological detectors. *Sens. Actuators B, Chem.* (Switzerland) B89 (1–2):33–9.

Tang, T., M.Y. Badal, G. Ocvirk, W.E. Lee, D.E. Bader, F. Bekkaoui, and D.J. Harrison. 2002. Integrated microfluidic electrophoresis system for analysis of genetic materials using signal amplification methods. *Anal. Chem.* 74:725–733.

Walther, I., B.H. van der Schoot, S. Jeanneret, Ph. Arquint, N.F. de Rooij, V. Gass, B. Bechler, G. Lorenzi, and A. Cogoli. 1994. Development of a miniature bioreactor for continuous culture in a space laboratory, *J. Biotechnology* 38:21–32.

Wang, J., A. Iba´n˜ez, M.P. Chatrathi, and A. Escarpa. 2001. Electrochemical enzyme immunoassays on microchip platforms. Anal. Chem. 73:5323–5327.

Woolley, A.T., and R.A. Mathies. 1994. *Proc. Natl. Acad. Sci. USA* 91:11348–11352.

Wu, G., R.H. Datar, K.M. Hansen, T. Thundat, R.J. Cote, and A. Majumdar. 2001. Bioassay of prostate-specific antigen (PSA) using microcantilevers. Nat. Biotechnol. 19:856.

Xu, J., X.B. Wang, E. Brooks, M. Li, L. Wu, A. Guia, and J.Q. Xu. 2001. Ion-channel assay technologies: Quo vadis? *Drug Discovery Today* 6 (24):1278–1287.

Yang, Y., H.F. Ji, and T. Thundat. 2003. Nerve agents detection using a Cu2+/L-cysteine bilayer-coated microcantilever. *J. Am. Chem. Soc.* 125:1124–5.

Zubritsky, E. 2000. E-noses keep an eye on the future. *Anal. Chem.* 72:421A-426 A.

CHAPTER 7

Information Systems for Biosensing

David J. Brady

INFORMATION SYSTEM CHALLENGES IN BIOSENSING

Biosensing integrates biochemistry, physical electronics, and information systems. The role of biochemistry is illustrated in many examples of the design or discovery of molecular recognition elements. The utility of physical electronics is illustrated in optoelectronic sensors and "lab-on-a-chip" projects. Careful information system design for biosensing systems is less developed, and clear examples are harder to find. Sophisticated algorithms and acquisition schemes have emerged from efforts to construct "artificial noses" (Gardner and Bartlett 1999; Snopok and Kruglenko 2002), but general methodologies for analysis of information processing and communication in biosensor systems are only beginning to emerge.

The challenge of developing a general methodology for information in biosensing systems derives from the subtlety and complexity of biological information. Biological information is the subtlest class of real-world data because wildly different biological systems may have the same gross physical properties and even basically the same chemical composition. Biological information may be geometric (as in protein function), contextual (as in epidemiology), and functional (as in microbial identity). Despite this complexity, most sensor technology measures gross physical features, such as temperature, time or mass, optical intensities, and spectra. (The obvious exceptions are cellular sensors, which may provide biological amplifiers.)

Biosensing is an interface between a biological state and a digital representation of that state. One may consider biosensing a process of shifting complex biological information back along the sensor hierarchy outlined above by translating structural and chemical complexity into digital numbers. The bio-digital interface consists of subinterfaces at

- transduction of the biological state into physical form via optical, mechanical, or electrical signals
- transduction of the physical state into electronic form
- analog to digital electrical conversion
- communication of digital signals
- digital signal processing for state estimation

This cascade of interfaces is in some cases considered as a static operator on a time-free state and in others is considered as a dynamic (and in some cases, bi-directional) communication channel from the biological to the digital world.

Information systems enter in biosensor design at every level, ranging from algorithms for rational design of molecular receptors (Looger et al. 2003; Looger and Hellinga 2001; Arnold 2001) through the decision theory for responses to sensor data. Conventionally, information challenges are considered discretely as they arise. In recent years, however, a structured methodology for integrated design of sensing and data processing has begun to develop (NSF 2002; DARPA 2001; OSA 2001; NIAAA 2002). Information system designers have long understood the dramatic advantages of wise algorithm design over naïve algorithms. Integrated sensing and processing (ISP) is critically enabling because it extends algorithm design into the transduction and sampling layers.

A methodology for biosensing information systems includes the following:

1. **Source and sensor state specification.** The source state may be a spatial and temporal distribution of chemical or biological species, or it may be statistical distribution or a morphological state, etc. The source state is typically transduced by molecular or cellular recognition or by spatial imaging. A precise model of the transduction process is necessary to develop information strategies.
2. **Sampling strategies**. Sampling consists of choosing locations in space and time for measurements and in choosing the physical form into which molecular recognition events are transduced.
3. **Inverse problem specification**. This provides a detailed description of the nature of input and output data. The input data consists of a source state and a model for how the source state is transformed into measured data. The output data may be a spatio-temporal description of the source state, as in the spatial density of a target toxin, or a localized or space-free description of the source state, as in the

presence of a toxin in a sample. In biosensing, problem specification includes choices of what measures to take. Considerable success has been achieved in developing sensors for specific pathogens or chemicals. Many sensor applications, however, require multifunctional sensitivity to a variety of targets. Development of data and computationally efficient systems capable of discriminating and analyzing multiple species requires coding and decoding of multiple information streams. A particular challenge involves matching the range of phenomena that can occur in low-level cellular or molecular recognition to the range of physical markers and the range of digital transducers.

4. **Algorithm specification**. Given a model for sensor data and a set of possible output states, algorithm specification transforms the sensor data into an output state. In many cases, one may know that it is possible to discriminate a set of output states from sensor data without necessarily knowing an inversion algorithm to perform this task. Even where one or more inversion algorithms are known, one may wish to compare the computational efficiency or estimation fidelity of different algorithms.
5. **Communications and display**. Sensors must be considered in the context of their utility. Sensor networks are used to improve the area coverage and robustness of sensing systems. Algorithms at the sensor node and network levels may be jointly optimized for improved performance, and human interfaces may be examined for high utility.
6. **Logistics**. Sensor system design includes deployment and operations. Information system analysis considers how reagents and probes may be deployed and replenished in an operating biosensor system.

A complete information theory of biosensors is beyond the level and extent of this report. In terms of information systems as applied to biosensing, the WTEC study has focused on identifying

- leading examples of ISP design in the United States, Europe, and Japan
- the current state of the art in the three regions
- potential shortcomings in current efforts
- opportunities for extending and integrating current work in emerging systems

This chapter also addresses how each example approaches the information system components described above. Generally, current systems consider only one or two of these components in detail; for those, the chapter briefly addresses how each example might be extended through more complete system integration.

BIOSENSING INFORMATION SYSTEMS IN THE UNITED STATES

While the United States has a long history of early leadership in integration of digital and physical systems, the main challenge in the coming decade will be to translate this leadership into deployed sensor systems. As a step in this direction, U.S. leadership is apparent in three important aspects of digital sensor systems development: geometric sensor systems, embedded sensor networks, and epidemiological systems.

Geometric Sensor Systems

While most biosensors target a single molecule or species, the most interesting systems from an information sciences perspective are more complex. Interesting examples include systems with multiple or unknown targets, such as electronic noses; systems with complex targets, such as cellular or DNA fragment recognition; and systems that target abstract states, such as functional and epidemiological studies.

Geometric sensors are an emerging class based on the creation of a vector space to describe the sensor state. The most established example from a biosensing perspective is DNA microarray technology (Schena et al. 1995; Chee et al. 1996; Lockart and Winzeler 2000). Microarrays have become extremely important for sequence identification and statistical analysis, too. Arrays provide a mechanism for mapping biological information on a space-time distribution; the spatial mapping enables data to be systematically recorded and analyzed. Following on the success of oligonucleotide arrays, microarray technology has recently developed in many directions. Protein chips are the leading example (Chen et al. 2003; MacBeath 2002; Mitchell 2002), but other examples abound. Many electronic nose systems have been developed based on array technology (Gardner and Bartlett 1999; Snopok and Kruglenko 2002, Dickinson et al. 1996; Drew et al. 2001; Drew, Janzen, and Mann 2002; Epstein, Stitzel, and Walt 2002; Karunamuni et al. 2001), and cellular arrays have recently begun to emerge (Biran et al. 2003; Biran and Walt 2002).

Microarrays can produce enormous quantities of data. Interpretation, logging, and databasing this data produce fascinating information science challenges. Even in the case of DNA arrays, methodologies and standards to address these challenges are undeveloped (Chee et al. 1996; Hariharan 2003; Irizarry et al. 2003; He et al. 2003; Valafar 2002; Butte 2002; Pan 2002; Brazma et al. 2001; Li and Wong 2001; Brown et al. 2000; Bassett, Eisen, and Boguski 1999). The term "geometric sensors" is used here to describe array technologies, because one can view these sensors as mechanisms for transforming complex data into a geometric space. Common methods for analyzing the data involve discovery of basis vectors to span the resulting hyperspace (Holter et al. 2000; Alter, Brown and Botstein 2000) or development of clustering and segmentation algorithms (Brown et al. 2000; Alon et al. 1999; Eisen et al. 1998).

Array technology translates abstract data into a physical sampling space. Analysis of this data typically involves creation of a reduced dimensional clustering space (Priebe 2001). At each level, the nature of sensing challenge translates abstract data onto a geometry. Unfortunately, the dimension of the sensing space is usually absurdly large. Over the past several years, mathematicians have developed profoundly enabling techniques for abstraction of exact or approximate low-dimensional embedding spaces for high-dimensional data (Achlioptas 2003; Hjaltason and Samet 2003; Cowen and Priebe 1997; Bourgain 1985). More recently still, mathematical methods for structure abstraction from the computer vision community have begun to be applied on unconventional spaces, although literature on this new approach is sparse. These approaches are just beginning to translate to biosensor array technology, but the match between the nature of the math and the complexity of the biosensing problem seems profound.

As a final comment on geometric sensors, consider how sensor design relates to the six components of the information systems methodology discussed in the introduction. Geometric sensors present interesting design and analysis options at every level, but most interestingly, they illuminate the significance of source and sensor state specification and data inversion. Arrays transform abstract source states, such as the presence, prevalence, or functional context of a target molecule or species, into a spatio-temporal sensor pattern. Many design choices are made in the nature and sampling strategy of the array. One must be sure that the entropy of the sensor state is sufficient to represent the range of potential source states. For electronic noses or other multiplex sensors, this means ensuring that the basis states of the potential responses span the range of potential chemical species. For DNA microarrays, this means ensuring sufficient variety in probes. Sampling strategies refers to how samples are prepared and presented to the microarray. Inverse problem specification refers to a precise description of the goal of analysis. Note that the inverse problem may be different from the source state specification. One often seeks, for example, to classify sources with less discrimination than the range of the underlying source state. Once one has determined the problem to be solved—for example, clustering the types of molecules present in terms of their functional response—one must develop an efficient algorithm based on the sensor state model to achieve this classification. Communication and display of the data may affect the definition of the inverse problem and the algorithm, especially if human analysis is intended. One seeks to match the source state description obtained from the inverse problem to a human-segmentable data structure.

Geometric sensing represents a profoundly new approach to sensor design arising from the capacity for abstract source and data models in digital systems. The leading communities for array technology especially for abstract data analysis are in the United States, although strong groups and leadership are also present in Europe. Geometric sensing is a particular example of the "integrated sensing and processing" paradigm mentioned in the chapter introduction.

Embedded Sensor Networks

Ubiquitous embedded digital biosensing systems have emerged as a central vision of future information systems. These systems continue a half-century trend of digital processor migration from central facilities to personal devices. The number of embedded microprocessors per individual in developed countries is in the 10-100 range. By the end of this decade, thousands of microprocessors will be deployed per person. Most of these microprocessors will drive sensor systems.

Embedded sensor networks are among the strongest areas of research in the academic computer science and computer engineering communities in the United States. After years of strong support by the Defense Advanced Research Projects Agency (DARPA), the National Science Foundation (NSF) has completed a second proposal round for integrated sensing systems. Literature on networking and platforms for distributed sensing indicates strong programs (Warneke, Atwood, and Pister 2001; Wang and Chandrakasan 2001; Wang and Jones 2001; Tsiatsis, Zimbeck, and Srivastava 2001; Sohrabi et al. 2000; Slijepcevic and Potkonjak 200; Schwiebert, Gupta, and Weinmann 2001; Qi, Iyengar, and Chakrabarty 2001; Pottie and Kaiser 2000; Nagel 2001). Nevertheless, these efforts have not been integrated with emerging computational sensor programs generally or with biosensing in particular.

Epidemiological Systems

Data abstraction from geometric sensors presents challenges for source and sensor state definition and algorithms, but at least one usually has a good model for data acquisition on these systems. One can view data analysis from well-characterized sensors deterministically. One would also like the capacity to analyze data from heterogeneous sensors in unexpected ways. The most common problem in biosensing with this characteristic is epidemiology, where it is necessary to identify an epidemic and both forward- and backward-track its origins and impact. DARPA's ENCOMPASS program (Williams et al. 2002; Graser et al. 2002) is an example of current research showing that it is possible to use information systems to opportunistically combine heterogeneous data. Intelligent design of human-in-the-loop systems at the heterogeneous analysis level may reduce data reduction requirements at the sensor level. While the capacity and impact of these systems is still unknown and sometimes controversial (especially with regard to privacy issues), the U.S. research community has shown substantial leadership in this area.

BIOSENSING INFORMATION SYSTEMS IN EUROPE

In addition to the excellence of the individual research observed on the WTEC biosensing panel's visits to laboratories in Europe, the level of coordination and cooperation across the community was extremely impressive and important. As a result of this coordination, European biosensing programs have tended to involve more fully developed systems than comparable U.S. projects. See site reports of the WTEC panel's visits to laboratories in Europe in Appendix B. Two examples are presented here.

Lab on a Chip

The "Laboratory-on-a-Chip" concept has broad appeal and has reached a well-developed level. Britain's Royal Society of Chemistry publishes a journal wholly dedicated to the concept, and considerable work in the United States, Europe, and Japan focuses on realizing the vision. From an information science perspective, laboratory-on-a-chip systems include embedded data processing and communications and digital interfaces. Sensors developed in the Physical Electronics Laboratory (PEL) at ETH are among the most complete versions of a molecule-to-bits lab on a chip (Hagleitner et al. 2001). PEL's work is notable for the full digital interface to chemical sensor chips. An image of a PEL sensor chip is shown in Figure 7.1.

The Laboratory of Biosensors at the University of Twente also illustrates the capacity for complete systems. Twente's MESA+ Institute (www.mesaplus.utwente.nl) provides a capacity for complete lab-on-a-chip system development. Indicating an emerging worldwide focus on cellular systems and biosensors, both MESA+ and PEL indicate strong current interest in extending chemical systems success to biosensors. The MESA+ biosensors lab has developed an extensive visualization of "lab-in-a-cell systems" based on nanorobotics and sophisticated probes. Robotic control and data analysis for such systems will present significant information science challenges.

Integrated lab-on-a-chip projects face similar information challenges to those outlined in the geometric sensors discussion for microarrays and artificial noses. The primary information challenges are specification of the nature of the target source information, specification of internal sensor models for representing sensed data, and algorithms for source estimation. These challenges become particularly fascinating as integrated analog/digital systems for analysis of complex biological systems, such as the cell, are developed.

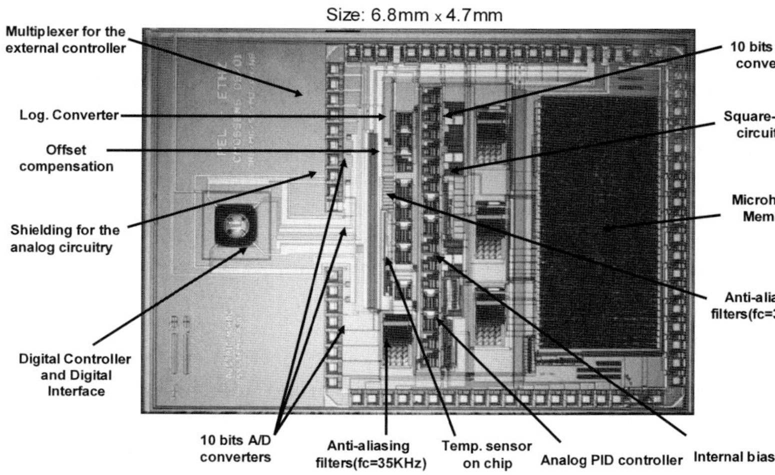

Fig. 7.1. Baltes group multifunctional chemical sensor on a chip. (Courtesy ETH, Physical Electronics Lab)

Deployment Logistics

The vertical integration of European research initiatives is particularly clear in programs at Linköping University. Biosensing initiatives at Linköping included several groups within the university, regional, and national healthcare institutions, and industry. Some of Linköping's programs focus on home healthcare. As an example of the cross-disciplinary flavor, computer scientists are collaborating with sensor array developers to imagine home-deployable biosensors. Filippini and Lundstrom (Filippini, Svensson, and Lundstrom 2003; Filippini, Svensson, and Lundstrom 2002) have demonstrated the use of a computer screen as a programmable spectral light source for bio-assays. While one may question both the capability of a RGB source for such applications and whether or not the illumination source is the most expensive component in such systems, the willingness of the Linköping group to think expansively about biosensors and their deployment in homes is impressive.

BIOSENSING INFORMATION SYSTEMS IN JAPAN

Japan has a history of large-scale home healthcare biosensor projects, as indicated, for example, by both commercial and R&D efforts at Matsushita Electric Company. It has also had a strong national commitment to sensors for counterterrorism. With the exception of the ambitious "Bionics" program under development at the Tokyo University of Technology, however, integrated sensing and information processing does not seem to fit naturally in to Japan's current academic culture. The strongest biosensing programs from an information science perspective are located in national laboratories. Bioelectronics is a research initiative at the National Institute of Advanced Industrial Science and Technology (AIST), with a primary focus on lab-on-a-chip development. See Appendix C for site reports of the WTEC panel's visits to laboratories in Japan.

Spatio-Temporal Dynamics

The Frontier Research Project on Spatio-Temporal Materials Function at RIKEN is one the most theoretically expansive integrated system analysis efforts observed by the WTEC panel. The Local Spatio-Temporal Functions Laboratory is attempting to scale from single molecule dynamics on surfaces up to the information science of complex systems. RIKEN has a history of imaginative combinations of biology, physics, and information, as illustrated most famously in the use of slime mold to solve mazes (Nakagaki, Yamada, and Toth 2000). These projects are not currently as closely integrated with mathematical and information theoretic studies as the integrated sensing and processing efforts in the United States, however.

OPPORTUNITIES

WTEC's survey of international research and development in the area of biosensing illuminates various perspectives, but the nexus of frontier research consists of discovery of molecular or physical recognition and transduction elements and system integration (as in lab on a chip or lab on a cell). System integration may consist of parallel testing, as in geometric sensors, or multifunctional components, as in the ETH and Twente systems. Analysis of system integration for biosensing is the heart of the integrated sensing and processing programs described in the chapter introduction.

The most fascinating opportunities for future research in biosensing information systems lie in the success of mathematical methodologies for sensor system generalization. Dimension reduction techniques for microarrays have successfully demonstrated clustering. Systemization of these techniques, development of standards, and integration with spatio-temporal analyses all promise to yield substantial improvements in the specificity and sensitivity of biosensors. Microarray and lab-on-a-chip development have successfully transformed abstract biological source states into "images." Dimension reduction, pattern recognition, and related projects are the digital image processing techniques for these biosensor impressions. Joint design of image transduction and image processing and display are the core of integrated sensing and processing. Initiatives centered on these goals are underway in the United States at NSF, DARPA, and the National Institutes of Health (NIH). The dramatic research community response to these initiatives (over 900 proposals in this 2003's NSF Integrated Sensing program, for example), suggests that substantial expansion may be needed to realize the full opportunity.

REFERENCES

Achlioptas, D. 2003. Database-friendly random projections: Johnson-Lindenstrauss with binary coins. *Journal of Computer and System Sciences* 66 (4):671–687.

Alon, U., N. Barkai, D.A. Notterman, K. Gish, S. Ybarra, D. Mack, and A.J. Levine. 1999. Broad patterns of gene expression revealed by clustering analysis of tumor and normal colon tissues probed by oligonucleotide arrays. *Proceedings of the National Academy of Sciences of the United States of America* 96 (12):6745–6750.

Alter, O., P.O. Brown, and D. Botstein. 2000. Singular value decomposition for genome-wide expression data processing and modeling. *Proceedings of the National Academy of Sciences of the United States of America* 97 (18):10101–10106.

Arnold, F.H. 2001. Combinatorial and computational challenges for biocatalyst design. *Nature* 409 (6817):253–257.

Bassett, D.E., M.B. Eisen, and M.S. Boguski. 1999. Gene expression informatics - it's all in your mine. *Nature Genetics* 21:51–55.

Biran, I., and D.R. Walt. 2002. Optical Imaging fiber-based single live cell arrays: A high-density cell assay platform. *Analytical Chemistry* 74 (13):3046–3054.

Biran, I., D.M. Rissin, E.Z. Ron, and D.R. Walt. 2003. Optical imaging fiber-based live bacterial cell array biosensor. *Analytical Biochemistry* 315 (1):106–113.

Bourgain, J. 1985. On Lipschitz Embedding of Finite Metric-Spaces in Hilbert-Space. *Israel Journal of Mathematics* 52 (1–2):46–52.

Brazma, A., P. Hingamp, J. Quackenbush, G. Sherlock, P. Spellman, C. Stoeckert, J. Aach, W. Ansorge, C.A. Ball, H.C. Causton, T. Gaasterland, P. Glenisson, F.C.P. Holstege, I.F. Kim, V. Markowitz, J.C. Matese, H. Parkinson, A. Robinson, U. Sarkans, S. Schulze-Kremer, J. Stewart, R. Taylor, J. Vilo, and M. Vingron. 2001. Minimum information about a microarray experiment (MIAME) - toward standards for microarray data. *Nature Genetics* 29 (4):365–371.

Brown, M.P.S., W.N. Grundy, D. Lin, N. Cristianini, C.W. Sugnet, T.S. Furey, M. Ares, and D. Haussler. 2000. Knowledge-based analysis of microarray gene expression data by using support vector machines. *Proceedings of the National Academy of Sciences of the United States of America* 97 (1):262–267.

Butte, A. 2002. The use and analysis of microarray data. *Nature Reviews, Drug Discovery* 1 (12):951–960.

Chee, M., R. Yang, E. Hubbell, A. Berno, X.C. Huang, D. Stern, J. Winkler, D.J. Lockhart, M.S. Morris, and S.P.A. Fodor. 1996. Accessing genetic information with high-density DNA arrays. *Science* 274 (5287):610–614.

Chen, G.Y.J., M. Uttamchandani, R.Y.P. Lue, M.L. Lesaicherre, and S.Q. Yao. 2003. Array-based technologies and their applications in proteomics. *Current Topics in Medicinal Chemistry* 3 (6):705–724.

Cowen, L.J., and C.E. Priebe. 1997. Randomized nonlinear projections uncover high-dimensional structure. *Advances in Applied Mathematics* 19 (3):319–331.

DARPA (Defense Advanced Research Projects Agency). 2001. Integrated Sensing and Processing. In DARPA DSO BAA01–30. Defense Advanced Research Projects Agency: *Commerce Business Daily*. www.darpa.mil/dso/solicitations/01/Baa01-30/s/index.htm .

Dickinson, T.A., J. White, J.S. Kauer, and D.R. Walt. 1996. A chemical-detecting system based on a cross-reactive optical sensor array. *Nature* 382 (6593):697–700.

Drew, S.M., D.E. Janzen, and K.R. Mann. 2002. Characterization of a cross-reactive electronic nose with vapoluminescent array elements. *Analytical Chemistry* 74 (11):2547–2555.

Drew, S.M., D.E. Janzen, C.E. Buss, D.I. MacEwan, K.M. Dublin, and K.R. Mann. 2001. An electronic nose transducer array of vapoluminescent platinum(II) double salts. *Journal of the American Chemical Society* 123 (34):8414–8415.

Eisen, M.B., P.T. Spellman, P.O. Brown, and D. Botstein. 1998. Cluster analysis and display of genome-wide expression patterns. *Proceedings of the National Academy of Sciences of the United States of America* 95 (25):14863–14868.

Epstein, J.R., S.E. Stitzel, and D.R. Walt. 2002. Randomly-ordered high-density fiber optic microsensor array sensors. *Microfabricated Sensors* 129–148.

Filippini, D., and I. Lundstrom. 2002. Chemical imaging by a computer screen aided scanning light pulse technique. *Applied Physics Letters* 81 (20):3891–3893.

Filippini, D., S.P.S. Svensson, and I. Lundstrom. 2003. Computer screen as a programmable light source for visible absorption characterization of (bio)chemical assays. *Chemical Communications* (2):240–241.

Gardner, J.W., and P.N. Bartlett. 1999. *Electronic noses: Principles and applications*. Oxford; New York: Oxford University Press.

Graser, T., K.S. Barber, B. Williams, F. Saghir, and K.A. Henry. 2002. Advanced consequence management program: Challenges and recent real world implementations. In *Sensors, and Command, Control, Communications, and Intelligence (C31) Technologies for Homeland Defense and Law Enforcement*, Apr 1–5 2002. Orlando, FL. The International Society for Optical Engineering.

Hagleitner, C., A. Hierlemann, D. Lange, A. Kummer, N. Kerness, O. Brand, and H. Baltes. 2001. Smart single-chip gas sensor microsystem. *Nature* 414 (6861):293–296.

Hariharan, R. 2003. The analysis of microarray data. *Pharmacogenomics* 4 (4):477–497.

He, Y.D.D., H.Y. Dai, E.E. Schadt, G. Cavet, S.W. Edwards, S.B. Stepaniants, S. Duenwald, R. Kleinhanz, A.R. Jones, D.D. Shoemaker, and R.B. Stoughton. 2003. Microarray standard data set and figures of merit for comparing data processing methods and experiment designs. *Bioinformatics* 19 (8):956–965.

Hjaltason, G.R., and H. Samet. 2003. Properties of embedding methods for similarity searching in metric spaces. *IEEE Transactions on Pattern Analysis and Machine Intelligence* 25 (5):530–549.

Holter, N.S., M. Mitra, A. Maritan, M. Cieplak, J.R. Banavar, and N.V. Fedoroff. 2000. Fundamental patterns underlying gene expression profiles: Simplicity from complexity. *Proceedings of the National Academy of Sciences of the United States of America* 97 (15):8409–8414.

Irizarry, R.A., B. Hobbs, F. Collin, Y.D. Beazer-Barclay, K.J. Antonellis, U. Scherf, and T.P. Speed. 2003. Exploration, normalization, and summaries of high density oligonucleotide array probe level data. *Biostatistics* 4 (2):249–264.

Karunamuni, J., K.E. Stitzer, D. Eastwood, K.J. Albert, D.R. Walt, S.B. Brown, and M.L. Myrick. 2001. Interference filter refinement for artificial nose fluorescence sensing. *Optical Engineering* 40 (6):888–895.

Li, C., and W.H. Wong. 2001. Model-based analysis of oligonucleotide arrays: Expression index computation and outlier detection. *Proceedings of the National Academy of Sciences of the United States of America* 98 (1):31–36.

Lockhart, D.J., and E.A. Winzeler. 2000. Genomics, gene expression and DNA arrays. *Nature* 405 (6788):827–836.

Looger, L.L., and H.W. Hellinga. 2001. Generalized dead-end elimination algorithms make large-scale protein side-chain structure prediction tractable: Implications for protein design and structural genomics. *Journal of Molecular Biology* 307 (1):429–445.

Looger, L.L., M.A. Dwyer, J.J. Smith, and H.W. Hellinga. 2003. Computational design of receptor and sensor proteins with novel functions. *Nature* 423 (6936):185–190.

MacBeath, G. 2002. Protein microarrays and proteomics. *Nature Genetics* 32:526–532.

Mitchell, P. 2002. A perspective on protein microarrays. *Nature Biotechnology* 20 (3):225–229.

Nagel, D.J. 2001. Wireless smart sensor networks. In *ISA TECH/EXPO Technology Update Conference Proceedings*. Houston, TX.

Nakagaki, T., H. Yamada, and A. Toth. 2000. Maze-solving by an amoeboid organism. *Nature* 407 (6803):470–470.

NIAAA (National Institute on Alcohol Abuse and Alcoholism). 2002. Integrated Alcohol Sensing and Data Analysis System, in Broad Agency Announcement No. BAA-02–01. www.niaaa.nih.gov/extramural/BAA1.pdf

NSF (National Science Foundation). 2002. Integrated Sensing, Computation, and Networked Systems for Decision and Action, in Program Solicitation NSF-02-039. Available online at www.nsf.gov/pubs/2002/nsf02039/nsf02039.html .

OSA (Optical Society of America). 2001. 2001 Integrated Computational Imaging Systems. Optical Society of America. www.osa.org/meetings/topicals/ICIS/

Pan, W. 2002. A comparative review of statistical methods for discovering differentially expressed genes in replicated microarray experiments. *Bioinformatics* 18 (4):546–554.

Pottie, G.J., and W.J. Kaiser. 2000. Wireless integrated network sensors. *Communications of the ACM* 43 (5):51–58.

Priebe, C.E. 2001. Olfactory classification via interpoint distance analysis. *IEEE Transactions on Pattern Analysis and Machine Intelligence* 23 (4):404–413.

Qi, H., S.S. Iyengar, and K. Chakrabarty. 2001. Distributed sensor networks - A review of recent research. *Journal of the Franklin Institute* 338 (6):655–668.

Schena, M., D. Shalon, R.W. Davis, and P.O. Brown. 1995. quantitative monitoring of gene-expression patterns with a complementary-DNA microarray. *Science* 270 (5235):467–470.

Schwiebert, L., S.K.S. Gupta, and J. Weinmann. 2001. Research challenges in wireless networks of biomedical sensors. In *Proceedings of the Annual International Conference on Mobile Computing and Networking*, MOBICOM. Rome.

Slijepcevic, S., and M. Potkonjak. 2001. Power efficient organization of wireless sensor networks. IEEE International Conference on Communications. Helsinki.

Snopok, B.A., and I.V. Kruglenko. 2002. Multisensor systems for chemical analysis: State-of-the-art in electronic nose technology and new trends in machine olfaction. *Thin Solid Films* 418 (1):21–41.

Sohrabi, K., J. Gao, V. Ailawadhi, G.J. Pottie. 2000. Protocols for self-organization of a wireless sensor network. *IEEE Personal Communications* 7 (5):16–27.

Tsiatsis, V., S.A. Zimbeck, and M.B. Srivastava. 2001. Architecture strategies for energy-efficient packet forwarding in wireless sensor networks. In *Proceedings of the International Symposium on Low Power Electronics and Design* (Huntington Beach, CA), *Digest of Technical Papers*.

Valafar, F. 2002. Pattern recognition techniques in microarray data analysis: A survey in techniques. *Bioinformatics and Medical Informatics* 41–64.

Wang, A., and A. Chandrakasan. 2001. Energy efficient system partitioning for distributed wireless sensor networks. In *Proceedings ICASSP, IEEE International Conference on Acoustics, Speech and Signal Processing,* Salt Lake City, UT.

Wang, I.J., and S.D. Jones. 2001. The scalability of a class of wireless sensor networks. In *Proceedings of SPIE* - The International Society for Optical Engineering, Denver, CO.

Warneke, B., B. Atwood, and K.S.J. Pister. 2001. Smart dust mote forerunners. In *Proceedings of the International Conference on Microelectromechanical Systems,* Interlaken, Switzerland, January 2001.

Williams, R.S., J.L. Brush, M.L. Heinrich, J.M. Mantock, B.E. Jones, and K.A. Henry. 2002. Advanced crisis response and consequence management: Enabling a coordinated response. In *Sensors, and Command, Control, Communications, and Intelligence (C31) Technologies for Homeland Defense and Law Enforcement*, Apr 1–5 2002. Orlando, FL, United States: The International Society for Optical Engineering.

APPENDIX A. PANEL BIOGRAPHIES

Jerome S. Schultz (Chair)

Distinguished Professor
College of Engineering
A 247 Bourns Hall
University of California
Riverside, CA 92521
jssbio@engr.ucr.edu

Professor Schultz enjoys a distinguished international reputation for his research initiatives in the areas of biosensors and synthetic membranes. His study of biosensors involves the utilization of biomolecules that have recognition functions (e.g., antibodies, membrane proteins, bioreceptors), to provide the selectivity capability of sensor probe devices. Professor Schultz has shown that these biological transducer molecules can be coupled with readout devices, such as fiberoptics, to result in biosensors that provide unique characteristics to measure biomolecules such as sugars, drugs, and toxic drugs in situ.

Professor Schultz has made several seminal contributions to the use of membranes in separation and purification. He proved the mechanism of selective separations in microporous membranes to be a function of hydrodynamic drag and partitioning of molecules. He also demonstrated and developed theories for carrier-mediated diffusion of gases through liquid membranes.

Professor Schultz received his B.S. and M.S. degrees in chemical engineering from Columbia University. He was awarded the Ph.D. degree in biochemistry by the University of Wisconsin and was subsequently employed for six years by Lederle Laboratories. While at Lederle, he was a group leader in the Research Division developing new antibiotics, enzymes, and steroids. He then joined the University of Michigan's Department of Chemical Engineering, where, in addition to his professorial responsibilities, he led research in applied microbiology, biomaterials, and membrane separations. Dr. Schultz served as chairman of the department from 1977 until 1985, where he championed the concept of molecular engineering. He then took a two-year leave of absence from Michigan to accept an assignment as Deputy Director for the Engineering Research Center Programs at the National Science Foundation.

In 1987, Dr. Schultz joined the University of Pittsburgh as director of the newly established Center for Biotechnology and Bioengineering. This interdisciplinary research center has programs in bioprocessing, biosensors, bioartificial organs, and gene therapy. He has been responsible for the establishment of the Bioengineering Department (and served as chairman) and its initiation of the B.S., M.S., and Ph.D. degrees. In 2004, he assumed his current position at the University of California, Riverside, as Distinguished Professor of Engineering and Director of the newly established Center for Bioengineering Research.

Dr. Schultz has served as chairman of the Biotechnology Division of the American Chemical Society and the Food, Pharmaceutical, and Bioengineering Division of the American Institute of Chemical Engineers. He is the editor of *Biotechnology Progress*, published jointly by these two societies. He helped organize the American Institute for Medical and Biological Engineering, of which he is a past president. Dr. Schultz was elected to the National Academy of Engineering and served on several NRC committees. He is also a Fellow of the American Association for the Advancement of Science. During 2002 he served as an advisor to NASA on fundamental biology program initiatives.

Milan Mrksich (Vice-chair)

Department of Chemistry
The University of Chicago
5735 South Ellis Avenue
Chicago, IL 60637
mmrksich@midway.uchicago.edu
http://mrksich-group.uchicago.edu/

Milan Mrksich is Associate Professor of Chemistry at the University of Chicago. He received degrees in Chemistry from the University of Illinois (B.S., 1989) and the California Institute of Technology (Ph.D., 1994). He was a postdoctoral fellow at Harvard University for two years and then joined the faculty at the University of Chicago. His research interests are in surface chemistry and tailored bio/materials interfaces. His research group is using model substrates that present peptide and carbohydrate ligands for mechanistic studies of cell adhesion and migration. His group has also developed routes towards dynamic substrates that can alter the presentation of ligands under electrochemical control and is applying these active substrates to chip-based systems. His many honors include the Searle Scholar Award, the A.P. Sloan Research Fellowship, the Camille Dreyfus Teacher-Scholar Award, the TR100 Young Innovator Award, and the American Chemical Society Arthur Cope Scholar Award. He serves on the Defense Sciences Research Council of DARPA, the Editorial Advisory Board of *Langmuir*, and as a frequent consultant and advisory board member to government and biotechnology companies.

Sangeeta N. Bhatia

Department of Bioengineering
Mail Stop 0412
University of California at San Diego
San Diego, CA 92037
sbhatia@ucsd.edu
http://www-bioeng.ucsd.edu/
http://mtel.ucsd.edu/

Dr. Sangeeta Bhatia is an Associate Professor of Bioengineering and Medicine at the University of California at San Diego. She received degrees at Brown University (B.S., Biomedical Engineering), Massachusetts Institute of Technology (M.S., Mechanical Engineering; Ph.D., Bioengineering), and Harvard Medical School (M.D.). She has been named a David and Lucile Packard Fellow (awarded to the nation's most promising young university professors in engineering), a Whitaker Foundation Fellow, Teacher of the Year, and she has been awarded the National Science Foundation's CAREER Award and the Biomedical Nanotechnology Society's Scientific Leadership Award. Dr. Bhatia has industrial experience in the areas of biotechnology, medical devices, and pharmaceutical drug development. She holds a number of patents for both clinical and biotechnological applications of engineering principles. She has extensive experience in the fields of biological microelectromechanical systems (BioMEMS), cell-based biosensing, and hepatic tissue engineering.

David J. Brady

Duke University
Electrical and Computer Engineering
Box 90291
Durham, North Carolina 27708
dbrady@duke.edu
http://www.disp.duke.edu/
http://www.fitzpatrick.duke.edu/

David J. Brady is Director of the Fitzpatrick Center for Photonics and Communication Systems and Brian F. Addy Professor of Electrical and Computer Engineering in the Pratt School of Engineering at Duke University. Brady's research focuses on computational sensor systems. He leads the DISP group (www.disp.duke.edu), which builds interferometric, tomographic, and coherence sensors for 3D microscopy, infrared tomography, multiplex spectroscopy, and heterogeneous sensor networks. Brady holds a B.A. in physics and mathematics from Macalester College and M.S. and Ph.D. degrees in applied physics from the California Institute of Technology. He was on the faculty of electrical and computer engineering at the University of Illinois in Urbana-Champaign from 1990 until joining the Duke faculty in 2001. Brady was a David and Lucile Packard Foundation Fellow from 1990 until 1995 and twice won the Xerox Award for Faculty Research at the University of Illinois.

Antonio J. Ricco

Consultant: bioMEMS, microfluidics, bio/chemical sensors
101 Bacigalupi Dr.
Los Gatos, CA 95032
ajricco@attbi.com

Antonio J. Ricco is Senior Director of Microtechnologies and Materials at ACLARA BioSciences. He received his B.S. and Ph.D. degrees in Chemistry from the University of California at Berkeley (1980) and the Massachusetts Institute of Technology (1984), respectively. He was a member of Sandia National Laboratories' Microsensor R&D Department from 1984–1998, focusing on chemical microsensor systems utilizing acoustic wave, optical, micromachined, electrochemical, and electronic platforms. In 1999 he joined ACLARA BioSciences, where his group develops core technologies for the commercialization of single-use plastic microfluidic array systems for bioanalytical applications, particularly genetic analysis, high-throughput pharmaceutical discovery, and proteomics. Dr. Ricco is the co-author of over 200 presentations, 140 publications, and a dozen patents. Dr. Ricco is a past chair of the Sensor Division of the Electrochemical Society, a Fellow of the Electrochemical Society, and a recipient of the Sensor Division's Outstanding Achievement Award. With Professor Richard Crooks, he co-founded the Gordon Research Conference on Chemical Sensors and Interfacial Design. He served on the Editorial Advisory Board of *Analytical Chemistry* and is presently an Associate Editor of the *Journal of Microelectromechanical Systems* and the *Sensors Update* series. He is a past chair (1998) of the Hilton Head Workshop on Solid-State Sensors, Actuators, and Microsystems and a trustee of the Transducers Research Foundation.

David R. Walt

Tufts University
Robinson Professor of Chemistry
Department of Chemistry
62 Talbot Avenue
Medford, MA 02155
david.walt@tufts.edu
http://ase.tufts.edu/chemistry/walt/index.htm

Dr. David R. Walt is a Professor of Chemistry at Tufts University. He received a B.S. in Chemistry from the University of Michigan and a Ph.D. in Chemical Biology from SUNY at Stony Brook. After postdoctoral studies at MIT, he joined the chemistry faculty at Tufts. Professor Walt served as Chemistry Department chairman from 1989 to 1996. His research interests are in the areas of sensors, arrays, artificial olfaction, and genomics. Dr. Walt serves on many government advisory panels and boards and chaired a National Research Council (NRC) panel on New Measurement Technologies for the Oceans and is a member of the NRC Committee on Waterborne Pathogens. He is Executive Editor of *Applied Biochemistry and Biotechnology* and serves on the editorial advisory board for numerous journals. Dr. Walt is the scientific founder of Illumina, Inc. He has received numerous national and international awards and honors recognizing his work, including a National Science Foundation Special Creativity Award, the Biosensors and Bioelectronics Award, and the Samuel R. Scholes Award in Glass Science. He was elected a fellow of the American Association for the Advancement of Science in 2000. Funding for his work has come from the Departments of Energy, National Science Foundation, National Institutes of Health, Office of Naval Research, DARPA, Environmental Protection Agency, as well as numerous foundations and corporations. Dr. Walt has published over 150 papers, holds over thirty patents, and has given hundreds of invited scientific presentations.

Charles L. Wilkins

Department of Chemistry and Biochemistry
115 Chemistry Building
University of Arkansas
Fayetteville, AR 72701
cwilkins@comp.uark.edu
http://www.uark.edu/depts/cheminfo/uarkchem/

Charles L. Wilkins is Distinguished Professor of Chemistry and Biochemistry at the University of Arkansas. Presently, he also serves as Director of the Center for Sensing Technology and Research at the

University of Arkansas. He received a B.S. in Chemistry from Chapman College and a Ph.D. in Chemistry from the University of Oregon in 1966. After postdoctoral studies at the University of California, Berkeley, he joined the chemistry faculty at the University of Nebraska as an assistant professor. He rose to the rank of professor and in 1981 moved to the University of California, Riverside. From 1982 to 1989, Wilkins served as Chemistry Department chairman. Subsequently, he served as Associate Dean, Physical and Mathematical Sciences, from 1994 to 1997. In 1997 he was named Distinguished Professor of Chemistry. He moved to the University of Arkansas in 1998. His research interests are in the areas of Fourier transform nuclear magnetic resonance, infrared and mass spectrometry, polymer analysis by mass spectrometry, fundamentals of gas phase ion-molecule reactions, sensors, and bioanalytical chemistry in the broadest sense.

Wilkins has served both as Chair of the Computers in Chemistry Division and of the Analytical Chemistry Division of the American Chemical Society. He also serves on many government advisory panels and boards. He is a contributing editor for *Trends in Analytical Chemistry* and serves on a number of editorial advisory boards, including those of the *Journal of the American Society for Mass Spectrometry*, *Mass Spectrometry Reviews*, *Computers in Chemistry*, and *Applied Spectroscopy Reviews*. He has received numerous national and international awards and honors recognizing his work, including the Pittsburgh Analytical Chemistry Award (1994), the American Chemical Society Franklin and Field Award for Outstanding Achievements in Mass Spectrometry (1997), and the Eastern Analytical Symposium Award for Outstanding Achievement in the Fields of Analytical Chemistry (2002). He was elected a fellow of the American Association for the Advancement of Science in 1996. Funding for his work has come primarily from the National Science Foundation, National Institutes of Health, Environmental Protection Agency, and the Petroleum Research Fund of the American Chemical Society. Dr. Wilkins has published over 230 papers, and has given hundreds of invited scientific presentations.

APPENDIX B. SITE REPORTS—EUROPE AND AUSTRALIA

Site: **Biacore Sweden**
Rapsgatan 7
SE-754 50 Uppsala, Sweden
Tel: +46 (0) 18 675700
Fax: +46 (0) 18 150110

Date Visited: 19 March 2003

WTEC Attendees: D. Brady (report author), A. Ricco

Host: Dr. Stephan Löfås, Chief Scientific Officer

BACKGROUND

Biacore was founded in 1984 as Pharmacia Biosensor AB. Building on research at the former Pharmacia, the Linköping Institute of Technology, and the Swedish National Defense Research Institute (FOA), the company pursued research and development of surface plasmon resonance (SPR) based biosensors from 1984 until 1990. In 1990 a product for biomolecular interaction analysis in a core instrument (Biacore), was introduced. The name of the company was changed to Biacore AB in 1996.

Biacore grew to 100 employees from 1984 to 1990 and today has 325 employees. Since the original instrument was introduced, Biacore has gone through several generations. The basic components of Biacore's systems are optical systems for measuring SPR, sensor chips, and microfluidic systems. Successive generations of instruments have improved chip and microfluidic interfaces and have increased automation. Biacore manufactures a variety of sensor chips; the most common are coated with dextran to improve molecular binding capacity.

The primary advantages of the Biacore instruments are realtime sensing and label-free and contact-free molecular recognition. SPR systems measure the mass of molecules bound onto functionalized sensor chips. The specificity of the measurement to the target molecular species depends on the specificity of the surface-binding agent.

Surface plasmon resonance couples light energy into a metalized interface between dielectrics. The resonance is very sharp and depends on the relative indices of refraction of the dielectrics. Molecules binding to the sensor surface change the refractive index at the surface and change the SPR angle. Biacore systems measure the angle of minimum reflected light intensity. Molecular sensitivities may be better than 1 pg/mm^2.

Biacore systems are used for research in pharmaceuticals, proteomics, cell biology, immunology, and food analysis. 44% of the market is in the Americas, 28% in Europe, and 28% in Asia.

Biacore has expanded into live cell analysis with the Procel fluorescence detection/microfluidic system.

REFERENCES

More than 2500 peer-reviewed papers have cited Biacore systems. This literature is accessible online at http://www.biacore.com/k2/references/core.lasso.

Site: **Cranfield University at Silsoe**
Institute of BioScience and Technology
Silsoe, Bedfordshire MK45 4DT, UK
http://www.silsoe.cranfield.ac.uk/

Date Visited: 18 March 2003

WTEC Attendees: C. Kelley (report author), D. Walt

Hosts:
Professor Anthony P.F. Turner, Head of Cranfield University at Silsoe and Chair in Biosensor Technology, Tel: +44 (0) 1525 863005; Email: a.p.turner@cranfield.ac.uk; Website: http://www.silsoe.cranfield.ac.uk/staff/apturner.htm
Professor Phil Warner, Head of Institute of BioScience and Technology, Tel: +44 (0) 1525 863532; Email: p.j.warner@cranfield.ac.uk
Professor Ruikang K. Wang, Chair in Biomedical Optics, Tel: +44 (0) 1525 863450; Email: r.k.wang@cranfield.ac.uk
Steve Setford, Ph.D., Diagnostics and Bioanalysis, Cranfield Centre for Analytical Science, Tel: +44 (0) 1525 863549; Email: s.j.setford@cranfield.ac.uk
Professor Seamus P.J. Higson, Chair in Bio- and Electroanalysis, Tel: +44 (0) 1525 863453; Email: s.p.j.higson@cranfield.ac.uk
Dr. David C. Cullen, Reader in Biophysics and Biosensors, Cranfield Biotechnology Centre, Tel: +44 (0) 1525 863538; Email: d.cullen@cranfield.ac.uk
Professor Sergey Piletsky

BACKGROUND

Cranfield University consists of three campuses: Cranfield, Silsoe, and Royal Military College of Science at Shrivenham. At the Silsoe campus the focus is on environmental, medical and life sciences. The University is a post-graduate research institution with approximately 3,500 students who are divided between the Masters (two thirds) and Ph.D. (one third) degree programs. The Silsoe campus has recently renovated 36 laboratories to provide state-of-the-art bioscience facilities. The University is well known for its commitment to industry and strategic applications of its research and has been rated second in the UK for research impact by the Higher Education Funding Council.

The organizational framework of the Silsoe campus consists of three academic departments: the National Soil Resources Institute, the Institute of Water and the Environment, and the Institute of BioScience and Technology. The National Soil Resource Institute was founded in 2001 as a center of scientific expertise and capable facilities for the efficient management of soil and land resources in the UK. The Institute of Water and the Environment was created in order to help sustain the world's environment and fresh water supplies. The Institute for Bioscience and Technology began in 1981 to provide new products, processes, consultancy, and skills for the biosciences. The key focus areas for the institute are biotechnology, biomedicine, ecochemistry, supramolecular technology, and bioinformatics and IT. It offers Masters courses with a 50% research element in Medical Diagnostics, Translational Medicine, Bioinformatics, Information Technology, Environmental Diagnostics, and Analytical Technology. While other parts of the university are involved in some biosensor research, the site visit was devoted to the activities of the Institute of BioScience and Technology.

With regard to biosensing in Europe, there exists a fledgling Network of Excellence in Sensing Technology (NEST) comprised of 120 biosensor labs selected from over 4,000 sensor labs in 24 countries. There are over 100 people at Cranfield working as a part of this sensor network. The predecessor to this new network was SENSPOL, http://www.cranfield.ac.uk/biotech/senspol/.

RESEARCH AND DEVELOPMENT ACTIVITIES

Upon commencement of the visit, Professor Anthony P.F. Turner, head of Cranfield University at Silsoe, presented an overview of the research portfolio of the university with additional comments from Professor Phil Warner, the head of the Institute of BioScience and Technology. Following this overview, four other hosts presented their relevant research interests and achievements.

Overview by Professor Anthony P. F. Turner, Head of Cranfield University at Silsoe

One of the great early successes of the biotechnology research program at Cranfield University was the invention of an at-home blood glucose monitoring system. In conjunction with Oxford University and MediSense,

Inc., researchers at Cranfield University supported the design and development of what is described as "the world's most successful biosensor." The blood glucose monitoring technology was patented worldwide between 1981 and 1984 and successfully launched as a commercial product in 1987.

Dr. Turner mentioned several research focus areas under examination by the Institute of BioScience and Technology, including electrochemical immunoassays, surface plasmon resonance, biosensors, molecularly imprinted polymers (MIPs), enzyme electrodes, and screen-printing. He briefly described a few other projects of interest to the institute. An optoelectronic technology dubbed the "sniffing" endoscope uses arrays of chemical sensors to detect infection. A device that can predict epileptic seizures through breath analysis has recently proven to be very promising, but is yet unpublished. The use of combinatorial ligands for analytes such as glycosylated hemoglobin and BSE have been explored, and use of synthetic receptors for high-density arrays is being investigated as a possible method for sensing extraterrestrial life on Mars. Finally, the use of imprinted ice as a recognition material is being investigated for applications such as chromatographic columns for enantiomer resolution.

Dr. Turner then commented on ongoing projects at the National Soil Resources Institute. This Institute is currently involved in an innovative sensor technology project concerning precision farming. Researchers have designed tractors that do not require human interaction to determine chemical application rates. These "smart" tractors, through the use of various biosensors, can deliver the appropriate amount of fertilizer to a field based upon data from previous years and can determine the adjustments necessary to improve the harvest. Moreover, the Sports Surface Technology Program at the institute has been utilizing sensors to study an assortment of surfaces, such as golf courses and running tracks. In one program, up to 246 different measurements of soil and soil structure, testing for nearly any type of analyte, i.e., lead or sulfur, were performed, producing a 5 km digital grid map of England and Wales. Although in this latter program all of the sampling was done manually, investigators are working to automate the system.

Professor Ruikang K. Wang, Chair in Biomedical Optics

Dr. Wang's research efforts are primarily focused on enhancing the imaging depths for Optical Coherence Tomography (OCT) and characterizing fluid flow using confocal microscopy. His approach is to chemically alter the optical properties of tissues so that the visual depth is greater than 1.5 mm—the effective working depth without chemical treatment. The chemicals make the tissue transparent, causing less light scattering and greater photon penetration. He has filed for a patent on the technology and has expressed interest in using nanoparticles to enhance contrast.

In Dr. Wang's laboratory we spoke with an investigator who explained his research on sonodynamic chemiluminescence for cancer diagnosis. The process involves the reaction between a fluorescent chemiluminescent agent (FCLA) and a reactive species of oxygen (O_2^-), the product of which is luminescence. The investigator uses the technique of ultrasound sensitizing to localize tumors and to generate the reactive oxygen. He claims to have shown that the system works in vivo.

Professor Seamus P.J. Higson

Dr. Higson, whose expertise lies in analytical biochemistry, began his presentation with a short tutorial addressing biosensors. He upheld that biosensor technology is in its infancy but is growing exponentially. To reiterate his point, he cited the blood glucose sensor as an anomalous example. In the case of this sensor, the sample is a drop of blood, which is held in place by surface tension. However, other samples are not as compliant with such a method, leading Dr. Higson to stress the need for a commercial biosensor similar in structure to a pH electrode.

Dr. Higson continued by describing in great detail his patented sensor production system, which he hoped to have published in *Nature Biotechnology*. The sensor is made of microelectrode arrays on a commercially viable platform. In manufacturing the platform, a carbon/gold composite conduction surface is mounted on a ceramic substrate. An electropolymerizing insulation film is then placed on top of the conduction surface. Using a technique similar to ultrasound lithography, small pores or holes of reproducible size are blown into the insulating layer, exposing the conductive surface. Conductive polymers carrying antibodies can then be adsorbed on the conduction surface in each of the pores. According to Dr. Higson, the advantages of his system include inexpensive mass production, a very high pore density, and a three-fold increase in the signal-to-noise ratio.

Dr. Higson also spoke of research on electronic nose technology. The eNose system, a commercial instrument, employs sensors that detect the chemicals associated with certain odors. The group is evaluating the technology for analysis of transformer oils, edible oils, and heavy metals.

Dr. David C. Cullen, Reader in Biophysics and Biosensors

The research focus areas outlined by Dr. Cullen were optical evanescent wave systems, scanning probe microscopy, bio-interface science, optical sensors, array sensors, microsystems, and nanotechnology. His research using scanning probe microscopy primarily deals with elucidating material surface properties pertaining to biocompatibility. In studying biointerfaces, Dr. Cullen has been analyzing the adsorption of antibodies to synthetic materials, such as polystyrene, in order to discover fundamental biological knowledge of molecular-level interactions. Furthermore, he has been attempting chemical modifications of surfaces to achieve micro-heterogeneity, as he hopes to be able to integrate MIP ultrathin films into biosensors and diagnostic devices. For much of his bio-interface analysis, Dr. Cullen uses surface plasmon resonance (SPR) and atomic force microscopy (AFM).

In terms of sensor technology, Dr. Cullen discussed several research projects. He is working on the astrobiology project that is developing sensor array concepts for identifying biomolecular markers in space during planetary exploration. These extraterrestrial biosensors incorporate both optical and electrochemical transducers on the same device. Similar sensor arrays are being investigated for remote environmental analysis. In the realm of biomedical applications, Dr. Cullen expressed the institute's interest in commercializing a microsystem for optical glucose measurement, as well as an online optical sensor for endotoxin detection. He also mentioned SPR arrays for genomics and proteomics, photochromic systems for addressed optical control of biomolecules, and novel acoustic sensor arrays, all of which are currently under development at the institute.

Professor Sergey Piletsky

Professor Piletsky is working in the area of molecularly imprinted polymers. He uses molecular modeling to find monomers and polymerization conditions for optimal MIP preparation. The modeling phase of the work takes approximately one week and is followed up by the preparative effort. This part of the visit was cut short, but it was clear that the approach being taken in the Piletsky laboratory was broadly applicable to solving many recognition problems.

SUPPORT

The University of Cranfield at Silsoe is primarily funded by competitive funds totaling approximately 18 million/year, approximately half of which is provided by international governments. Within the Institute of Bioscience and Technology, the source of funding varies with the application areas. In healthcare applications, a majority of the funding is from industry. In environment applications, funding is procured through government contracts. Food safety is very conservative and regulated, as it is usually supported by government Euros.

In general, funding from the government scientific research councils is aimed at academically responsive research. The research council that generates the most support for Cranfield is the EPSRC, the engineering council. Funding from government departments such as Defense and Environment is more needs-driven and appears to be akin to U.S. government contract work.

Site: **DiagnoSwiss**
(Representatives met with WTEC panel at EPFL)

Date Visited: 22 March 2003

WTEC Attendees: D. Brady (report author), C. Kelly, A. Ricco, D. Walt

Hosts: Hubert Girault
Fredertc Reymond, DiagnoSwiss
Joel Rossier, DiagnoSwiss

SUMMARY

DiagnoSwiss is a spin-off of the Laboratory for Physical and Analytical Electrochemistry at EPFL. The company was founded in 1999 and now has 6 employees. Its headquarters are in Monthey, Switzerland.

DiagnoSwiss manufactures plasma-etched microchips for biochemical analysis and partners with several large biotech companies in Switzerland in pursuing applications. Manufacture of enzyme-linked immunosorbent assay or enzyme-linked oligosorbent assay systems in portable formats is a primary goal. The chips are intended for "lab-on-a-chip" systems and can be mass produced with integrated control electrodes.

In collaboration with EPFL, DiagnoSwiss has also developed and patented "off-gel" electrophoresis technology, which separates proteins according to their charge. Protein solutions flow under an immobilized pH gradient gel (IPG) coupled to a dynode array. Proteins with an isoelectric point (pI) close to pH of the gel are not extracted by the electric field and stay in solution. Off-gel technology is an attractive integrated protein purification system for proteomic analysis.

REFERENCES

Gobry, V., J. Van Oostrum, M. Martinelli, T. Rohner, F. Reymond, J.S. Rossier, and H.H. Girault. 2002. Microfabricated polymer injector for direct mass spectrometry coupling. *Proteomics* 2:405–412.

Roberts, M.A., J.S. Rossier, P. Bercier, and H.H. Girault. 1997. UV laser machined polymer substrates for the development of microdiagnostic systems. *Anal. Chem.* 69:2035–2042.

Ros, A., M. Faupel, H. Mees, J. Van Oostrum, R. Ferrigno, F. Reymond, P. Michel, J.S. Rossier, and H.H. Girault. 2002. Protein purification by off-gel electrophoresis. *J. Proteomics* 2:151–156.

Rossier, J.S., F. Reymond, and P.E. Michel. 2002. Polymer microfluidic chips for electrochemical and biochemical analyses. *Electrophoresis* 23:858–867.

Rossier, J.S., M.A. Roberts, R. Ferrigno, and H.H. Girault. 1999. Electrochemical detection in polymer microchannels. *Anal. Chem.* 71:4294–4299.

Rossier, J.S., R. Ferrigno, and H.H. Girault. 2000. Electrophoresis with electrochemical detection in a polymer microdevice. *J. Electroanal. Chem*. 492:15–22.

Rossier, J.S., and H. H. Girault. 2001. Enzyme linked immunosorbent assay on a microchip with electrochemical detection. *Lab Chip* 1:153–157.

Schwarz, A., J.S. Rossier, M.A. Roberts, H.H. Girault, E. Roulet, and H. Mermod. 1998. Micro-patterning of biomolecules on polymer substrates. *Langmuir* 14:5526–5531.

Schwarz, A., J.S. Rossier, F. Bianchi, F. Reymond, R. Ferrigno, and H.H. Girault. 1998. Micro-TAS on polymer substrates micromachined by laser photoablation, In D.J. Harrson and A. van den Berg, ed., *Proceedings of the μTAS'98 Workshop,* 241–244. (Held in Banff, Canada, 13–16 October 1998.) Dordrecht: Kluwer Academic Publishers.

Site:	**Dublin City University** **National Centre for Sensor Research** **Dublin 9, Ireland** **Tel: 353-1-7005299** **Fax: 353-1-7008021** **http://www.ncsr.ie**
Date Visited:	19 March 2003
WTEC Attendees:	J. Schultz (report author), D. Walt
Host:	Prof. Brian MacCraith, Director, Email: brian.maccraith@dcu.ie
Other Presenters:	Dermot Diamond, Vice President for Research Prof. Robert J. Forster Prof. Richard O' Kennedy

OVERVIEW

The National Centre for Sensor Research is a large-scale multidisciplinary research organization focused on the development of chemical sensors and biosensors. Centre development is assisted by an enhancement program in Ireland that is devoting about 1.5% of its GDP to science (€2.5 billion/year). This group was established at Dublin City University in 1999 and is a partnership between sensor-related researchers in a variety of departments, e.g., chemistry, biology, physics, etc. The Centre has 27 academic members, 130 full-time researchers, a facility of 20,000 ft^2, and had a budget of €5 million for initial setup.

Their research program is based on eight clusters of expertise:

- Synthesis and molecular recognition
- Biomolecular interactions
- Deposition and surface characterization
- Electrochemical sensors
- Photonic sensors and devices
- Separation science
- Microsystems and instrumentation
- Nanomaterials

Center facilities include class-100 cleanrooms, mask aligner, photolithographic station, microinjection molder, hot embosser, laser ablation, UHV-STM, AFM, SECM XPS, LEED, LC-MS/GC-MS.

In addition to education and research programs, the Centre has a strong commitment to commercialization. Some examples of commercial outcomes are "ClearCense" water color and turbidity sensor system; custom synthesis and supply of calixarene-based ionophores; Metohm ISE for Na^+; nondestructive sensor for monitoring the integrity of packaged foods; and a joint venture agreement with Growcorp Group.

In the field of photonics, the Centre has developed a library of luminescent indicators and labels based on ruthenium and osmium polypyridyl complexes. These materials have been incorporated into polymers and sol-gel devices. A process has been developed to manufacture printable Ormosils from sol-gel materials. Another development has been the perfection of printable sensors for oxygen and carbon dioxide that will be used to test the integrity of food packaging. The Centre developed fluorophores for these two gases that could be excited and measured simultaneously by a single-wavelength light source in a lifetime phase fluorimetry approach. Its researchers also perfected a new readout device based on a new theory of dual luminophore referencing that allows for self-calibration of the measurements over long periods of time. This system is the basis of a new product called "Intellipak."

Some other photonic developments include a single-reflection-based pH sensor that has a resolution of 0.0007 pH units (Polerecky et al. 2002), and a biochip/bio-MEMS system (Rowe et al. 1999). By a thorough theoretical lightwave analysis of light emission from fluorescent spots on a planar surface they have developed a software program that can be used to improve light capture by 15 fold of the emitted signal from multiarray-type devices (PCT Patent Application WO 02/059583 A1). This theory has allowed them to produce a very efficient multianalyte sensor chip that has eight waveguides of 50 μm width and 200 μm separation (Feldstein, et al., U.S. Patent 6,137,117, 200).

Richard O' Kennedy has perfected a rapid and efficient process for producing single-chain Fv antibodies by a combination of monoclonal techniques, combinatorial phage display libraries, and protein engineering. Utilizing this methodology, the Centre has produced highly effective biorecognition proteins for coumarin, warfarin, Aflatoxin, illicit drugs, listeria, immunosensors, and atrazine. The Centre is working closely with Xenosense and Biacore to develop reagents for their devices.

An extensive program in nanoelectrodes and nanoparticles in the Centre is led by Prof. Robert J. Forster. The primary foci of his group are nanode and nanofiber sensors; functionalization of carbon nanotubes and metal/semiconductor nanoparticles; and novel hybrid materials.

For example, nanotubes have been functionalized with fluorophores for nanosensor applications. Microsensors on the order of 160 nm have been prepared; time constants for electrodes of this nature can be on the order of nanoseconds (Forster et al. 2000). Methods have been developed to synthesize ordered monolayers with different electroactive substances with spacings on the order of 10 Å. This type of structure has been used to produce prototyped molecularly switched immunosensors. Methods have been developed at the Centre to produce functionalized nanoparticles with a variety of detection elements. For example, nanoparticles have been labeled with different fluorophores to produce optical barcodes. These have been applied to produce multiplexed optical systems for immunoassays. Nanoparticles are being modified with controlled release formulations to be applied to surfaces to provide antimicrobial properties.

The Centre is associated with the Virtual Centre for Supramolecular Nanoscale Science that is led by Prof. Han Vos of Dublin City University. The researchers have extensive collaborations with scientists in Europe and the United States.

Site:	**Eberhard Karls University Tübingen** **Institute for Physical Theoretical Chemistry** **Auf de Morgenstelle 8** **72076 Tübingen, Germany** **http://barolo.ipc.uni-tuebingen.de/**
Date Visited:	21 March 2003
WTEC Attendees:	J. Shultz (report author), M. Mrksich
Host:	Prof. Dr. Günter Gauglitz

BACKGROUND

The university was founded in 1477 and now comprises 20,000 students, 16 faculties, 200 institutes, 450 professors, and 2000 scientific staff. Its total budget is about $1 billion, and research funds are about $70 million. The Institute for Physical and Theoretical Chemistry has two department heads, a staff of about 100 (60 are funded by external research funds), and about 25 graduate students.

Prof. Gauglitz' research group is focused on

- Examination of (bio-)molecular interactions
- Characterization of thin films
- Multivariate data analysis

Technical themes of the groups include spectroscopy, kinetics, analytics (process control, data acquisition, simulation, and chemometrics), computer applications, and sensors. The laboratory predominantly utilizes optical techniques in its research. These include

- Direct Optical Detection
  - Reflectometric interference spectroscopy
  - Ellipometry
  - Mach-Zehnder integrated optics
  - Surface plasmon resonance
- Detection of Labeled Systems
  - Fluorescence intensity/lifetime
  - Fluorescence anisotropy
  - Fluorescence correlation
  - Total Internal Reflection Fluorescence (TIRF)
  - Fluorescence Energy Transfer (FRET)

In recent years a major effort has been undertaken to utilize the principles of Reflectometric Interference Spectroscopy (RIfS) to devise instruments for the detection and monitoring of a wide range of analytes without the use of any labels. The principle of this methodology is illustrated in Figure B.1 below. The transducer element consists of a thin inert transparent film (shown in green in Figure B.1) placed on the surface of another (thicker) transparent material with a different refractive index. If a third layer of material (again of different refractive index, the red layer) is on the top of the inert layer, then there will be interference between the light reflected from the green and red layers. The effect of this interference will be to shift the spectrum of the reflected light towards higher wavelengths, as illustrated in Figure B.2. The shift in the spectrum (measured in nm) is directly proportional to the thickness of the red layer. If instead of a polymer film, a ligand is immobilized to the inert (green) layer, a fixed wavelength shift in the reflected light will also occur that is related to the amount of ligand and its refractive index. Further, if the slide is exposed to a protein that binds to the immobilized ligand, a further shifting in the reflected spectrum will occur that is proportional to the amount of protein adsorbed. A typical response is shown in Figure B.3. This is a powerful new technique that will compete with Surface Plasmon Resonance (SPR) methods for biosensing.

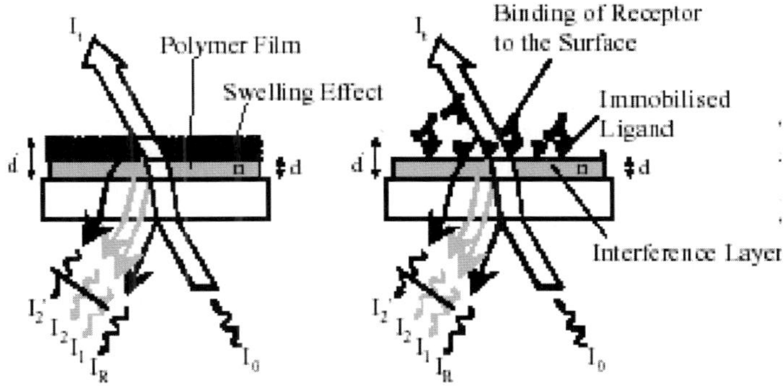

Fig. B.1. Illustration of using principles of reflectometric interference spectroscopy to detect and monitor a wide range of analytes. *Left*: The transducer element consists of three layers of material, each with different refractive indexes; a shift in the spectrum of the reflected light towards higher wavelengths is directly proportional to the thickness of the top layer. *Right*: The element contains an immobilized ligand that, when exposed to a protein that binds to it, produces a shifting in the reflected spectrum that is proportional to the amount of protein adsorbed.

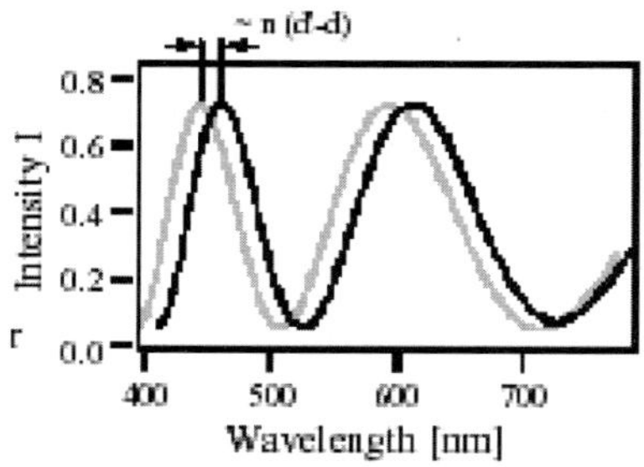

Fig. B.2. Illustration of the shift in wavelength to higher frequency due to the interference of the red layer shown in the element on the left in Fig. B.1 above.

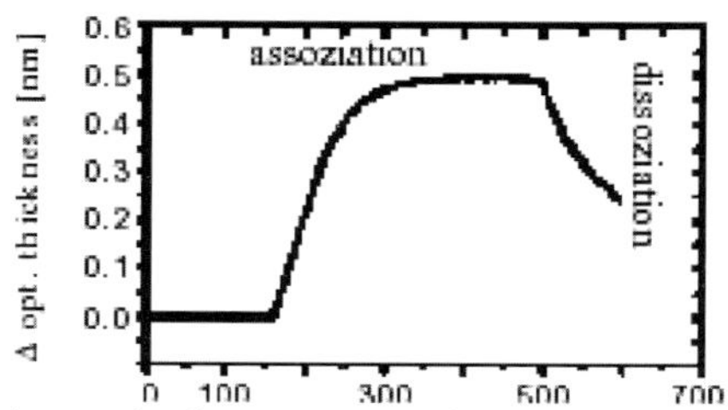

Fig. B.3. A typical response from the shifting of the reflected spectrum, illustrated in Figures B.1 and B.2 above.

The Gauglitz laboratory has exploited these phenomena to develop a multitude of assay systems. It has instruments that have a resolution of optical thickness of the assay layer (the red layer) of about 20 pm. For their instrumentation its researchers can obtain sufficient data in 10–20 sec to provide a good estimate of the amount of material adsorbed. The method has been validated for a variety of applications, including typical immunoassays, binding of antisense oligonucleotides, monitoring of fermentation processes, gas analysis (discrimination of refrigerants R22 and R134a), kinetics of antibody-antigen interactions, antibody binding constants, volatile organic carbon (VOC) pollutants, and epitope mapping.

The Institute for Physical Theoretical Chemistry has developed an inexpensive robust laboratory instrument for routine laboratory analyses with an expected cost of less than $2000. This type of apparatus is shown in Figure B.4 below.

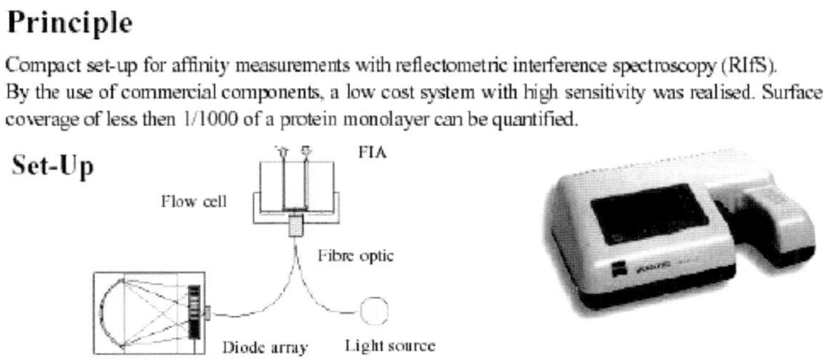

Fig. B.4. Institute for Physical Theoretical Chemistry's compact, low-cost RIfS instrument for routine assay measurements.

Surface Chemistry. Methods have been developed to functionalize surfaces to immobilize a variety of biological recognition elements, e.g., antibodies. These immobilization techniques can support a surface loading of about 20 ng protein/mm^2 and have maintained the protein for over 300 cycles of use.

Applications of the analysis techniques under development in this laboratory include immunoassays; receptor-ligand interactions for drug screening and diagnostics; DNA-diagnostics; functional proteome analytics; and surface binding of antibodies and bacteria.

Mach-Zehnder Integrated Optics. This system is used for the continuous measurement of the refractive index of thin films for the detection of receptor ligand interactions, hydrocarbons in air and water, and characterization of new anisotropic sensor layers. Miniature flow-through cells on the order of 35 mm in length have the potential to resolve changes in refractive index on the order of 10^{-5}. This system has the capability for parallel sampling and for integration for immunoassays and on-line control.

Total Internal Reflection Fluorescence (TIRF). A portable river water analyzer (RIANA) has been developed for the monitoring of multiple organic pollutants simultaneously with minimum sample preparation. The system consists of a modular flow system that periodically alternates samples, labeled antibodies, and regeneration fluids. One application of this technique is the measurement of estrogens in wastewater. The limits of detection were in the range of 0.03 and 0.16 ppb with an assay period of 15 min. This project being developed with a consortium that includes the Central Research Labs (London), Siemens (Germany), CSIC (Spain), TZW (Germany), EI (Slovakia), Optoelectronics Research Centre (Southhampton), and King's College (London).

Various techniques for high-throughput screening (HTS) have been developed. Recently a nanotiterplate system has been perfected. The plate is about 2x2 mm and contains 625 wells that can handle test volumes between 10–60 nl. The piezoelectric dosing device can produce drop sizes from 15–500 pl with an error of < 2%. The limit of detection of a label fluorophore is about 250 amol, and the consumption of antibody is about 5000 times less than current microtiterplate format systems. Results of over 600 data points can be obtained in 30 minutes.

The laboratory has developed a variety of chemometric techniques to improve and optimize data acquisition and accuracy. Statistical methods (PCA, MLR, PCR, PLS, and QPLS) are used in conjunction with artificial neural networks and classification schemes (SOM, LCQ, etc).

REFERENCES

Barzen, C., A. Brecht, and G. Gauglitz. 2002. Optical multiple-analyte immunosensor for water pollution control. *Biosensors & Bioelectronics* 17:289–295.

Belge, G., D. Beyerlein, C. Betsch, K.-J. Eichhorn, G. Gauglitz, K. Grundke, and B. Voigt. 2002. Suitability of hyperbranched polyester for sensoric applications: Investigation with reflectometric interference spectroscopy. *Anal. Bioanal. Chem.* (ABC) 374:403–411.

Beyerlein, D., G. Belge, K.-J. Eichhorn, G. Gauglitz, K. Grundke, and B. Voit. 2001. Preparation and properties of thin films of hyperbranched polyesters with different endgroups. *Macromol. Symp.* 164:117.

Birkert, O., and G. Gauglitz. 2002. Development of an assay for label-free high-throughput screening of thrombin inhibitors by use of reflectometric interference spectroscopy. *Analytical and Bioanalytical Chemistry* 372:141–147.

Birkert, O., H.-M. Haake, A. Schütz, J. Mack, A. Brecht, G. Jung, and G. Gauglitz. 2000. A streptavidin surface on planar glass substrates for the detection of biomolecular interaction. *Analytical Biochemistry* 282:200–208.

Birkert, O., R. Tünnemann, G. Jung, and G. Gauglitz. 2002. Label-free parallel screening of combinatorial triazine libraries using reflectometric interference spectroscopy. *Analytical Chem.* 74:834–840.

Braun, M., X. Camps, O. Vostrowsky, A. Hirsch, E. Endreß, T.M. Bayerl, O. Birkert, and G. Gauglitz. 2000. Synthesis of a biotinated lipofullerene as a new type of transmembrane anchor. *Eng. J. Org. Chem.* 2000:1173–1181.

Bühler, B., D. Fröhlich, H.-M. Haake, A. Brecht, and G. Gauglitz. 2001. Structure and characterization of a micro flow system for the study of modulated concentrations. *Trends in Analytical Chemistry* 20(4):186–194.

Coille, I., S. Reder, S. Bucher, and G. Gauglitz. 2002. Comparison of two fluorescence immunoassay methods for the detection of endocrine disrupting chemicals in water. *Biomolecular Engineering* 18:273–280.

Dieterle, F., D. Nopper, and G. Gauglitz. 2001. Quantification of butanol and ethanol in aqueous phases by reflectometric interference spectroscopy: Different approaches to multivariate calibration. *Fresenius J. Anal. Chem.* 370:723–730.

Gauglitz, G. 2000. Optical sensors. In *Biosensors for environmental monitoring*, pgs. 28–51, ed. U. Bilitewski and Anthony P.F. Turner. City: Harwood Academic Publishers.

Gauglitz, G. 2000. Optical detection methods for combinatorial libraries. *Current Opinion in Chemical Biology* 4:351–355.

Gauglitz, G., J. Piehler, and U. Bilitewski. 2000. Affinity sensor systems. In *Biosensors for environmental monitoring*, pgs. 150–165, ed. U. Bilitewski and Anthony P.F. Turner. City:Harwood Academic Publishers.

Haake, H.-M., A. Schütz, and G. Gauglitz. 2000. Label-free detection of biomolecular interaction by optical sensors. *Fresenius' J. Anal. Chem.* 366:576–585.

Haake, H.-M., R. Tünnemann, A. Brecht, V. Austel, G. Jung, and G. Gauglitz. 2002. Online monitoring of solid-phase peptide syntheses on glass-type surfaces using white light interference. *Analytical Biochemistry* 300:107–112.

Hänel, C., and G. Gauglitz. 2002. Comparison of reflectometric interference spectroscopy with other instruments for label-free optical detection. *Analytical and Bioanalytical Chemistry* 372:91–100.

Jung, A. 2002. DNA chip technology. *Analytical and Bioanalytical Chemistry* 372:41–42.

Kieser, B., C. Fiethzek, R. Schmidt, G. Belge, U. Weimar, V. Schurig, and G. Gauglitz. 2002. Use of a modified cyclodextrin host for the enantioselective detection of a hologenated diether as chiral guest via optical and electrical transducers. *Analytical Chemistry* 74:3005–3012.

Kieser, B., D. Pauluth, and G. Gauglitz. 2001. Nematic liquid crystals as sensitive layers for surface plasmon resonance sensors. *Analytica Chimica Acta*. 434:231–237.

Kieser, B., F. Dieterle, and G. Gauglitz. 2002. Discrimination of methanol and ethanol vapors by the use of a single optical sensor with a microporous sensitive layer. *Analytical Chemistry* 74:4781–4787.

Killenberg-Jabs, M., G. Gauglitz, and C. Hänel. 2001. Markierungsfreie detektion biomolekularer reaktionen. *LaborPraxis* 10:26–29.

Kröger, K., A. Jung, S. Reder, and G. Gauglitz. 2002. Versatile biosensor surface based on peptide nucleic acid with label free and total internal reflection fluorescence detection for quantification of endocrine disruptors. *Analytica Chimica Acta*. 469:37–48.

Kröger, K., J. Bauer, B. Fleckenstein, J. Rademann, G. Jung, and G. Gauglitz. In press. Epitope-mapping of transglutaminase with parallel label-free optical detection.

Länge, K., G. Griffin, T. Vo-Dinh, and G. Gauglitz. 2002. Characterization of antibodies against benzo[a]pyrene with thermodynamic and kinetic constants. *Talanta*. 56:1153–1161.

Mallat, E., C. Barzen, R. Abuknesha, G. Gauglitz, and D. Barceló. 2001. Part per trillion level determination of isoproturon in certified and estuarine water samples with a direct optical immunosensor. *Analytica Chimica Acta*. 426:209–216.

Mallat, E., D. Barceló, C. Barzen, G. Gauglitz, and R. Abuknesha. 2001. Immunosensors for pesticide determination in natural waters. *Trends in Analytical Chemistry* 20(3):124–132.

Mutschler, T., B. Kieser, R. Frank, and G. Gauglitz. 2002. Characterization of thin polymer and biopolymer layers by ellipsometrie and evanescent field technology. *Anal. Bioanal. Chem.* 374:658–664.

Piehler, J., A. Brecht, R. Valiokas, B. Liedberg, and G. Gauglitz. 2000. A high-density poly(ethylene glycol) polymer brush for immobilization on glass-type surfaces. *Biosensors & Bio-electronics* 15:473–481.

Raitza, M., M. Herold, A. Ellwanger, G. Gauglitz, and K. Albert. 2000. Solid-state NMR and ellipsometric investigations of C-30 chains bonded to SiO2 surfaces. *Macromol. Chem. Phys.* 201:825–829.

Rathgeb, F., D. Reichl, M. Herold, O. Mader, and T. Mutschler. 2000. Dyeless optical detection of ammonia in the gaseous phase using a pH-responsive polymer. *Fres. J. Anal. Chem.* 368:192–195.

Rathgeb, F., G. Belge, M. Herold, C. Bogenschütz, and G. Gauglitz. 2000. Enhancement of the selectivity of sensor arrays by the use of a size selective sensitive layer. *Eingereicht*. Vol/pages?

Rathgeb, F., and G. Gauglitz. 2000. Optical gas sensors in analytical chemistry: Applications, trends and general comments. In *Encyclopedia of Analytical Chemistry*, pgs. 2189–2203, ed. R.A. Meyers. Chichester: John Wiley & Sons.

Reichl, D., R. Krage, C. Krummel, and G. Gauglitz. 2000. Sensing of volatile organic compounds using a simplified reflectometric interference spectroscopy setup. *Appl. Spectr.* 54:583–586.

Schobel, U., C. Barzen, and G. Gauglitz. 2000. Immunoanalytical techniques for pesticide monitoring based on fluorescence detection. *Fresenius' J. Anal. Chem.* 366:646–658.

Schobel, U., H.-J. Egelhaaf, D. Fröhlich, A. Brecht, D. Oelkrug, and G. Gauglitz. 2000. Mechanisms of fluorescence quenching in donor-acceptor labeled antibody-antigen conjugates. *Journal of Fluorescence* 10(2):147–157.

Schobel, U., I. Coille, A. Brecht, M. Steinwand, and G. Gauglitz. 2001. Miniaturization of a homogeneous fluorescence immunoassay based on energy transfer using nanotiter plates as high-density sample carriers. *Analytical Chemistry* 73:5172–5179.

Tünnemann, R., M. Mehlmann, R. D. Süssmuth, B. Bühler, S. Pelzer, W. Wohlleben, H.-P. Fiedler, K.-H. Wiesmüller, G. Gauglitz, and G. Jung. 2001. Optical biosensors. Monitoring studies of glycopeptide antibiotic fermentation using white light interference. *Analytical Chemistry* 73:4313–4318.

PATENTS BY THE RESEARCH GROUP GAUGLITZ

Patentanmeldung: Reversibles chemisches Aktinometer, Nr. 1825828 Gauglitz, Günter; Hubig, Stephan

Patentanmeldung: Sensor zum Stoffnachweis: DE 3 832 185 A1 N. Fabricius, G. Gauglitz, J. Ingenhoff

Patentanmeldung: Interferenzsensor: P 42 00 088.2 W. Nahm, A. Brecht, Biochem, G. Gauglitz United States Patent: Optical Sensors...: 5,262,842 G. Gauglitz, J. Ingenhoff, N. Fabricius

Patentanmeldung: Kapillarelektrophorese: P 42 18 721.4 H. Bauer, B. Wulf, W. H. Hoffmann, W. Nahm, J. Nagel, G. Gauglitz

Patentanmeldung: Optisches Parallel High Through-put Screening: P 196 15 366.2 A. Brecht, G. Gauglitz, D. Gräfe, Fuchs (C. Zeiss)

Patentanmeldung: Mikroreflektometrie: 197 29 259.3 G. Gauglitz, J. Seemann, D. Reichl, A. Brecht

Patentanmeldung: Optische Messwandler auf der Grundlage von flüssigkristalli-nen Phase: DE 199 31 754 A1 D. Paluth, J. Krause, G. Gauglitz, B. Drapp.

Site: **École Normale Supérieure (ENS)**
24, rue Lhomond
F-75231 Paris Cedex 05, France
Fax: 33+ (0)1 44 32 38 63

Date Visited: 21 March 2003

WTEC Attendees: D. Walt (report author), C. Kelley, C. Wilkins

Host: Professor Christian Amatore, Département de chimie,
École normale supérieure,
Email: christian.amatore@ens.fr,
Website: http://www.helene.ens.fr/w3amatore

BACKGROUND

Structure: Chemistry at ENS is under the auspices of France's Centre National de la Recherche Scientifique (CNRS). Its funding is allotted every four years and is based primarily on the previous term's output. (All universities in France are autonomous, belong to the government, and are funded by the government.) Personnel are funded through permanent lifetime positions, and their funding is not included in the operational budgets of the institution, as they get paid by CNRS directly. Therefore, it is always a struggle to secure additional faculty. The university's organization is hierarchical, with one professor and his/her subordinates. The recurrent funding pays to maintain lab operations. In addition, there are annual requests for proposals that will fund specific projects of contemporary need and interest. These projects tend to be in applied areas of research. Postdoctoral associates are also funded directly by the ministries, with salary going to the individual. For Ph.D. candidates, the head of the lab applies for a certain number of positions per year. ENS requests 25–30 positions per year and gets the very best students, as it is considered to be among France's top institutions.

Support for sensors and biosensors has been difficult to obtain, as the subject is considered to be part of the Department of Analytical Chemistry, which suffered a significant decline in support in the last 15–20 years, ascribed to the nature of succession of laboratory heads. Analytical Chemistry was producing data, not new methods, and was viewed as a service and not a science. Amatore is spearheading a NEW Analytical Methodologies and Sensors Initiative and has encountered resistance. While biologists recognize the power of new analytical tools to help solve their scientific problems, they expect chemists to work for them in a service capacity. The new initiative is patterned after Cornell's

Nanotechnology Center, with chemistry to be integrated with microfluidics.

RESEARCH ACTIVITIES

The laboratory at ENS Paris is focusing on single cell measurements, new methods for separating and analyzing components, and microfluidics. Amatore is well known for developing the basic theory of microelectrodes. Electrochemistry is a kinetic method and does not perturb the system. Both cells and electrochemistry work by kinetics and, in principle, are matched to one another. The problem has been the lack of selectivity of electrochemical methods, and methods have been developed to couple electrodes to biological systems. Ideally the transducer will amplify the signal and the substrate electrode can supply electrons or detect a side product of the reaction. One can also use enzymes as the sensing element coupled to the electrode and measure their inhibition by a toxin for food and water monitoring.

Microfluidics. Using a very clever design strategy based on fundamental understanding of mass transport within microfluidics capillaries, the group has developed more efficient and more sensitive designs for electrochemical detection.

Artificial neurons have been designed in which coupled redox cycles are employed in which separated source and detector electrodes are used to generate and detect an electroactive redox pair. The system has been used as molecular logic gates in which coincidence of signals is required to provide a signal, thereby making the system fault tolerant.

Microelectrodes. The laboratory is interested in mapping concentration profiles within the transient or steady state diffusion layer using a microelectrode. The resolution is determined by the probe diameter. Diffusion layers have been detected within single molecules (dendrimers). In addition, studies of electron transport in DNA molecules are being undertaken using electron transport probes attached to the DNA target molecule that hybridizes to the electrode surface. By temporal measurements, one can measure the rates at which hopping occurs and can possibly detect the location of a single base mismatch.

Exocytosis. The Amatore laboratory has had a long-standing collaboration with the Wightman laboratory at UNC-Chapel Hill in which they have been studying the use of microelectrodes to measure single vesicle neurotransmitter release. The Amatore laboratory has extended this study to understand exocytosis and the mechanism of vesicular fusion with the plasma membrane. The study involves both theoretical analysis of the energies involved as well as experimental verification of the calculation. Excellent agreement between theory and experiment has been obtained.

Oxidative Stress. Oxidative stress is in balance with defense mechanisms. When oxidative stress exceeds the rate of defense mechanisms, damage occurs. The Amatore laboratory researchers are performing amperometry measurements in oxidation mode only. They have determined the electrochemical behavior of the oxidation signal appears to be due to four different compounds: hydrogen peroxide, peroxynitrite, nitrite ions, and nitric oxide. They have made the first analytical characterization of the so-called "ROS" (reactive oxygen species) oxidative bursts.

Site: **École Polytechnique Fédérale de Lausanne (EPFL)**
Institute of Biomolecular Sciences
ISB-VO
CH-1015 Lausanne, Switzerland
http://icpsg3.epfl.ch/

Date Visited: 21 March 2003

WTEC Attendees: D. Walt (report author), D. Brady, C. Kelley, and T. Ricco

Host: Prof Horst Vogel, Rund Hovius
Office: CHB B3 494; Tel: +41 2169 33155;
Fax: +41 2169 36190;
Email: horst.vogel@epfl.ch;
Website: http://icpsg3.epfl.ch/Personalpages/HorstVogel.htm

RESEARCH OVERVIEW

The Vogel Laboratory is involved in biosensing and employing nanobiotechnology for elucidating biochemical networks. The laboratory has 35 researchers with expertise ranging from molecular biology to surface physics. One of the major interests is in understanding what is going on inside the cell. The WTEC team was given a tour of the spacious state-of-the-art facilities and laboratory equipment.

SPECIFIC PROJECTS

G Protein Coupled (GPC) receptors are complicated to study in living cells. The laboratory is micropatterning GPC receptors using microcontact printing (*Nature Biotechnology* 1999). It uses SPR to detect binding.

In an exciting new methodology, the laboratory is patterning vesicles onto a surface. First, the researchers print 100 x 400 nm linelet features on a surface that bind cell surface ligands (e.g., streptavidin printed on the surface binds to a biotin ligand on the vesicle). Two different types of vesicles have been prepared containing red and green dye labels. When the vesicular solution is brought into contact with the patterned surface, a single vesicle attaches to each of the nanofeatures. Of the spots, 80–90% are occupied with a randomized distribution of red and green vesicles. The researchers have also employed laser tweezers to guide vesicles to individual spots. The laboratory has also developed a technique for converting native cells into vesicles by treating them with detergent and a toxin. The resulting vesicles are 0.5 to several microns in diameter and contain both cell surface and cytosolic contents representative of the cell from which they were derived. The long-term goal is to create a library of vesicles containing all the putative olfactory receptors (also GPC receptors) and to array the vesicles so as to create an artificial olfactory system.

Patch Clamp Technology. The laboratory reconstitutes membranes and inserts channel proteins. Immunosensing is performed by synthetic ligand-gated ion channels. Mellitin, a pore-forming protein, is used in conjunction with antibody epitopes such as malaria.

Planar patch clamps are also being fabricated. A Si chip with small holes is treated to modify the surface. Vesicles fuse and form an intact membrane over each hole in the Si chip. In this manner, single channels are created.

These systems can be used to correlate both electrical and fluorescence signals.

The laboratory is also involved in fluorescence correlation spectroscopy for single molecule detection, bio-functionalized quantum dots, and suppressor tRNA to introduce fluorescent labels into specific positions in proteins.

Site: **École Polytechnique Fédérale de Lausanne (EPFL)**
Institute of Molecular and Biological Chemistry
Laboratory of Physical and Analytical Electrochemistry
Lausanne CH-1015, Switzerland

Date Visited: 21 March 2003

WTEC Attendees: A.J. Ricco (report author), D. Brady, D. Walt, C. Kelly

Host: Professor Hubert Girault, Tel: +41 (21) 693 3151;
Fax: +(41) 21 693 3667;
Email: hubert.girault@epfl.ch

INTRODUCTION

The research topics under investigation in Professor Girault's laboratory include the following, as described in the laboratory website, dcwww.epfl.ch/le/:

Applied Electrochemistry

Electrochemical Microreactors/Integrated Chemical Microsystems

Interfacial Physical Chemistry

- Structure and reactivity of polarizable liquid/liquid junctions
- Self-assembly and specific adsorption at electrified interfaces
- Nucleation, optical properties and reactivity of nanoparticles
- Light energy conversion at molecular interfaces

Miniaturized Analytical Systems

- Theoretical and practical studies of mass transport in miniaturized total analysis systems (μ-TAS)
- Interfacing polymer microanalytical devices to mass spectrometers
- Theoretical and practical approach of the isoelectric separation of proteins by "off-gel" electrophoresis
- Enzyme-linked-ImmunoSorbent-Assays (ELISA) in polymer microchips
- Microchromatography

Finite Element Simulations

Analytical Experimental Methods

- Electrochemical
- Spectroscopic
- Bioanalytical

RESEARCH HIGHLIGHTS

Plastic Microfluidic Chip Microfabrication

The Laboratoire d'Electrochimie has well-recognized, long-standing expertise in the design and fabrication of plastic microfluidic chips, particularly using laser ablation and plasma processing to fabricate the microchannels. It also has specific expertise in sample injection methods, separation, and very notably in electrochemical detection. A current area of intense activity is the use of nanospray chips to couple fluidic devices, including separation systems, to mass spectrometry; this is discussed in more detail below.

Theoretical and Practical Studies of Mass Transport in μ-TAS

Significant effort is devoted to the understanding of mass transport in miniaturized analytical systems. Both simulation and experimental analyses are undertaken for optimization of mass transport by diffusion, convection, and migration.

Purely diffusional phenomena. Recently examples of unexpected diffusional behavior in microchannels have been observed during electrochemical detection. Microelectrodes inserted in a microchannel exhibit a nonconventional linear diffusion response due to the depletion of the volume along the microchannel direction. The electrochemical response and simulated behavior now match perfectly.

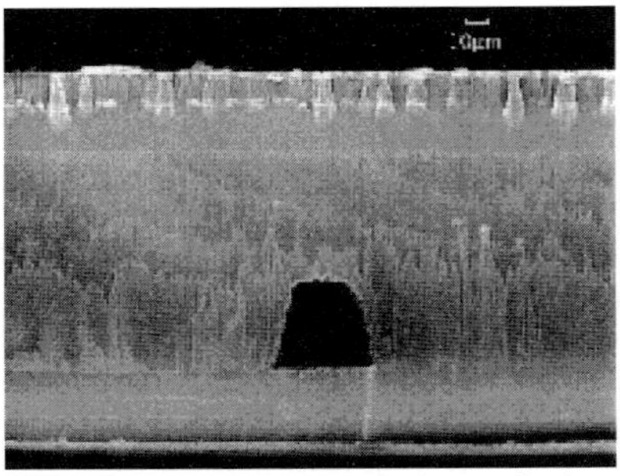

Fig. B.5. SEM view of a cross section of photoablated PET nanospray outlet used for MS analysis (channel dimension: 40 x 40 μm2).

Convection-migration.. Electroosmotic flow in polymer microchannels has been analyzed and calculated with 2D or 3D simulation tools. The effect of inhomogeneous microchannel surface charge density on the local electroosmotic flow velocity or Taylor dispersion has been examined. In order to avoid these effects, homogeneous microchannels were fabricated by means of plasma bonding instead of thermal lamination. This treatment enables enhanced performance for electrophoretic separations by at least one order of magnitude.

Interfacing polymer microanalytical devices to mass spectrometers

Polymer microchips have been shown to enable easy connection to mass spectrometers thanks to the capability for electrospray ionization. Photoablated or plasma-etched chips have been adapted to enable electrospray, with and without supporting external flow. Integrated microelectrodes can be placed in the microchannel in order to study the electrochemical behavior of this ionization source. Special attention has been paid to the spray conditions to reduce chemical noise as well as decreasing the limit of detection using disposable polymer nanoelectrospray chips. Preliminary experiments on the coupling of electrophoresis microdevices with mass spectrometry detection have been performed, showing promising results for direct spray from these microdevices.

Theoretical and practical approach to the isoelectric separation of proteins using "Off-Gel" electrophoresis

This technology for protein purification is described in the report on DiagnoSwiss, which is in the process of commercializing this development from Professor Girault's laboratories.

Enzyme-Linked-ImmunoSorbent-Assays in Polymer Microchips

Enzyme-Linked-ImmunoSorbent Assays (ELISA) with electrochemical detection has been developed in polymer-based microchannel devices in which an electrochemical detector is incorporated. Both photoablated and plasma-etched structures are used in a single-use immunoassay cell where antigens can be detected below 1 pM. The sandwich-type immunoassay process begins with a 1–5 min. incubation (rather than as much as one hour for standard immunoassays) in the microchannel, the short time being a result of the very short diffusional distance. Assays are realized by labeling the antibody that is the outer part of the sandwich with alkaline phosphatase, horseradish peroxidase, or β-galactosidase.

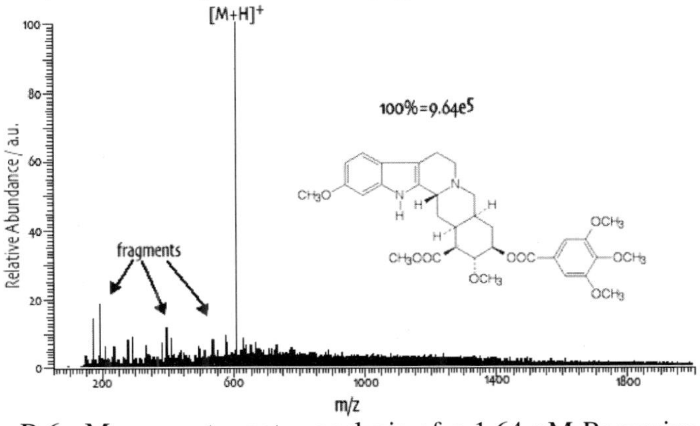

Fig. B.6. Mass spectrometry analysis of a 1.64 pM Reserpine sample sprayed with the chips presented in Figure B.5 with imposed flow rate. The spray voltage is applied through an embedded microelectrode located at 2.5 mm from the outlet (spray voltage: 7kV, pressure flow: 100 nl/min; sample in 50/49/1% methanol/water/acetic acid [v/v]).

Fortunately, many antibodies are commercially available already conjugated to one or another of these enzymes. Biotin-conjugated enzymes are also available, and these are readily bound to any enzyme labeled with avidin or streptavidin. The presence of this enzyme following sandwich formation and washing converts a substrate to an electrochemically detectable product; the rapidity of turnover for such enzymes yields an effective amplification of the signal, by generating several orders of magnitude more detectable redox species than the number of enzyme-labeled target species captured. In the case of alkaline phosphatase labeling, the substrate typically used is *p*-aminophenyl phosphate, which is converted rapidly to *p*-aminophenol by the enzyme. As little as 10^{-18} mole of target has been detected in this way, and the current best-case limit of detection is 35 fM (35 x 10^{-15} M).

Microchromatography

A novel mechanical injection method was developed to inject sub-nL sample plugs into microchannels by means of a multiport valve and syringe pumps. This injection scheme has the advantage of being independent of the composition of the sample and/or of the wall properties of the channel. Its development enables the integration of injector, separation column, and detector on the same chips for chromatographic purposes such as ion chromatography. Generally speaking, a first position of the multiport valve allows the preparation of the injection plug by pinching the sample within the intersection region of two channels, whereas the second position allows the sample injection, including a push-back effect in the side arms that enables very precise injection of nL volumes. The system can be used to prepare a sample plug by pressure in order to perform chromatography with a broad range of solutions that can be buffered or not. Fluorescent molecules were injected and detected in a photoablated planar polymer device to demonstrate the feasibility of the injection concept. With the pinching effect from two side channels in the microchannel network, this injection method is time-independent.

Finite-Element Simulation with Flux-Expert Software

The following systems and phenomena are under study using this software approach for FEM:

- Relief electrodes in flow channels
- Interdigitated band electrodes
- Liquid-liquid interfaces
- Electroosmotic flow at a T-junction
- Laminar micro-mixing
- Off-gel™ electrophoresis buffering
- Pinched pressure driven flow injection

REFERENCES

Bai, X., J. Josserand, H. Jensen, J.S. Rossier, and H.H. Girault. 2002. Finite element simulation of pinched pressure-driven flow injection in microchannels. *Anal. Chem.* 74:6205.

Bai, X., H.J. Lee, J.S. Rossier, F. Reymond, H. Schafer, M. Wossner, and H.H. Girault. 2002. Pressure pinched injection of nanolitre volumes in planar micro-analytical devices. *Lab on a Chip* 2:45.

Bianchi, F., Y. Chevolot, H.J. Matthieu, and H.H. Girault. 2001. Photomodification of polymer microchannels induced by static and dynamic excimer ablation: Effect on the electroosmotic flow. *Anal. Chem.* 73:3845.

Bianchi, F., H.J. Lee, and H.H. Girault. 2002. Ionode detection and capillary electrophoresis integrated on a polymer micro-chip. *Journal of Electroanalytical Chemistry* 523:40.

Bianchi, F., F. Wagner, P. Hoffmann, and H.H. Girault. 2001. Electroosmotic flow in composite microchannels and implications in microcapillary electrophoresis systems. *Anal. Chem.* 73:829.

FermÌn, D.J., H. Jensen, J.E. Moser, and H.H. Girault. 2003. Organisation and reactivity of nanoparticles at molecular interfaces. Part II. Dye sensitisation of TiO_2 nanoparticles assembled at the water I 1,2-dichloroethane interface. *ChemPhysChem.* 1:85.

Gobry, V., J. van Oostrum, M. Martinelli, T.C. Rohner, F. Reymond, J.S. Rossier, and H.H. Girault. 2002. Microfabricated polymer injector for direct mass spectrometry coupling. *Proteomics* 2:405.

Lion, N., V. Gobry, H. Jensen, J.S. Rossier, and H. Girault. 2002. Integration of a membrane-based desalting step in a microfabricated disposable polymer injector for mass spectrometric protein analysis. *Electrophoresis* 23:3583.

Mengeaud, V., J. Josserand, and H.H. Girault. 2002. Mixing processes in a zigzag microchannel: Finite element simulations and optical study. *Anal. Chem.* 74:4279.

Rohner, T.C., J.S. Rossier, and H.H. Girault. 2001. Polymer microspray with an integrated thick-film microelectrode. *Anal. Chem.* 73:5353.

Ros, A., M. Faupel, H. Mees, J. van Oostrum, R. Ferrigno, F. Reymond, P. Michel, J.S. Rossier, and H.H. Girault. 2002. Protein purification by Off-Gel electrophoresis. *Proteomics* 2:151.

Rossier, J.S., N.Youhnovski, N. Lion, E. Damoc, S. Becker, F. Reymond, H.H. Girault, and M. Przybylski. 2003. Thin-chip microspray system for high-performance Fourier-transform ion-cyclotron resonance mass spectrometry of biopolymers. *Angew. Chem. Int. Ed.* 42:53.

Rossier, J.S., C. Vollet, A. Carnal, G. Lagger, V. Gobry, H.H. Girault, P. Michel, and F.Reymond. 2002. Plasma etched polymer microelectrochemical systems. *Lab on a Chip* 2:145.

Rossier, J.S., F. Reymond, and P.E. Michel. 2002. Polymer microfluidic chips for electrochemical and biochemical analyses. *Review in Electrophoresis* 23:858.

Rossier, J.S., G. Gokulrangan, H.H. Girault, S. Svojanovsky, and G.S. Wilson. 2000. Characterisation of protein adsorption and immunosorption kinetics in photoablated polymer microchannels. *Langmuir* 16:8489.

Rossier, J.S., R. Ferrigno, and H.H. Girault. 2000. Electrophoresis with electrochemical detection in a polymer microdevice. *J. Electroanal. Chem.* 492:15.

Rossier, J.S., and H.H. Girault. 2001. Enzyme linked immunosorbent assay on a microchip with electrochemical detection. *Lab on a Chip* 1:153–157.

Roussel, C., T.C. Rohner, H. Jensen, and H.H. Girault. 2003. Mechanistic aspects of on-line electrochemical tagging of free L-cysteine residues during electrospray ionisation for mass spectrometry in protein analysis. *ChemPhysChem.* 4:200.

Schwarz, A., O. Bagel, and H.H. Girault. 2000. A sensitive electrochemical protein quantification method. *Electroanalysis* 12:811.

Site: **Griffith University, Gold Coast Campus**
PMB 50, GCMC
Queensland 9726, Australia

Date Visited: 12 April 2003

WTEC Attendees: David Walt (report author)

Host: Dr. Richard John, Chair, Electrochemistry Division; Royal Australian Chemical Institute and Senior Lecturer, Environmental Chemistry,
Tel: +61 7 555 28260; Fax: +61 7 555 28067;
Email: r.john@griffith.edu.au

BACKGROUND

Griffith University, located on Australia's Gold Coast, hosts nearly 9,000 students. The university is home to several major research facilities including the national Heart Foundation Research Centre and the Centre for Drug Discovery and Design, headed by Prof. Mark von Iztstein, developer of the of the influenza vaccine Relenza.

RESEARCH OVERVIEW

The University has three core faculty members whose main focus is on biosensors. The primary application is for environmental monitoring, primarily directed to water quality monitoring.

SPECIFIC PROJECTS

Chemical Oxygen Demand (COD) sensors have been developed that combine photocatalysis at TiO_2 surfaces with electrochemical detection. This approach allows the rapid determination of organics in waters and wastewaters by direct electrochemical measurement of the oxidation process.

Biological Oxygen Demand (BOD) biosensors are a major innovation of the laboratory. In this project, ferricyanide mediated microbial reactions are employed for the development of a rapid BOD assay. Multiple species of microorganisms are mixed with a water sample and ferricyanide, which acts an artificial terminal electron acceptor. The mixed culture enables close to full metabolism of the organic components in the sample in a very short period of time (< 3hrs). After a suitable incubation time, an electrode then measures the extent of the ferricyanide reduction product. By adding ferricyanide to the sample, the cells are relieved of the rate limitation imposed by transferring electrons (from the organics) to poorly soluble oxygen. The result is a much more rapid test for BOD compared to the conventional 5-day assay.

An exciting new project employs artificial neural nets to determine water quality. The researchers have made significant progress and have demonstrated the ability to determine overall water quality parameters rapidly using ANN pattern recognition coupled to simple physical (e.g., turbidity, pH, etc) and spectrophotometric measurements.

Finally, the group has been involved extensively in developing theoretical models to understand kinetic rate limitations of various biosensors. Its models involve both mass transport limitations as well as enzyme kinetic limitations to put biosensor performance on a sound theoretical footing.

REFERENCES

Catterall, K., K. Morris, C. Gladman, H. Zhao, N. Pasco, and R. John. 2001. The use of microorganisms with broad range substrate utilization for the ferricyanide-mediated rapid determination of biochemical oxygen demand. *Talanta* 55:1187–1194.

Jiang, D, H. Zhao, Z. Jia, J. Cao, and R. John. 2001. Photoelectrochemical behavior of methanol oxidation at nanoporous TiO_2 film electrodes. *Journal of Photochemistry and Photobiology A: Chemistry* 144:197–204.

Killard, A., S. Zhang, H. Zhao, R. John, E. Iwuoha, and M. Smyth. 1999. Development of an electrochemical flow injection immunoassay (FIIA) for the real-time monitoring of biospecific interactions. *Analytica Chimica Acta* 400:109–119.

Morris, K., K. Catterall, H. Zhao, N. Pasco, and R. John. 2001. Ferricyanide mediated biochemical oxygen demand: Development of a rapid biochemical oxygen demand assay. *Analytica Chimica Acta* 442:129–139.

Zhang, S.,H. Zhao, and R. John. 2000. Development of a generic microelectrode array biosensing system. *Analytica Chimica Acta* 421:175–187.

———. 2001a. Development of a quantitative relationship between inhibition percentage and both incubation time and inhibitor concentration for inhibition biosensors: Theoretical and practical considerations. *Biosensors & Bioelectronics* 16:1119–1126.

———. 2001b. A theoretical model for immobilized enzyme inhibition biosensors. *Electroanalysis* 13:1528–1534.

Site: **Institute for Chemical and Biochemical Sensors (ICB)**

Mendelstr. 7
48149 Münster, Germany
http://www.icb-online.de

Date Visited: 19 March 2003

WTEC Attendees: M. Mrksich (report author) and F. Heineken

Hosts: Professor Karl Cammann, Institute for Chemical and BioSensors
Dr. Torsten Börchers, Head of DNA Technology,
Tel: +49(0) 251 980-2880;
Fax: +49(0) 251 980-2890;
Email: t.boerchers@icb-online.de
Dr. Andreas Gorschluter, Group Leader of Microreactors, Institute for Chemical and BioSensors,
Tel: +49(0)2 51 980-2891;
Fax: +49(0)2 51 980-2891;
Email: a.gorschlueter@uni- muenster.de

BACKGROUND

The Institute for Chemical and BioSensors (ICB) was founded in 1991 by Professor Karl Cammann, a professor of analytical chemistry at Munster University. The ICB was established as a not-for-profit institute that was associated with the University. The ICB and University shared research infrastructure, and provided for Ph.D. student training in the Institute. A building was constructed for the ICB in 1994 at a cost of €15 million, with funding from the federal (90%) and state (10%) governments.

The institute focused on select areas of chemical and biochemical sensor research, with a strong emphasis on developing prototype sensors for commercialization. In prioritizing research efforts, the institute pays special attention to market opportunities for the proposed technology and patent positions for protecting developments, both of which are important for later transitions to spin-out companies. The philosophy is that the university research activity would undertake goal-oriented basic research in sensors, and the ICB would undertake a market-oriented applied program for research and development. In 2002, the institute staff included approximately 30 postdoctoral associates, 20 engineers, and 40 Ph.D. students, with an annual operating budget of nearly DM 10 million. Approximately 40% of the budget is derived from licensing income, with the balance divided equally between contract research, peer-reviewed federal funding, and public sources.

The ICB has an extremely impressive record of patenting and licensing of its technology. It has obtained more than 150 German patents—at a rate of about 1 per month—and 50 international patent families, of which 23 have been licensed to 15 companies. The licensing contracts generated several million DM of revenue, and had commitments for an additional several million DM. The worldwide market slowdown of 2002 resulted in the termination of deals with several partners and the loss of important operating funds. As a result, the ICB filed for insolvency at the end of 2002 and reorganized as a GMBH company co-owned by the Applied University of Gelgenkirchen and the City of Munster. The ICB retains a strong emphasis on transitioning basic research to commercial opportunities and expects programs to yield spin-out companies.

R&D ACTIVITIES

The site visit began with an overview of the ICB by Dr. Börchers. Professor Cammann then gave an overview of the platform technologies that have been developed at ICB. Dr. Börchers gave a technical presentation on the development of low- to mid-density DNA arrays for realtime hybridization monitoring. Dr. Gorschluter presented a basic science innovation based on electromagnetic beads for extremely sensitive detection. This formal session was followed by a lab tour to observe demonstrations on cell-scan optical devices, sensor systems, assay development for immuno-optical sensors, DNA sensor chips, and electromagnetic bead sensors. Following lunch, Dr. Degen described the background and future plans for gasBeetle, a spin-out company from the ICB. The panel then had a final discussion with the hosts and ended our tour with a reception hosted by the Rektor of the University.

Containment Platform Technologies

The ICB has developed and patented a containment platform technology, which integrates microfluidics and microelectrode-based assays for sampling and assaying on a single platform. The major application of this technology has been to a minimally invasive glucose sensor. In this program, a patch in contact with the skin contains a bubble that projects a microneedle. The pressure is reduced in the bubble (through the clever use of a vacuum container, and therefore without mechanical pumping), which results in a minor blistering of the skin and penetration of the skin by the needle. The interstitial fluid is then processed for an electrode-based measurement of glucose. This technology has been placed in a spin-out company, NIMOS, and has brought several million Euro of funding to the ICB. The distinguishing characteristic of this technology is its simplicity. The prototype sensor captures the mission of the ICB, which is to develop reliable and patent-protected technologies for real-world sensing.

DNA Sensor Chips

Dr. Torsten Börchers described the Institute's efforts to develop DNA chips for applications that include animal species differentiation (for validation of food imports), microorganisms, and point-of-care diagnostics. The ICB has focused on DNA arrays having less than 400 sequences, because these low-density arrays are outside the scope of the patent landscape and provide an entry to a commercial scale technology. Again, the emphasis has been on simple devices that integrate sample preparation, fluidics, and analysis. The devices comprise a simple flow cell that is designed to uniformly deliver a sample to an array. The detection is based on evanescent field fluorescence using Cy-5 labeled DNA (generated from PCR extension) and the use of a CCD camera to image the array. The strategy is also distinguished in that it employs a kinetic measurement, which offers the primary benefit that measurements can be made without calibration of the sensor.

An early application of the DNA arrays has been to differentiating animal species. The mitochondrial DNA of an unknown sample is PCR-amplified with primers that identify fingerprint regions for different species. An array of approximately 30 DNA sequences is sufficient for typing the sample. Another example uses an array having 132 elements to probe the 16s ribosomal RNA to determine bacterial levels. The ICB sees the greatest potential for this technology in the food industry. The next-generation versions of these technologies will move to a lateral scanning mode, since the cost of photomultiplier tubes is significantly less than the CCD cameras.

Cell Scan Technology

The ICB has developed a simple device for monitoring cell viability in microtiterplate cultures. The strategy is based on measuring molecular oxygen in the cell culture; it uses a polymer film that contains a chromophore whose fluorescence is quenched by oxygen to make the measurement. The top microtiterplate has been modified so that a cylindrical optical element having the polymer grafted at the end resides in the medium of each well. The optical path allows light to be coupled to the photoactive polymer layer and each well to be read in a scanning mode. This technology has been patent-protected and is now at the stage of identifying partners for commercialization.

Electromagnetic Bead Sensors

Dr. Gorschluter described a basic science effort in the company to develop a new sensor for detection of low numbers of analytes. The strategy is based on the binding of a magnetic bead to an electrode by way of ligand-receptor interactions, analogous to sandwich immunoassays. The binding of the bead to the microelectrode obstructs the flux of a soluble redox-active species to the electrode and results in a reduction in the steady state current. Further, trapped particles can be removed from the surface with a magnetic field, and the magnitude of the field required gives information on the strength of the molecular interaction. This technology, while still at a basic stage, offers many advantages over the more traditional AFM-based methods for measuring small numbers of interactions. Primarily, it can be performed with a small investment in instrumentation.

Technology Transfer

Intellectual property is of top importance to the ICB. The Institute recognizes that strong patents with worldwide coverage are essential to attracting investment for spin-outs and for ensuring the success of these ventures. The ICB has licensed nearly 40% of its international patents—a very impressive record. Further, a few key technologies have resulted in very large deals, including a €6 million initial payment for the noninvasive glucose-monitoring project in 2001. The panel also learned that the ICB offers a strong incentive to inventors of technology, which more closely resembles the policies of U.S. research institutions.

Assessment

It was clear that the ICB is entirely focused on developing technologies that can be successfully translated to companies for commercialization. This group—more than any other site visited—applies a sophisticated analysis to potential projects to ensure that the program, if successful in meeting the scientific goals, will have the combination of performance, cost structure, and patent position that will make it market-competitive. The ICB does not confuse its mission; it does not, for example, invest heavily in basic research. It also favors the simplicity of established science over the complexity of many of today's most innovative research programs.

The record of patenting and licensing are outstanding. The panel recognizes that the economic failure of the original institute reflects an ill-timed economic downturn and not a flaw in the operating model. Rather, this institute serves a unique mission in biosensor development and validates the role for institutes that border between academic units and start-up companies. It also reveals that there is a need for government support to weather the downturns of business cycles.

Site: **Linköping University**
Institute of Technology
Department of Physics and Measurement Technology, Biology, and Chemistry (IFM)
SE-581 83 Linköping, Sweden
http://www.ifm.liu.se

Date Visited: 19 March 2003

WTEC Attendees: A.J. Ricco (report author), D. Brady

Hosts: Professor Ingemar Lundström, Tel: +46 13 281200;
Fax: +46 13 288969;
Email: ingemar@ifm.liu.se
Professor Bo Liedberg

INTRODUCTION AND FUNDING SITUATION

Linköping University has three faculties: Philosophical, Technical, and Medical. The university's new president has a keen interest in biomedicine and biotechnology; at Linköping's Institute of Technology, he found solid expertise in biomedical engineering, particularly measurement technology, but recognized a need for more support in chemistry and biology if the university were to make a major impact in this field. Therefore, new faculty members were recruited to add synthetic chemistry and fundamental biology expertise to the applied physics/engineering faculty.

The Department of Physics and Measurement Technology, Biology, and Chemistry (known as IFM) is comprised of six research areas: Applied Physics; Biology; Chemistry; Environmental Technique & Management; Material Physics; and Theory & Modeling. Each of these research areas comprises a number of research groups, some being relatively diverse in their technical focus areas. For example, Applied Physics includes individual groups in Biomaterials, Catalytic Reactions, Biomolecular and Organic Electronics, Sensor Science, Sensor Science and Molecular Physics, The Swedish Sensor Centre (SSENCE), Scanning Probe Microscopy, and Applied Optics, as well as the Division of Biotechnology. Also included are two multi-institutional programs, Forum Scientium and Biomimetic Materials Science.

Professor Lundström emphasized the importance of *multi*disciplinary, as opposed to *inter*disciplinary, teams. The difference, reflected to a significant degree in the diversity of disciplines within the Applied Physics Area at Linköping, is that of closely linked involvement of many disciplines to achieve a single technical outcome, rather than carving a large program into separate funding blocks driving largely independent efforts in separate departments.

In an effort to develop effective, useful new technology, a concept known as the "triple helix" was described, representing the intertwining of academia, industry, and society in defining needs, then devising and implementing solutions. A recent Swedish initiative, called "New Tools for Life," will fund over a dozen new projects at a level of about $1.2 million per year each (total $16 million/year), each project for a total of ten years. The key goal of Linköping's project is to address the needs of the aging Swedish population by designing measures of health, not illness. A second goal is to develop primary care diagnostics that are better controlled and more accurate than those in use today. At the academic level, such technical targets can be very effective "grease" for the innovative system.

The "county" medical council in the Linköping area distributes funds for the social healthcare system, and also supports strategic research in technology and medicine at a level of about $4 million/year. The goal of this research is to save healthcare funds in the long run by adding to knowledge and understanding. One project at Linköping University funded by this agency is "Biomimetic Measurement Systems for Medical Diagnosis," targeted at new diagnostic arrays and systems for primary care as well as home use.

FORUM SCIENTIUM

The Forum Scientium is a multidisciplinary graduate school, established in 1996, that spans the Faculties of Health Sciences, Art and Sciences, and the Institute of Technology at Linköping. It encompasses biology, physics, chemistry, medicine, life science technologies, and biomedicine. A governing concept is that the research should be relevant to the advancement of Swedish industry. This program is supported by a number of the sub-councils of the Swedish Foundation for Strategic Research, including those in Electronics, Biomimetic Materials Science, and the Materials Research Programs in Low-Temperature Thin-Film Synthesis and in Nano-optical Systems.

The research topics being pursued by Forum Scientium participants include artificial proteins, biomimetic materials, distributed physics and chemistry, multisensing surfaces for proteomics, chemical sensing and brain computing, hydrogen extraction using anaerobic microorganisms, next-generation biosensors (diagnostic implants, artificial receptors, transduction cascades), protein-protein interactions, and soft lithography.

SWEDISH SENSOR CENTRE (SSENCE)

This Swedish national center of excellence for biological and chemical sensor technology is directed by Prof. Lars-Gunnar Ekedahl. Its central goal is to develop and evaluate chemical and biological sensors for industrial applications. SSENCE is one of 30 governmental centers of excellence of the Swedish National Board for Technical and Industrial Development (NUTEK). SSENCE is financed by NUTEK, Linköping University, and 12 industrial partners: AppliedSensor, AssiDomän Carton, Biacore, Duni/Finess, Volvo, Otre, Vinnova, Vattenfall, Asko Cylindra, Global Hemostasis Institute, Tekniska Verken, and Tetra Pak. More than 70 students, faculty, and staff are involved in seven interrelated projects, including senior researchers and Ph.D. students from both university and industry. Of these, 25 (including 13 Ph.D. students) are at Linköping University and another 45 are onsite at the industrial partners. The budget for three years is about $9 million. SSENCE has four main focus areas: gas-phase applications, including high temperature silicon carbide electronic devices for automotive applications; liquid-phase applications, notably the "electronic tongue"; biosensor applications, focused to a significant degree on surface plasmon resonance (SPR); and evaluation technology.

RESEARCH HIGHLIGHTS

Electronic Tongue

The basis for this exotic sounding sensor system is cyclic voltammetric measurement of electroactive species at multiple electrodes with different patterns of response; the electrode materials selected include Au, Ir, Pd, Pt, Re, and Rh. SSENCE researchers have demonstrated the capability to determine breakthrough of microbial contamination in a water quality treatment plant using this approach.

Sensor Science and Molecular Physics

Prof. Liedberg gave an overview of projects in his group in self-assembly, soft lithography, and molecular gradients; together, he referred to these projects as "surface science directed at sensing." Included in this research are projects in molecular design, synthesis, and self-assembly; biosensing chips, arrays, and imaging; and a molecular wires collaborative effort with Prof. David Allara at Pennsylvania State University. Included as well is a $600–700 thousand/year program (involving Chalmers and Lund Universities as well as Linköping) in biomimetic materials science, with focus on artificial cells, biofunctional materials and coatings, lipid-based surface modification, and liposomes on surfaces. One of the recent accomplishments of this program is fabrication of single carbon nanotube-connected networks of lipid vesicles, shown in Figure B.7. This project is exploring the possibilities of using lipid bilayer vesicles and lipid nanotube-vesicle-networks (NVNs) for encapsulation and support of reconstituted biological functions such as receptors, synaptic vesicles, and signal-transduction systems/pathways. A second theme is to integrate these structures into bioelectronic systems by use of soft and porous polymer electrodes. The integrated biological functions within lipid bilayer structures are complex biomimetic systems, approximating higher-order cellular structures. A key goal is to design such systems to respond to complex stimuli, affecting multiple receptor systems, and also to store and process this information.

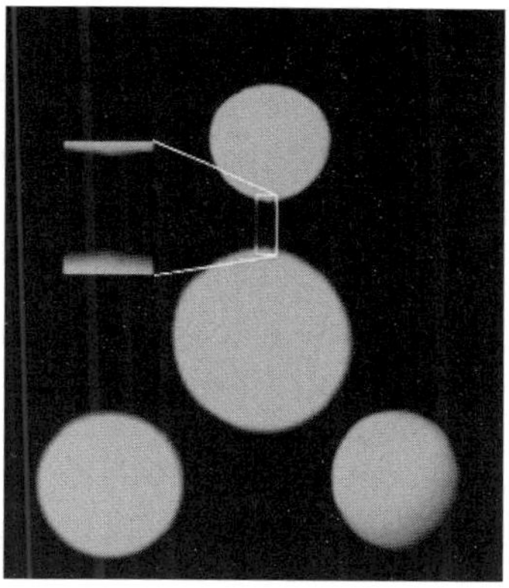

Fig. B.7. Networks of lipid vesicles comprised of single carbon nanotubes.

Molecular Design, Synthesis, and Self-Assembly

The group of Prof. Liedberg is using these experimental approaches in biosensors, biochips, arrays, and imaging. Helix-loop-helix polypeptides, oligosaccharide-specific receptors, and attachment strategies using ethylene glycol self-assembled monolayers are among the materials being examined. Decoration of the synthetic polypeptides with receptors and ligands has been demonstrated, including incorporation of lysine and histidine residues. Several specific examples are described below.

Quantitative Fluorescence Protein Affinity Assay

As in many enzymatic drug discovery assays, a fluorophore is attached near the ligation/binding region to make its fluorescence sensitive to the binding event. This has been used in a sort of affinity-based quantitative assay, wherein spots on a microarray were created with a range of binding constants from 10^{-4} to 10^{-9} M; the particular subset of spots changing their fluorescence in response to the presence of analyte thus indicates its concentration.

Urinary Tract Infection Sensor

Using surface plasmon resonance (SPR), the group of Prof. Lars Baltzer developed an assay for urinary tract infections. A trisaccharide (globotrios) that interacts selectively with an antibody specific to the infection was immobilized on the gold-coated SPR sensing surface using an alkane thiol tail. Binding of the antibody gives a measurable SPR response. The optimal response was found for a surface alkane thiol coverage of about 1%. A company, ModPro (modular protein), spun out of Linköping University to commercialize this assay.

Quartz Microbalance Explosive Detector

Quartz crystal microbalances were used to detect TNT via displacement binding to surface-attached antibodies, yielding a significant mass change as a result of displacing the antibodies from the device surface upon binding to the TNT; this provides a mass amplification in proportion to the ratio of masses of the antibodies relative to the TNT. The amount of signal was further optimized through the use of mixed alkane thiol monolayers for the attachment of antibodies to the gold-coated QCM surface. This technology led to the spin-out of a company called Biosensor Applications, focusing on detection of narcotics and explosives.

Collaborations: Molecular Wires and Nanoparticles for DNA Detection

In collaboration with Prof. David Allara of Pennsylvania State University, Prof. Liedberg's group is making molecular wires based on diacetylenic alkane thiol monolayers. A single Pt nanoparticle attached to a scanning probe microscope tip is used to make contact with the surface-attached molecular wires one by one.

In collaboration with Prof. Chad Mirkin, the details of the nucleotide-gold interaction are being studied by thermal desorption; the work is jointly funded by the (U.S.) National Science Foundation and its Swedish counterpart.

Photoluminescent Readout of Biomolecular Interactions

Work on this project by the group of Prof. Olle Inganäs focuses on uses for conjugated polymers in biochips. They have synthesized a derivative of the photoluminescent organic conducting polymer polythiophene that has zwitterionic peptide sidechains, enabling it to interact directly with various biomolecules; it is known as POWT. They study electron transfer, for example with redox enzymes, and the effects of geometric changes in the polymer chains, on the luminescent properties. Mixing this material with 12-base-long single-stranded (ss) DNA, for example, results in an observable shift from red to green in the luminescence spectrum. Adding the complementary ss-DNA then causes the spectrum to shift back to its original peak in the green due to the different structural changes imposed by single- and double-stranded DNA within the polymer matrix. Prof. Inganäs's group has shown the differentiation of 0, 1, 2, or 3 single-base mismatches between two complementary DNA strands according to the ratio of luminescent intensities. One picomole of DNA has been measured in 10 μL of solution. A DNA microarray concept has also been developed in which ss-DNA oligomers are immobilized on a surface, POWT is present in solution, and the binding of the complementary strand of ss-DNA affects the luminescence signal in the corresponding surface location.

This group has also made synthetic 42-amino acid polypeptides, incorporating 4 helices, designed (and verified) to complex with POWT.

Computer Screen Photo-Assisted Technologies

In a novel concept, an engineering project is under way to use the multispectral, two-dimensionally patterned, time-varying, intensity-controlled light emission capabilities of a computer screen as a light source for a range of optically based assays. This effort, led by Dr. Daniel Fillippini, aims to take advantage of such measurement approaches as light-assisted potentiometric spectroscopy (LAPS, pioneered by Molecular Devices and Stanford), light pulse techniques, and absorbance spectroscopy. An example application for which this approach is being examined includes the antibiotic resistance of different types of cells at different concentrations. This approach is touted as having some of the capabilities of a multiwell microplate reader, but of course, the approach will give up many orders of magnitude in limits of detection, limiting its utility to carefully designed assays for which sensitivity is not paramount.

An example of the biological systems that might be used effectively with this measurement approach are specific types of pigmented cells that either aggregate or disperse pigment particles to effect a color change in response to a stimulus (the mechanism by which an octopus changes color, for example). By engineering such cells to recognize target analytes, the burden of specificity and, to a major degree, sensitivity as well, is transferred to the biological system, leaving a fairly simple detection task for the computer screen and simple detector. Immortal frog cells were transfected to yield sensitivity to morphine at the concentrations typical of blood levels characteristic of the medical use of this opiate. It is claimed that detection of some toxins at femtomolar concentrations is possible with this approach.

Electric field-induced surface photovoltaic spectroscopy utilizes a two-dimensional readout in which excitation wavelength varies along one axis while the electric field applied to a 2D semiconductor detector varies along an orthogonal axis (producing a varying surface potential along the surface).

Meat freshness is being assayed using a scanning light pulse technique, where transient photocurrents are produced by the 85 Hz refresh rate of the computer screen. An MOS capacitive sensor using patches of Pd, Pt, or Au film is being combined with a heater based on a projector lamp to provide a thermal gradient that offers a variable parameter for response mapping to identify analytes. With the addition of an inexpensive Fresnel lens, lateral resolution of 200 μm can be obtained.

A new company in a building under construction during our visit will house a company charged with commercializing the computer screen-based technology. This technology incubator building is partially funded by the Linköping community of 170,000 residents, which donated the land where the building is sited.

REFERENCES

A series of detailed reports, including extensive lists of references, is available at www.ifm.liu.se/ar/.

Edwin, W., H. Jager, O. Inganäs, and I. Lundström. 2000. Microrobots for micrometer-size objects in aqueous media: Ppotential tools for single cell manipulation. *Science* 288:2335–2338.

Enander, K., G. T. Dolphin, L. K. Andersson, B. Liedberg, I. Lundström, and L. Baltzer. 2002. Designed, folded polypeptide scaffolds that combine key biosensing events of recognition and reporting. *Journal of Organic Chemistry*. 67 (9): 3120–3123.

Holmin, S., C. Krantz-Rülcker, I. Lundström, and F. Winquist. 2001. Drift correction of electronic tongue responses. *Measurement Science and Technology*. 12:1348–54.

Holmin, S., P. Spångeus, C. Krantz-Rülcker, and F. Winquist. 2001a. Compression of electronic tongue data: A comparative study. *Sensors and Actuators B*. 76:454–463.

———. 2001b. Compression of electronic tongue data: A comparative study. *Sensors and Actuators B*. 76:454–463.

Immerstrand, C., K. Holmgren Peterson, K.E. Magnusson, E.W.H. Jager, M. Krogh, M. Skoglund, A. Selbing, and O. Inganäs. 2002. Conjugated-polymer micro- and milliactuators for biological applications. *Materials Research Society Bulletin*. 27 (6):461–464.

Ivarsson, P., S. Holmin, N-E. Höjer, C. Krantz-Rülcker, I. Lundström, and F. Winquist. 2001. Discrimination of tea by means of a voltammetric electronic tongue and different applied waveforms. *Sensors and Actuators B*. 76:454–463.

Ivarsson, P., Y. Kikkawa, F. Winquist, C. Krantz-Rülcker, N-E. Höjer Hayashi, K. K. Toko, and I. Lundström. 2001. Comparison of a voltammetric electronic tongue and a lipid membrane taste sensor. *Analytica Chimica Acta*. (449): 59–68.

Jager, E.W.H., C. Immerstrand, K. Holmgren Peterson, K.E. Magnusson, I. Lundström, and O. Inganäs. 2002. The cell clinic: Closable microvials for single cell studies. *Biomedical Microdevices.* 4 (3):177–187.

Jager, E.W.H., E. Semla, and O. Inganäs. 2000. Microfabricating conjugated polymer actuators. *Science* 290:1540–1545.

Jungar, C., and C.-F. Mandenius. 2001. Neoglycoconjugates as affinity ligands in surface plasmon resonance. *Analytica Chimica Acta.* 449:51–58.

Jungar, C., M. Strandh, S. Ohlson, and C.-F. Mandenius. 2000. Analysis of carbohydrates using liquid chromatography-surface plasmon resonance immunosensing systems. *Analytical Biochemistry* 281:151–158.

Karlsson, A.M., K. Bjuhr, M. Testorf, P. Å. Öberg, E. Lerner, I. Lundström, and S. P. S. Svensson. 2002. Biosensing of opioids using frog melanophores. *Biosensors & Bioelectronics* 17:331–335.

Karlsson, L.M., P. Tengvall, I. Lundström, and H. Arwin. Adsorption of human serum albumin in porous silicon gradients. *Physica Status Solidi (a)* 197 (1/2).

Noort, D. V., J. Rumberg, E.W.H. Jager, and C.F. Mandenius. 2000. Silicon based affinity biochips viewed with imaging ellipsometry. *Measurement Science & Technology* 11:801.

Ohlson, S., Ch. Jungar, M. Strandh, and C-F. Mandenius. 2000. Continuous weak affinity immunosensing. *Trends in Biotechnology* 18 (2):49–52.

Svedhem, S., C.-Å. Hollander, J. Shi, B. Liedberg, P. Konradsson, and S.C.T. Svensson. 2000. Synthesis of a series of oligo(ethylene glycol) terminated alkanethiol amides designed to address structure and stability of biosensing interfaces. *Journal of Organic Chemistry* 66 (13):4494–4503.

Svedhem, S., K. Enander, M. Karlsson, H. Sjöbom, B. Liedberg, S. Löfås, L.-G. Mårtensson, S.-E. Sjöstrand, S. Svensson, U. Carlsson, and I. Lundström. 2001. Subtle differences in dissociation rates of interactions between destabilized human carbonic anhydrase II mutants and immobilized benzenesulfonamide inhibitors probed by a surface plasmon resonance biosensor. *Analytical Biochemistry*. 296:2 188–196.

Winquist, F., S. Holmin, C. Krantz-Rülcker, P. Wide, and I. Lundström. 2000. A hybrid electronic tongue. *Analytica Chimica Acta.* 406 (2):147–157.

Site: **Oxford Glycosciences (UK), Ltd.**
The Forum 86 Milton Park
Abingdom Oxon OX14 4RY
United Kingdom

Date Visited: 17 March 2003

WTEC Attendees: C.L. Wilkins (report author), C. Kelley, D. Walt

Hosts: Peter K. Zagorin, Vice President, Marketing,
Tel: +44 (0) 1235 208000;
Fax: +44 (0) 1235 208230;
Email: peter.zagorin@ogs.co.uk
Christian Pohlff, Director of Proteome Research
Jim Bruce, Director of Protein Separations

INTRODUCTION

Oxford Glycosciences (OGS) is a company specializing in the general area of glycobiology and glycoproteins, with principal emphasis on applications to proteome research. The primary tool used for this work is 2D gel-based mass spectrometry, employing highly automated 2D-gel ICAT techniques to do high-throughput protein analysis, The company has approximately 200 employees, of whom about one-third are devoted to proteomics applications. Through an agreement with the Institute for Systems Biology (Seattle, WA), the patented ICAT technology is licensed to OGS, and joint patents are being developed for intellectual property developed as a result of collaborative research.

BIOMARKER STUDIES

An important goal of many of the studies Oxford Glycosciences conducts is to identify biomarkers that can be used for clinical studies. A recent example of this activity is a contract study of Alzheimer's disease carried out for Pfizer and the National Institute of Mental Health (U.S.), where clinical samples from a 1000-patient group were investigated. As mentioned above, automated processing of 2D gels has been developed, and techniques that allow archival storage of the 2D separated samples have resulted. As currently implemented, the methodology allows ICAT mass spectral processing of about 150 2D gels per week. However, the equipment is available and the long-term goal is to implement the capability to process about 150 gels per day, allowing large-scale clinical studies, which will ultimately contribute to development of a human protein atlas that will become an important tool in human proteome research. It is intended to expand clinical studies into studies of oncology and other disease states. The primary substances in previous studies have been IGG, imunoglobulin, haptoglobin, and transferin. It is expected that other proteins will be added to this list in the future. The primary analytical mass spectrometry approaches involve matrix-assisted desorption/ionization time-of-flight mass spectrometry (MALDI-TOF) and quadrupole time-of-flight mass spectrometry sequencing. There are plans to link proteomic information to determination of posttranslational modifications using Sequest software. Planned future clinical studies will soon focus on inherited glycolipid storage disorders.

Site:	**Potsdam University** **Karl-Liebknecht-Str. 24-25, Haus 25, B 2.13-2.14** **14476 Golm, Germany**
Date Visited:	18 March 2003
WTEC Attendees:	M. Mrksich (report author), F. Heineken, J. Schultz
Hosts:	Professor Frieder W. Scheller, Professor forAnalytical Biochemistry, Tel: +49(0)3 31 977-5121; Fax: +49(0)3 31 977-5050; Email: fschell@rz.uni-potsdam.de Dr. Frank F. Bier, Head of Molecular Bioanalytics and Bioelectronics Department, Fraunhofer Institute for Biomedical Engineering, Tel: +49(0)332 00/88-378; Fax: +49(0)3 32 00/88-452; Email: frank.bier@ibmt.fhg.de

BACKGROUND

Potsdam University is located 30 km southwest of Berlin. The University has a student enrollment of approximately 16,000 and offers courses in the broad range of disciplines. The University and surrounding area have benefited from a substantial investment in research infrastructure and activity over the past several years. This investment includes the establishment of Max Planck and Fraunhofer Institutes, and new University research buildings dedicated to the biological sciences. Professor Scheller leads the major biosensor program on campus. He is widely recognized both in Germany—where he holds the only chair in analytical biochemistry—and internationally. The University has several active collaborations with the neighboring Institutes, including one with Dr. Frank Bier, who heads the Molecular Bioanalytics and Bioelectronics Department within the Fraunhofer Institute for Biomedical Engineering.

R&D ACTIVITIES

The site visit began with an overview of Potsdam University and neighboring institutes by Professor Scheller. He then described the current research thrusts in his group. Dr. Frank Bier then described the missions of the Fraunhofer Society and of the Institute for Biomedical Engineering. The panel was given several laboratory demonstrations of current work and then finished with an open discussion during lunch.

The Scheller research group develops sensors for a variety of purposes, with an emphasis on electrochemical methods to measure enzyme-generated species. The group has a particular focus on the development and evaluation of biomimetic recognition elements, including RNA aptamers, peptide nucleic acids, synzymes, imprinted polymers, and ionophores.

Imprinted Polymers

The Scheller group has collaborated with Professor Mosbach (Lund) to develop imprinted polymers as recognition units in sensors. The team has explored a strategy wherein a transition state analog for ester hydrolysis is first immobilized to silica beads, and then treated with a solution of monomer and polymerized. The volume of monomer is kept small, so that the reagent is taken up within the pores of the bead and results in polymerization only in those regions. In this way, the beads remain dispersed following the polymerization. Etching with HF dissolves the silica beads and yields a powder of imprinted polymers. Early experiments with imprinted polymers prepared from styrene and an imidazole functionality point to better activity over conventional methods for preparing imprints. Because the haptens are immobilized to a solid phase prior to imprinting, this method is expected to give material with far higher binding capacity than do traditional approaches.

Aptamers

Dr. Bier and Professor Scheller have developed a next generation version of the SELEX method. This method—termed MONOLEX—recognizes that when a pool of RNA is applied to a chromatography column that is derivatized with the analyte, the tightest binding members of the pool will be localized at the front end of the column. Hence, instead of eluting all bound RNAs, as in SELEX, the column material is sliced into sections, and the topmost section is used to isolate sequences for subsequent rounds of amplification and selection. This method has been patented, making it attractive to industrial partners that seek alternatives to the SELEX method. Related work in Germany (in the group of Furste) has developed a clever method for creating enantiomeric RNAs, which are much more stable to enzymatic degradation. The method starts with the conventional process to isolate an active RNA sequence but is targeted towards the enantiomer of the target analyte. Chemical synthesis of the enantiomeric RNA sequence then gives a reagent that recognizes the desired analyte.

Single Cell Studies

This team is moving forward with a program to apply its long expertise in electrochemical sensors to measurement of analytes within a single cell. The strategies use electrical probes that are modified with the appropriate sensor surface chemistries that are inserted into live cells. This work relies on a collaboration with the nearby Max Planck Institute on Plant Physiology and is applying the sensors to the realtime measurement of metabolites in single cells. The motivation for using these approaches, as opposed to the current mass spectrometry methods, is that the former will give kinetic information on metabolite flux. The Scheller group is also exploring new opportunities for using redox manipulations in the cell to control signal transduction processes.

Clinical Applications

This group has developed an electrochemical sensor for measuring superoxide in tissue. The sensors are based on electrodes having an immobilized layer of cytochrome c. The iron group is oxidized by superoxide, which can subsequently be detected by an electrochemical reduction of the protein. This sensor is positioned on a probe that can be inserted into tissue and can provide a realtime measure for biological stress.

Education

Potsdam University offers excellent courses in analytical biochemistry for undergraduate students. As many as 40 students are enrolled annually in a research course that introduces students to the experimental methods of biosensors. As is traditional in Germany, many of the graduate students who apply to Potsdam do so in response to the research program of a specific faculty member.

Funding

Professor Scheller's group includes 25 coworkers. The local government in Brandenburg provides funding for five of these coworkers. The majority of funding comes from peer-reviewed grants to the German agencies (Ministry and basic research) and approximately 10% from the European Union.

Technology Transfer

The environment for commercialization of university research is quite good. A recent system-wide change in Germany now directs that each university patent and own its intellectual property. This arrangement is beneficial to the creation of professor-led companies. It has in some cases, however, made it difficult to interact with industry, which has generally required ownership of patents. Nonetheless, there are several examples of start-up companies that were founded on the basis of university research. As part of the campus planning, Potsdam University is building a technology park that will provide space to start-up companies. Overall, there is a reasonably clear route for professor-entrepreneurs to transfer research to a commercial setting. The panel also noted that the technology transfer activities in Germany are somewhat understated: the ratio of substance to hype is quite high.

FRAUNHOFER INSTITUTE

Dr. Frank Bier described the programs at the nearby Fraunhofer Institute. The Fraunhofer Society supports 57 institutes across Germany, with a total of 12,000 employees. Unlike the Max Planck Institutes, the Fraunhofer Institutes emphasize industrial research and must raise approximately 70% of their budgets from industry (half of which represent industrial contracts for specific tasks). The German government provides approximately 30% of the Institutes' budgets. Dr. Bier belongs to the Institute for Biomedical Engineering and is in the Medical Biotechnology branch, which has programs in nanobiotechnology; cryobiotechnology; bioanalysis and bioelectronics; and biochips. A common thread in these programs is that they emphasize bio/hybrid technologies, with particular attention to integration of component technologies and greater complexity.

One set of projects uses standard microarraying technologies to create arrays of DNA, small molecules (steroids), and antibodies. These programs are product-oriented; the goal is to utilize existing methods to develop assays for designated panels of analytes and not to develop novel scientific approaches to these problems. In the antibody arrays, for example, the antibodies are synthetically labeled with biotin and immobilized to glass slides coated with a layer of streptavidin. The team does invest some effort in developing new surface chemistries, for the dual purposes of reducing nonspecific adsorption of protein and generating patents (the latter is a necessity for industry buy-in).

Dr. Bier and his group have developed a useful method for printing microarrays. They adapt the TopSpot method, which is based on a stamping structure that allows several source solutions to be placed in wells that wick the solutions to a microarray of channels. In this way, once filled, the structure can be used to stamp multiple copies of an identical array. The team has also formatted fluorescence-based assays for measuring DNA hybridization, restriction endonuclease activity on the chip, and PCR amplification with immobilized primers. The team is particularly interested in replacing end-point assays with kinetic assays that monitor ligand-receptor interactions. This interest has motivated much work with lab-on-a-chip systems, and the integration of assays, microfluidics, and fluorescent detection methods.

A final interest of the group is in the area of nanobiotechnology; its aim is to immobilize DNA strands to surfaces with control at the single-molecule scale. This effort is still in the basic science and exploratory stage, and the group is developing methods for immobilizing and characterizing single DNA molecules. The longer-term goals are to develop bioanalytical systems that are precisely ordered at the nanometer scale, but the form these opportunities will take is still unclear.

Assessment

The Scheller group provides a center of expertise in Germany for biosensor development. This group has a particular expertise in enzyme-linked electrical detection of biological analytes, and it leverages these platforms in two ways. In the first, it provides model systems for the development and evaluation of nontraditional recognition motifs that would replace antibodies. The group has a program to develop recognition units, but in addition has many collaborations with outside groups that provide recognition reagents. The strength of this program is that it provides an effective environment for evaluating this class of biosensor components; it does not lead the way in early innovations of new classes of recognition units.

The second way in which this group leverages its expertise is to partner with collaborators that have specific applications. The integration of a sensor for measurement of superoxide has a clear motivation in the surgical sciences. The Potsdam group has the expertise to devise and implement the appropriate sensing designs in prototype devices. In this way, this group provides an important contribution to sensor development in Germany.

The path for translating University research to sensor development within small companies is very good at Potsdam. Further, this group is well recognized by large companies, and serves as an important collaborator and provides well-trained employees to the industry. The major mode for commercialization relies on spin-out companies that mature the technology. Compared to other sites in Germany, the sensor technologies are at a less mature stage when transferred to private industry.

The establishment of a Fraunhofer Institute in Potsdam provides an important partner that can aid in the further development of biosensors before commercialization. Although just beginning, this Institute has a strong focus on commercially viable and market-driven projects. Indeed, the significant investment in research infrastructure in the greater Berlin area will make this region highly competitive in biosensor research and development.

Site:	**Ruprecht-Karls University Heidelberg** **Physical Chemistry Institute (APC)** **Im Neuenheimer Feld 253** **D-69120 Heidelberg, Germany** **http://www.pci.uni-heidelberg.de/apc/**
Date Visited:	20 March 2003
WTEC Attendees:	A.J. Ricco (report author), D. Brady
Hosts:	Professor Dr. Michael Grunze, Tel: +49 6221 548461; Fax: +49 6221 546199; Email: michael.grunze@urz.uni-heidelberg.de Professor Joachim P. Spatz Dr. Reiner Dahint

INTRODUCTION

Professor Grunze's research group investigates the chemistry and physics of surfaces and interfaces. This group is very strong in the detailed study of fundamental physical/chemical interactions that are key to the development of effective biosensing strategies, including the origins of biocompatibility of surfaces and the root causes of protein adsorption.

As noted on Dr. Grunze's APC website, the Institute's research is focused on the synthesis and the physical and chemical characterization of ultrathin organic films. Work includes the use of organic monolayers and polymer films in chemical and biochemical sensors in basic research as well as in applied projects in collaboration with industry, with work in the sensors area being led by Dr. Dahint. The development of novel spectroscopic and microscopic methods is a major aspect of the research. In more applied projects, organic films are examined for use as biocompatible coatings for medical implants, in corrosion inhibition, and in high-resolution nanolithography. Work presently includes the following topics:

- Self-assembled monolayers
- Polymer-metal(oxide) interfaces
- Low-energy electron point source (LEEPS) microscopy
- Chemical nanolithography
- Interaction of biological species with artificial surfaces, including application to chemical and biochemical sensors
- Coating of medical implants

- Materials chemistry
- Low-pressure synthesis of diamond films
- Electrochemical properties of self-assembled monolayers
- Mechanical properties of organic monolayers
- Surface science and analysis
- Industry collaborative projects: corrosion, biosensors, nano lithography

The group of Professor Joachim Spatz, with expertise in chemistry, physics, and biology, applies unusual but complementary combinations of the techniques of biology and materials science, including custom polymer manipulations, to address various physical and biologically inspired questions. In particular, they are interested in the use of nanotechnology to study biological systems and their underlying mechanisms, with emphasis on signal transduction. A key project area is the nature and manipulation of cell adhesion and activation.

RESEARCH HIGHLIGHTS

Polyphosphazene Coating for Medical Implants

For long-term implantation of biosensors, device surface biocompatibility is a critical issue. The objective of this project is to develop methods for the coating of implants with poly([bistrifluorethoxy] phosphazene), a coating material developed initially by a Russian group for coating vascular implants, and to investigate these coatings in vitro and in vivo. This coating has shown major success in the coating of cardiac stents that have been implanted in many patients. The stents show no encapsulation after months of implantation. They are found to coat initially with a native protein layer, which is then covered by epithelial cell growth to a maximum thickness of 100–200 μm; there is no restenosis. Use in dental implants results in no open pockets and perfect closure of gums around the implants. Surface analytical tools such as XPS and AFM are used to characterize the coatings, and investigation is ongoing to understand how these particular phosphazene coatings prevent thrombus formation, which is measurably lower than on PTFE or polyethylene surfaces. ELISA-like immunological methods are used to measure protein adsorption from blood plasma onto polyphosphazene-coated and reference samples. The coating material is found to have a relatively high, negative surface charge; essentially, it is an electret material. The high molecular weight ($\sim 10^7$) and all-linear structure of the Russian/Heidelberg version of this coating are thought to be key to its biocompatibility. Further, it remains relatively elastic from -70°C–260°C. Techniques have recently been developed to form pores of controlled sizes in this material for controlled-release applications. This project is conducted in collaboration with external groups and industrial partners, notably Polyzenix, a company now numbering 5 employees, which has commercialized cardiac stents with this coating.

Dr. Reiner Dahint: Surfaces for Biosensors

The research activities of Dr. Dahint's Biosensors and Biomaterials group (see APC website) are focused on a detailed understanding of the interaction of proteins and cells with artificial surfaces, which is important for many medical and biochemical applications. Investigations are conducted into the specific detection of molecules in solution and the relationship between physical/chemical surface parameters and nonspecific adsorption of proteins and cells. Not only homogeneous, but also chemically microstructured surfaces are of major interest. By the use of acoustic wave-based and optical diffraction-based sensors, adsorption and binding processes are monitored *in situ* without the necessity for labeling the molecules of interest. Derivatized surfaces are characterized by surface analytical techniques such as infrared spectroscopy (sum-frequency generation), X-ray photoelectron spectroscopy, neutron reflectivity measurements, and atomic force microscopy, as well as by standard immunoassays. Investigations are directed towards precise control of adsorption and binding processes via specific variations of surface properties.

For biosensing applications, protein adsorption can block desired receptors and/or bind nonspecifically, causing false or irreproducible signals. Answers to this problem include deliberate pre-adsorption of a well-characterized protein layer, such as bovine serum albumin or protein A; the formation of so-called mixed monolayers, where about 2% of the surface is terminated in the functional group of interest and the remaining 98% in an "inert" surface like poly(ethylene glycol); or functionalizing the surface with a polymer that contains "occasional" functional groups. Protein-resistant surfaces are found to have generally repulsive interactions—apparently ionic in nature—as characterized by AFM (atomic force microscopy).

Interactions of Water with Self-Assembled Monolayers (SAMs); Implications for Protein Adsorption

The relatively biocompatible ethylene glycol-terminated SAMs studied are found to have a 3–5 nm-thick layer of water adjacent to the surface that is some 85–90% of the density of bulk water; this particular surface has a contact angle of 65°; neutron reflectivity shows that this thin layer is depleted in protein concentration relative to bulk solution. Hydrophobic surfaces, such as a C18-terminated alkane thiol monolayer with a contact angle of 120°, have a similar low-density water layer that is about 10 nm thick; but proteins adsorb to a greater extent on this surface. Highly hydrophilic surfaces have no such low-density layer, and in general also exhibit more protein adsorption than the glycol-terminated SAM.

A related finding is that for a broad range of different types of surfaces, the solvation energy for OH^- is about 1/2 that of the energy for H_3O^+. The consequence is that, at neutral pH, most surfaces are negatively charged. In the case of the favorable ethylene glycol-terminated surface, a relatively low lateral density of hydroxyl groups allows water to penetrate some way into the surface of the film, which in turn helps prevent the protein from "squeegeeing away" the negatively charged hydroxyls as it approaches the surface. The consequence is that proteins do not adsorb. This condition is disrupted by high ionic strength or low pH, which thus results in greater protein adsorption.

Low-Energy Electron Point-Source (LEEPS) Microscopy

The understanding of biological systems requires an in-depth knowledge of the structural, mechanical and electrical properties of the involved macromolecular entities. Holographic imaging with LEEPS microscopy is a novel method to characterize single molecules. Work of this group (see APC website) is centered on construction and improvement of microscopes as well as the development of strategies for the preparation and identification of macromolecular entities. In the current instrument, they can image and reconstruct single polymer strands like DNA or nanotubes and observe ~2–3 nm-size features. Simulations performed in close collaboration with the LEEPS theory group at Dalhousie University (Halifax, Canada) show that the resolutions of features below 1 nm should be feasible.

Chemical Nanolithography

The effective fabrication of chemically defined surface structures is an important objective in nanoscience. A promising strategy towards this goal is the combination of well-established "top-down" lithographic techniques with novel self-assembling materials. In the lithography group (see APC website), electron beams generate "chemical nanostructures" in self-assembled monolayers. Terminal nitro groups in aromatic self-assembled monolayers are locally converted into amino groups while the underlying aromatic substrate is dehydrogenated and cross-linked. Other molecules may be coupled to the surface amino groups in a second step, producing molecularly distinguished nanostructures. The procedure can be multiply repeated so that nanopatterns with different chemical characteristics may be generated on one surface. For patterning, proximity printing with stencil masks is used, as well as direct writing with electron beams. The method has significant potential in future device fabrication, in particular for applications in electrochemical nanotechnology, molecular electronics, and biotechnology.

Industrial Collaborations: Corrosion, Biosensors, and Nanolithography

Heidelberg's Applied Physical Chemistry Institute has several collaborative projects with industrial partners (see APC website). Besides the application of surface science to "real-life" problems (chemical composition, morphology of surfaces), they also work on technology transfer, in particular in the areas of novel corrosion inhibitors based on self-assembled monolayers, application of ultrathin organic films for biosensors, and application of ultrathin organic films as resists in nanolithography.

Professor Joachim Spatz: Nanofabrication Applied to Cell Adhesion and Activation

One current area of interest and activity in the Spatz group is shown schematically below. They seek to study and control the process of cellular adhesion at the molecular scale using micro-patterned surfaces of bio-functionalized nanoparticles, focusing particularly on the various scales of protein organization. The adherent cells under study bind to fibronectin via actin filaments; the fibronectin binds to an integin "head" on the cell surface. Spatz and colleagues have fabricated nanometer structures small enough so that just one integrin binds on each artificial "spot" on the tailored surface.

The Spatz group displays an impressive ability to control the size, density, and overall geometry of patterns of nanometer-size "patches" of specific surface chemical functionalities relevant to biological attachment and interactions. This is a powerful tool for the study of the mechanism and the general nature of cellular adhesion. Further, since key cellular processes including growth and communication involve surface adhesion (for many cell types), local control of attachment morphology and chemistry provides an outstanding means to understand these phenomena at a fundamental level.

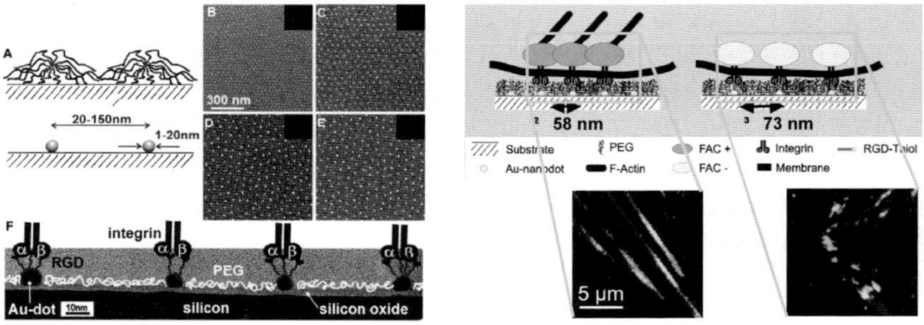

Fig. B.8. Studying the process of cellular adhesion as a function of the scale of protein organization: *Left*: micro-patterned surfaces of bio-functionalized nanoparticles; *Right:* the resulting effect of inter-particle spacing upon the binding of integrins, showing that binding sites must be <60 nm apart for activation. (Courtesy Dr. J. Spatz).

REFERENCES

Professor Grunze; Dr. Dahint (See www.pci.uni-heidelberg.de/apc/APC_Inst_Publikationen.htm for full list of this group's publications.)

APC (Ruprecht-Karls University Heidelberg, Physical Chemistry Institute) website: www.pci.uni-heidelberg.de/apc/APC_Inst_AG_English.htm.

Bender, F., F. Meimeth, R. Dahint, M. Grunze and F. Josse. 1997. Mechanisms of interaction in acoustic plate mode immunosensors. *Sensors and Actuators B* 40:105–110.

Bender, F., R. Dahint, M. Grunze, F. Josse, A.J. Ricco, and S.J. Martin. 1997. Investigation of high-sensitivity acoustic plate mode biosensors. *Electrochem. Society Proceedings* 97 (19):165–169.

Eisert, F., M. Gurka, A. Legant, M. Buck, and M. Grunze. 2000. Detection of molecular alignment in confined films. *Science* 287:468–470.

Feldman, K., G. Hähner, N.D. Spencer, P. Harder, and M. Grunze. 1999. Probing resistance to protein adsorption of oligo(ethylene glycol)-terminated self-assembled monolayers by scanning force microscopy. *J. Am. Chem. Soc.* 121:10134–10141.

Gölzhäuser, A., W. Eck, W. Geyer, V. Stadler, Th. Weimann, P. Hinze, and M. Grunze. 2001. Chemical nanolithography with electron beams. *Advanced Materials* 13:11.

Gölzhäuser, A., B. Völkel, B. Jäger, M. Zharnikov, H.J. Kreuzer, and M. Grunze. 1998. Holographic imaging of macromolecules. *J. Vac. Sci. Technol. A.* 16 (5):3025–3028.

Grunze, M. 1999. Driven liquids. *Science (Perspective)* 293 (41):98–14

Grunze, M., and A. Pertsin. 2000. Molecular conformations in organic monolayers affect their ability to resist protein adsorption. In *Chemistry at the beginning of the third millennium*, ed. L. Fabbrizzi and A. Poggi, Berlin, Heidelberg: Springer-Verlag.

Grunze, M., A. Welle, and D. Tur. 1998. Blood compatibility of poly[bis(trifluoroethoxy) phosphazene]. *Soc. Plast. Eng. Ann. Tech. Conf.* 44:2713–2717.

Harder, P., M. Grunze, R. Dahint, G.M. Whitesides, and P.E. Laibinis. 1998. Molecular conformation in oligo(ethylene glycol) terminated self-assembled monolayers on gold and silver surfaces determines their ability to resist protein adsorption. *J. Phys. Chem. B.* 102 (2):426–436.

Harder, P., M. Grunze, and J.H. Waite. 2000. Interaction of the adhesive protein Mefp-1 and fibrinogen with methyl and oligo(ethylene glycol)-terminated self assembled monolayers. *J. Adhesion* 73:161–177.

Herrwerth, S., T. Rosendahl, C. Feng, J. Fick, W. Eck, M. Himmelhaus, R. Dahint, and M. Grunze. 2003. Covalent coupling of antibodies to self-assembled monolayers of carboxy-functionalized poly(ethylene glycol): Protein resistance and specific binding of biomolecules. *Langmuir* 19 (5):1880–1887.

Morhard, F., J. Pipper, R. Dahint, and M. Grunze. 2000. Immobilization of antibodies in micropatterns for cell detection by optical diffraction. *Sensors and Actuators B* 70:232–242.

Morhard, F., R. Dahint, and M. Grunze. 1998. In-situ detection of cells and biochemical reactions by optical diffraction. In *Proceedings of the μTas ´98 Workshop,* ed. D.J. Harrison and A. van den Berg, 469–472. Dordrecht: Kluwer Academic Publishers. .

Morhard, F., J. Schumacher, A. Lenenbach, T. Wilhelm, R. Dahint, M. Grunze, and D.S. Everhart. 1997. Optical diffraction: A new concept for rapid on-line detection of chemical and biochemical analytes. *Electrochem. Society Proceedings* 97 (19):1058–1065.

Pertsin, A.J., and M. Grunze. 2000. Computer simulation of water near the surface of oligo(ethylene glycol) terminated alkanethiol self-assembled monolayers. *Langmuir* 16:8829–8841.

Pertsin, A. J., M. Grunze, and I.A. Garbuzova. 1998. Low energy configurations of methoxy tri(ethylene glycol) terminated alkanethiol self-assembled monolayers and their relevance to protein adsorption. *J. Phys. Chem B.* 102 (25):4918–4926.

RosSeigel, R., P. Harder, R. Dahint, M. Grunze, F. Josse, M. Mrksich, and G.M. Whitesides. 1997. Online detection of nonspecific protein adsorption at artificial surfaces. *Anal. Chem.* 69:3321–3328.

Schumacher, J., M. Ranft, T. Wilhelm, R. Dahint, and M. Grunze. 1998. Chemical analysis based on environmentally sensitive hydrogels and optical diffraction. In *Proceedings of the μTas ´98 Workshop,* ed. D.J. Harrison and A. van den Berg, 61–64. Dordrecht: Kluwer Academic Publishers.

Schwendel, D., R. Dahint, S. Herrwerth, M. Schlörholz, W. Eck, and M. Grunze. 2001. Temperature dependence of the protein resistance of poly- and oligo(ethylene glycol)-terminated alkanethiolate monolayers. *Langmuir* 17:5717–5720.

Wang, R.L.C., H.J. Kreuzer, and M. Grunze. 1997. Molecular conformation and solvation of oligo(ethylene glycol) terminated self-assembled monolayers and their resistance to protein adsorption. *J. Phys. Chem. B.* 101 (47):9767–9773.

Wang, R.L.C., H.J. Kreuzer, M. Grunze, and A.J. Pertsin. 2000. The effect of electrostatic fields on an oligo(ethylene glycol) molecule: Dipole moments, polarizabilities and field dissociation. *Phys. Chem. Chem. Phys.* 2:1721–1727.

Welle, A., M. Grunze, and D. Tur. 1998. Plasma protein adsorption and platelet adhesion on poly[bis(trifluoroethoxy)phosphazene] and reference material surfaces. *J. Colloid Interf. Sci.* 197:263–274.

Welle, A., and M. Grunze. 2000. Blood compatibility of poly[bis(trifluoroethoxy) phosphazene]. *JAMP* 4(1):6–10.

Professor Spatz (See www.pci.uni-heidelberg.de/bpc/publications.html for full list of the group's publications.)

Boyen, H.-G., G. Kästle, K. Zürn, Th. Herzog, F. Weigl, P. Ziemann, O. Mayer, Ch. Jerome, J.P. Spatz, M. Möller, M.G. Garnier, and P. Oelhafen. 2003. A micellar route to ordered arrays of magnetic nanoparticles: From size-selected pure cobalt dots to cobalt-cobaltoxide core-shell systems. *Advanced Functional Materials* 13:359–364.

Chan, V.Z.H., S.L. Codd, M.J. van der Helm, J.P. Spatz, C. Röcker, G.U. Nienhaus, S.I. Levi, F.C.J.M. van Veggel, D.N. Reinhoudt, and M. Möller. 2001. Sub-10 nm gold nanoarrays for tethering single molecules. In *Synthesis, Functional properties and applications of nanostructures*, ed. H.W. Hahn, D.L. Feldheim, C.P. Kubiak, R. Tannenbaum, and R.W. Siege, Y4.4.1-Y4.4.6. Materials Research Society Symposium Proceedings Vol. 676, l. Pittsburgh PA: MRS Publications.

Gorre-Talini, L., J.P. Spatz, and P. Silberzan. 1998. Dielectrophoetic ratchets. *Chaos* 8:650.

Haupt, M., S. Miller, K. Bitzer, K. Thonke, R. Sauer, J.P. Spatz, S. Mössmer, C. Hartmann, and M. Möller. 2001. Polymer masks on semiconductors: A novel way to nanostructures. *Phys. Stat. Sol. B.* 224 (3):867.

Haupt, M., S. Miller, R. Glass, M. Arnold, K. Thonke, M. Möller, and J.P. Spatz. 2003. Formation of nanoporous gold films by templating self-assembly structures of inorganic-block-copolymer micelles. *Advanced Materials* 15:829–831.

Möller, M., J.P. Spatz, M. Moessmer, P. Eibeck, P. Ziemann, and B. Kabius. 1999. Formation of chemical nanopattern by means of block copolymers. *Polymeric Materials Science and Engineering* 80 (1):3.

Semenov, A., J.P. Spatz, M. Möller, J.-M. Lehn, B. Sell, D. Schubert, C.H. Weidl, and U.S. Schubert. 1999. Controlled arrangement of supramolecular metal coordination arrays on surfaces. *Angew. Chem.* 111:2701–2705; *Angew. Chem. Int. Ed.* 38:2547–2550.

Sönnichsen, C., S. Geier, N.E. Hecker, G. von Plessen, J. Feldmann, H. Ditlbacher, B. Lamprecht, J.R. Krenn, F.R. Aussenegg, V. Chan, J.P. Spatz, and M. Möller. 2000. Spectroscopy of single metallic nanoparticles using total internal reflection microscopy. *Applied Physics Letters* 77:2949–2951.

Spatz, J.P. 2003. Cell - nanostructure interactions. In *Nanobiotechnology*. Weinheim: Wiley-VCH.

Spatz, J.P., S. Mößmer, F.-M. Kamm, A. Plettl, P. Ziemann, and M. Möller. 2002. A combined top down/bottom up approach for nanolithography. *Advanced Materials* 14:1827.

Site: **Swiss Federal Institute of Technology (ETH) Zürich**
Department of Chemistry
CH-8093 Zürich, Switzerland

Date Visited: 18 March 2003

WTEC Attendee: C.L. Wilkins (report author)

Hosts: Professor Dr. Renato Zenobi, Tel: +41 1 632 43 76;
Fax: +41 1 632 12 92;
Email: zenobi@org.chem.ethz.ch
Professor Dr. Ernö Pretsch

INTRODUCTION

The Chemistry Department at the Swiss Federal Institute of Technology [Eidgenössische Technische Hochschule (ETH)], Hönggerberg, has approximately 550 four-year chemistry undergraduate students, 400 graduate students, and 35 full professors. In addition, the department also employs approximately 35 additional senior scientific staff, many of whom run independent research groups. Research support is provided by a combination of ETH support funds and funds provided by granting agencies (e.g., the Swiss National Science Foundation).

LASER-ASSISTED ANALYTICAL CHEMISTRY, NEAR-FIELD SCANNING MICROSCOPY, AND MASS SPECTROMETRY

Professor Zenobi's group is not directly involved in what is traditionally viewed as biosensing research. However, his interests certainly have significant potential future implications in that regard, even if primarily in the role of development of the required new analytical tools to support biosensor development. He has summarized his diverse analytical chemistry research in an article in *Chimia* (Bakker and Pretsch 2002). With respect to the biological implications of some of his present mass spectrometry (MS) research, it is clear that that his group's work in the application of soft ionization mass spectrometry analysis to the study of noncovalent interactions is significant (Daniel et al. 2000; Friess and Zenobi 2001; Friess et al. 2002). The recent work has emphasized a new strategy that has been developed for obtaining topological information about biomolecules (Friess and Zenobi 2001; Friess et al. 2002).

Additionally, a good deal of effort in Zenobi's laboratory has been devoted to analysis of water (Bucheli et al. 2000) and aerosol samples

(Morrical and Zenobi 2002), both of which represent applications with direct implications as environmental sensor techniques that could clearly be adaptable to biosensor needs. The approach advocated by Zenobi involves a method he has termed two-step laser mass spectrometry, employing an infrared laser in the first step for ablation of the sample and a tunable ultraviolet laser in the second step for ionization. He has exchanged comments (Haefliger, Bucheli, and Zenobi 1999) in the literature with others (Reilly et al. 1999) who advocate the single-step laser desorption approach to aerosol particle analysis. This debate exemplifies another of the major interests in the laboratory, the development of MS instrumentation and sample preparation methods that has extended from the theory of matrix-assisted laser desorption/ionization (MALDI) (Zhang et al. 2002) to development of a MALDI sample preparation method applicable to insoluble polymers (Skelton, Dubois, and Zenobi 2000) and more recently, to construction of an atmospheric pressure nanosampling interface for mass spectrometry based on near-field laser ablation (Stöckle et al. 2001). The latter development represents a linkage of Zenobi's mass spectrometry interests with his interest in the development of the combination of scanning near-field optical microscopy (Zenobi and Dekert 2000) with optical spectroscopy. Furthermore, he and his students have accomplished important achievements in the area of near-field Raman spectroscopy measurements, further establishing the feasibility of this new technique.

OPTIMIZED ION-SELECTIVE ELECTRODES

Polymeric membrane electrodes are well established for their robust properties and the ability to use them for quantitative and highly selective analyses as ion-selective electrodes. Recent research in the Pretsch laboratory (Pretsch et al. 2003) has established successful new approaches to achievement of low detection limits and increased sensitivity of potentiometric measurements in various applications. A series of *Analytical Chemistry* papers (Bakker, Pretsch, and Bühlmann 2000; Ceresa, Pretsch, and Bakker 2000; Qin, Zwicki, and Pretsch 2000) outline the basis for the new approach, which was most recently the topic of a presentation at the 2003 Pittsburgh Conference on Analytical Chemistry and Applied Spectroscopy. The first paper from the Pretsch laboratory outlining this method described the picomolar detection of lead (Zenobi 2001). Subsequently, a report of the first trace-level potentiometric measurements was published (Ceresa et al. 2001). It seems very likely that the new methodology holds promise for rapid and sensitive future potentiometric biosensor applications. A recent A-page article in

Analytical Chemistry summarizes these developments (Bakker and Pretsch 2002).

REFERENCES

Bakker, E., and E. Pretsch. 2002. The new wave of ion-selective electrodes, *Anal. Chem.* 420A-426A.

Bakker, E., E. Pretsch, and P. Bühlmann. 2000. Selectivity of potentionmetric ion sensors. *Anal. Chem.* 72:1127–1133.

Bucheli, T.D., O.P. Haefliger, R. Dietiker, and R. Zenobi. 2000. Analysis of water contaminants and natural water samples using two-step laser mass spectrometry. *Anal. Chem.* 72:3671–3677.

Ceresa, A., E. Pretsch, and E. Bakker. 2000. Direct potentionmetric information on total ionic concentrations, *Anal. Chem.* 72:2050–2054.

Ceresa, A., E. Bakker, B. Hattendorf, D. Günther, and E. Pretsch. 2001. Potentiometric polymer membrane electrodes for measurement of environmental samples at trace levels: New requirements for selectivities and measuring protocols and comparison with ICPMS. *Anal. Chem.* 73:343–351.

Daniel, J. M., S.D. Friess, S. Rajagopalan, S. Wendt, and R. Zenobi. 2000. Quantitative determination of noncovalent binding interactions using soft ionization mass spectrometry. *Inter. J. Mass Spectrom.* 216:1–27.

Friess, S.D., and R. Zenobi. 2001. Protein structure information from mass spectrometry? Selective titration of arginine residues by sulfonates. *J. Am. Soc. Mass Spectrom.* 12:810–818.

Friess, S.D., J. M. Daniel, R. Hartmann, and R. Zenobi. 2002. Mass spectrometric noncovelent probing of amino acids in peptides and proteins, *Inter. J. Mass Spectrom.* 219:269–281.

Haefliger, O.P., T.D. Bucheli, and R. Zenobi. 1999. Comment on "Real-time characterization of the organic composition and size of individual diesel engine smoke particles." *Environ. Sci. Technol.* 33:3932.

Morrical, B.D., and R. Zenobi. 2002. Detection of polycyclic aromatic compounds at jungfraujoch high-alpine research station using two-step laser mass spectrometry, *Inter. J. Environ. Anal. Chem.* 82:377–385.

Pretsch, E., R.E. Gyurcsanyi, L.A. Muslinkina, J. Sutter, Z. Szigeti, T. Vigassy. 2003. Orlando, FL, March 9–14, 650–653.

Qin, W., T. Zwicki, and E. Pretsch. 2000. Improved detection limits and unbiased selectivity coefficients obtained by using ion-exchange resins in the inner reference solution of ion-selective polymeric membrane electrodes, *Anal. Chem.* 72:3236–3240.

Reilly, P.T.A., R.A. Gieray, W.B. Whitten, and J.M. Ramsey. 1999. Response to comment of "Real-time characterization of the organic composition and size of individual diesel engine smoke particles." *Environ. Sci. Tech.* 33:3933–3934.

Skelton, R., F. Dubois, and R. Zenobi. 2000. A MALDI sample preparation method suitable for insoluble polymers, *Anal. Chem.* 72:1707–1710.

Sokalski, T., A. Ceresa, T. Zwickl, and E. Pretsch. 1997. Large improvement of the detection limit of ion-selective polymer membrane electrodes. *J. Am. Chem. Soc.* 119:11347–11348.

Stöckle, R., P. Setz, V. Dekert, T. Lippert, A. Wokaun, and R. Zenobi. 2001. Nanoscale atmospheric pressure laser ablation-mass spectrometry. *Anal. Chem.* 73:1399–1402.

Stöckle, R., V. Dekert, C. Fokas, D. Zeisel, and R. Zenobi. 2000. Sub-wavelength Raman spectroscopy on isolated silver islands, *Vibr. Spec.* 22:39–48.

Zenobi, R. 2001. Laser-assisted analytical chemistry and mass spectrometry, *Chimia* 55:773–777.

Zenobi, R., and V. Dekert. 2000. Scanning near-field optical microscopy and spectroscopy as a tool for chemical analysis. *Angew. Chem. Int. Ed.* 39:1746–1756.

Zhang, J., T.-K. Ha, R. Knochenmuss, and R. Zenobi. 2002. Theoretical calculation of gas-phase sodium binding energies of common MALDI matrices. *J. Phys. Chem. A.* 106:6610–6617.

Site: **Swiss Federal Institute of Technology (ETH) Zürich**
Physical Electronics Laboratory
ETH Zürich, HPT-H6
CH-8093 Zürich, Switzerland
Tel: +41-1-633 20 90
Fax: +41-1-633 10 54
http://www.iqe.ethz.ch/pel

Date Visited: 17 March 2003

WTEC Attendees: D. Brady (report author), A. Ricco

Hosts: Prof. Dr. Henry Baltes,
Email: baltes@iqe.phys.ethz.ch
Dr. Andreas Hierlemann

BACKGROUND

For the past decade the Physical Electronics Laboratory (PEL) has focused on integration of transducers with integrated silicon microcircuits. The laboratory has developed a methodology for integrated sensor system design and manufacturing. A typical project begins with integrated commercially fabricated CMOS circuits with sensor interfaces fabricated at ETH. PEL typically receives raw processed dies designed internally and does CMOS post-processing, packaging, and testing in-house.

The lab has integrated many sensor interfaces with silicon, including cantilever (AFM), electronic (FET), calorimetric (Al/poly-Si thermocouple), and micro hotplates. A particularly impressive demonstration was presented in a 2001 *Nature* article, "Smart single-chip gas sensor microsystem" (Hagleitner et al. 2001), which describes a single-chip resonant cantilever, capacitive, and calorimetric system. PEL's results are particularly important as a demonstration of the added functionality and performance that can arise in careful sensor system integration. PEL has successfully manufactured complete sensor chips with standardized digital outputs. Such chips are easily inserted into digital platforms.

PEL collaborates directly and extensively with major microsystems companies and emphasizes attention to manufacturability in sensor research. Demonstrating practical integration of sensors and control systems is a primary goal of the lab.

PEL's work has mostly focused on chemical sensors, but work has begun on biosensors over the past couple of years. Resonant microcantilever sensors could be immediately applied to biosensing applications if biological receptors were coated on the devices, although standard lifetime and stabilization challenges would arise. PEL is currently working on two biosensing projects: BioFinger and Neurochip.

BIOSENSING PROJECTS

BioFinger

The goal of the BioFinger project (www.biofinger.org) is to build a microcantilever-based molecular interaction sensor. The project focuses on tools for clinical diagnosis, with specific reference to the detection of tumor-associated protein and the detection of viruses. The project is led by the Centro Nacional de Microelectrónica in Barcelona and includes collaborators at Cork University and the Cork Institute of Technology, in addition to PEL. BioFinger is supported by IST in the EU's Fifth Framework program and by the Swiss Federal Office for Eduation and Science.

PEL is responsible for micro- and nanocantilever fabrication and system integration for BioFinger. Molecular receptors for the project are developed at Cork University Hospital. For submicron cantilevers in particular, direct integration of signal processing circuitry with the cantilever will be essential to signal extraction. Rather than focus on mass-based changes to the cantilever itself, the goal of nanocantilever systems is to directly measure binding forces between a functionalized tip and a single substrate-bound molecule.

The BioFinger project aims to develop both handheld field diagnostic tools with microcantilever-based disposable sensor chips and high sensitivity laboratory instruments with nanocantilevers. In demonstration systems, the specific target of the microcantilever system is prostate-specific antigen. The target of the nanocantilever system is equine herpes virus.

Neurochip

The neurochip project produces CMOS chips that both stimulate and record cellular responses. Microfluidic systems are integrated with the chip to support neuronal cells. Several groups worldwide have developed neural interfaces over the past decade; the ETH project is most notable for its complete integration of drive circuitry, signal processing, and controllers on a single chip. The current chip interfaces with chicken neuronal cells drawn from eggs. The cells are adhered to Pt electrodes on a 250-micron pitch. The current effort focuses on a 32 by 32 electrode array.

REFERENCES

Hagleitner, C., A. Hierlemann, D. Lange, A. Kummer, N. Kerness, O. Brand, and H. Baltes. 2001. Smart single-chip gas sensor microsystem. *Nature* 414 (6861):293–296.

Hierlemann, A., and H. Baltes. 2003. CMOS-based chemical microsensors. *Analyst* 128 (1):15–28.

Site: **University of Cambridge**
Institute of Biotechnology
Tennis Court Road
Cambridge CB2 1QT, UK
http://www.dcu.ie/~ncsr/index_home.html

Date Visited: 18 March 2003

WTEC Attendees: C. Kelley (report author), D. Walt,

Host: Dr. Christopher Lowe, Director,
Tel: +44 (0)1223334160;
Email: crl1@biotech.cam.ac.uk

BACKGROUND

The University of Cambridge was established around 1209. Currently, there are 16,500 full-time students and nearly 7,000 faculty members at the university. The Institute of Biotechnology was founded in 1988 to meet growing demands for highly skilled research personnel and to uncover new knowledge for continuing the expansion of the science platform upon which technological innovations are based. The institute occupies modern, well-equipped laboratories representing a wide range of scientific areas, including biochemistry, molecular and cellular biology, plant growth, optics, analytical technology, electronics design and instrumentation, and nuclear magnetic spectroscopy. The Institute of Biotechnology at the University of Cambridge maintains significant industry contacts as a part of its mission to ensure the rapid and effective transfer of ideas and developments in biological science between academia and industry. Seven companies have been spun out of the institute. Moreover, the institute serves as a biotechnology resource center for surrounding universities and researchers.

RESEARCH AND DEVELOPMENT ACTIVITIES

Dr. Lowe, the director of the Institute of Biotechnology, has been involved in biosensor research for the past 25 years and obtains the most funding for sensing research in the United Kingdom. His biosensor research group consists of 30 members and is focused primarily on optical, acoustic, and microamperometric sensors, as well as novel microbial enzymes for biosensors. Dr. Lowe is interested in preparing very inexpensive sensors with high functionality—"sophisticated but cheap." Dr. Lowe has most recently led the development of holographic sensors for efficient analyte targeting.

The holographic sensors consist of a polymeric hydrogel, which deforms when in contact with a specific chemical or biological reagent; a hologram written into the hydrogel reports the hydrogel's changing volume. The holographic component provides both the analyte-specific polymer matrix and the optical detection mechanism. Furthermore, the holographic sensors do not require any additional electronic processing, as the visible hologram image serves as the test result. Sample holographic sensor test results can be seen in Figure 2.1, Chapter 2 of this report. As described by Dr. Lowe, biosensors of this nature hold great potential as inexpensive, noninvasive, and reliable means of monitoring whole-cell dynamics.

Dr. Lowe was also eager to discuss his research involving miniature acoustic wave devices for use as biosensors. A new acoustic sensor technology, the magnetic acoustic resonant sensor (MARS), was recently developed at the institute in which an acoustic pulse generates an electric current. The MARS device uses a frequency range of up to 4 GHz, making it capable of "fingerprinting" the conformational changes of single proteins. It is also possible to develop the device so that it can detect multiple analytes through the emission of multiple frequencies. Another advantage of the novel biosensor is it contains no electronic components, thus yielding a product that can be manufactured quickly and cheaply.

In addition to its innovative research, the Institute of Biotechnology boasts a proficiency in translating discoveries and developments into commercial ventures. Dr. Lowe reported that the institute maintains a strong interface with area industry. Furthermore, he alone has overseen the inauguration of seven spin-off companies ranging in scope from biosensors to medical devices to proteomics; three are described below.

ProMetic BioSciences, Ltd., established in 1987, was the first biotechnology company to be initiated as a result of innovative research at the Institute of Biotechnology. The company currently employs 130 people, making it the largest of the spin-offs. ProMetic Biosciences specializes in the development of separation technologies, medical devices, and biopharmaceuticals and therapeutics.

Cambridge Sensors, Ltd., was founded in 1992 to facilitate biosensor research and development and the manufacturing of low-cost environmental and biomedical diagnostic sensors. The company's activities focus on lab-on-a-chip technologies for genomics and proteomics, point-of-care diagnostics, and drug discovery. Presently, Cambridge Sensors holds patented technology for the fabrication of heavy metal sensors, chlorine sensors, and glucose sensors.

Smart Holograms, Inc., was established in 2001 to pursue some of the holographic sensor research at the Institute of Biotechnology. The company retains three patents for its developments based on the integration of interactive holograms into thin film sensors. These holographic sensors have potential applications in various markets, according to Dr. Lowe. Most significant is the promise the sensors show in the areas of drug discovery, pathogen detection (bioterrorism), and noninvasive biosensing.

SUPPORT

Funding and support for the Institute of Biotechnology come from an array of sources. UK research councils and government departments sustain a majority of the Institute of Biotechnology's budget, while the European Union and charities provide additional funding. Dr. Lowe disclosed the institute's funding for fiscal year 2002 to be around 2.5 million (~$4.0 million). Worldwide industry links generate the necessary support to facilitate and finance any transfer of technology from the institute.

REFERENCES

Gizeli, E. Acoustic wave devices. Available online 2003–4 at www.biot.cam.ac.uk/~eg/eg2.html.

Lowe, C.R., J. Blyth, A. Marshall, et al. 2003. Holograms that react to biological substances offer new diagnostic tools. *OE Magazine* 3 (3):20–23.

Lowe, C.R. Holographic biosensors. Available online 2003–4 at www.biot.cam.ac.uk/~crl/crl6.html.

Mayes, A.G., J. Blyth, R.B. Millington, et al. 2002. *Anal. Chem.* 74:3649–3657.

Sindi, H.S., and A.C. Stevenson. 2001. *Anal. Chem.* 73:1557–1586.

Site:	**The University of Manchester Institute of Science and Technology (UMIST)** **Department of Chemistry** **P.O. Box 88, Sackville Street** **Manchester M6O 1QD** **United Kingdom**
Date Visited:	18 March 2003
WTEC Attendees:	C.L. Wilkins (report author), C. Kelley
Hosts:	Dr. Simon J. Gaskell, Director, The Michael Barber Centre for Mass Spectrometry, Tel: +44 (0)161 200 4474; Fax: +44 (0)161 200 4421; Email: Simon.Gaskell@umist.ac.uk Professor Peter Fielden, Department of Instrumentation and Analytical Sciences Professor Douglas B. Kell, Department of Chemistry

INTRODUCTION

University of Manchester Institute of Science and Technology (UMIST) is a small university with about 4,500 undergraduates and approximately 1,500 graduate students. There are approximately 1,900 staff members, of whom 1,400 are classified as academic or academic-related. Plans are underway to administratively consolidate the campus with the University of Manchester. A unique feature of the present institution is the policy that any intellectual property developed as a result of research is vested in the faculty. This has caused UMIST to be a particularly productive source of start-up companies that become successful at facilitating technology transfer. It is not clear that this intellectual property policy will survive the merger.

METABOLOMICS AND MACHINE LEARNING

Professor Kell presented an overview of his research program. Among the goals of his research is the application of a technique called "genomic computing" to understanding the category of post-genomic studies called metabolome studies (Kell 2002a). The goal of this research is to understand certain aspects of the metabolism of various organisms. This work directly relates to the future development of biosensing technology in the broadest sense. For example, the tools and approaches employed are generally applicable to a very wide range of data and chemical analysis problems (Kell, Darby, and Draper 2001; Kell 2002b). For example, applications as diverse as quantitative detection of microbial spoilage of meat based on machine learning analysis of Fourier transform infrared spectroscopy (FT-IR) data (Ellis et al. 2002) and detection of *Bacillus* spores based on FT-IR and pyrolysis mass spectrometry data have been reported in the recent past (Goodacre et al. 2000).

MASS SPECTROMETRY

Professor Gaskell discussed his research interests in the context of broadly defined biosensing. His interests concern the development and application of state-of-the-art mass spectrometry, principally in the context of biological research. Individual research projects cover a broad range, including instrument development, aspects of gas-phase ion chemistry, and structural characterizations of large biomolecules.

Extended efforts have been devoted to understanding the fragmentations of gas-phase peptide ions, with a view to improving analytical capabilities. It is now well understood that extensive diagnostic fragmentation is promoted by a precursor ion population heterogeneous with respect to site of charge. Equivalently, "mobilization" of ionizing protons facilitates fragmentation. Furthermore, the extent and direction of peptide ion fragmentation is influenced by intra-ionic interactions; salt bridge formation, for example, may promote concomitant peptide bond and side-chain cleavage via a low energy process.

The improved understanding of peptide fragmentations has facilitated the productive application of tandem MS to the characterization of structurally modified proteins. Thus, for example, recent research provided the first direct evidence for the formation of lipid/protein conjugates following oxidation of low-density lipoprotein.

In the context of the characterization of biomolecules such as proteins and peptides, the compelling advantages of mass spectrometry are those of high sensitivity and a capability for mixture analysis. The analysis of peptides associated with molecules of the Major Histocompatibility Complex (MHC) provides an extraordinary challenge in both respects. Professor Gaskell's research in this area has included collaboration with several immunology research groups within the UK and outside.

Research in proteomics also provides a substantial challenge with respect to sensitivity and mixture characterization. Work in this area benefits critically from collaboration with colleagues in the UMIST Department of Biomolecular Sciences and in other biological science departments in Manchester. "Conventional" biochemical techniques and mass spectrometry are frequently of complementary value; thus, for example, Gaskell has developed (with Dr. J. Brookman of the University of Manchester) the combination of immunoaffinity adsorption and mass spectrometry for the characterization of minor components of complex cell lysates.

DIAS MINIATURIZATION GROUP RESEARCH

The Department of Instrumentation and Analytical Sciences research group currently consists of 3 faculty, 2 visiting faculty, 9 research assistants, and 7 Ph.D. students. A variety of research activities are underway, most of which are related to biosensor development. Among the more interesting developments are leaky waveguide biosensors that are adaptable as high sensitivity chemical sensors and have been shown to be capable of single bacterium detection. Micromoulded grating arrays capable of providing thousands of individual grating sensors have been developed. In addition, polymer-based micro-isotachtophoresis chips have been successfully developed. Genotoxicity testing methods have been investigated recently, as have the applications of conducting polymer devices for chemical sensing. Integrated waveguides for microfluidics devices as well as flow cytometry have also been the subject of recent research. Overall, an impressively diversified set of analytical goals has been pursued, with some significant accomplishments achieved.

REFERENCES

Ellis, D.I., D. Broadhurst, D.B. Kell, J.J. Rowland, and R. Goodacre, R. 2002. Rapid and quantitative detection of the microbial spoilage of meat by Fourier transform infrared spectroscopy and machine learning. *Appl. and Environ. Microbiology* 68:2822–2828.

Goodacre, R., B. Shann, R.J. Gilbert, E.M. Timmins, A.C. McGovern, B.K. Alsberg, D.B. Kell, and N.A. Logan. 2000. Detection of the dipicolinic acid biomarker in *bacillus* spores using Curie-point pyrolysis mass spectrometry and Fourier transform infrared spectroscopy., *Anal. Chem.* 72:119–127.

Kell, D.B. 2002a. Metabolomics and machine learning: Explanatory analysis of complex metabolome data using genetic programming to produce simple, robust rules. *Mol. Biol. Rep.* 29:237–241.

——. 2002b. Defense against the flood. *Bioinformatics World* Jan./Feb.:16–18.

Kell, D.B.,R.M. Darby, and J. Draper. 2001. Genomic computing. Explanatory analysis of plant expression profiling data using machine learning. *Plant Physiology* 126:943–951.

Site: **University of Neuchâtel**
Institute of Microtechnology (IMT)
Sensors, Actuators and Microsystems Laboratory (SAMLAB)
Rue Jaquet-Droz 1
Neuchâtel 2007, Switzerland
http://www-imt.unine.ch
http://www-samlab.unine.ch

Date Visited: 17 March 2003

WTEC Attendees: A.J. Ricco (report author), D. Brady

Hosts: Professor Nico de Rooij, Tel: +41 32 720 5303;
Fax: +41 32 718 3641;
Email: nico.derooij@unine.ch
Dr. Sabeth Verpoorte
Dr. Milena Koudelka-Hep

KEY INFORMATION

Part of the Faculty of Science, University of Neuchâtel.

Date of founding: 1975.

Faculty, staff, students at IMT: ~140 total; of faculty, 5 are full professors; of students, ~50 are Ph.D. students.

Degrees offered: Masters, Physical Electronics; advanced degrees in engineering (awarded by EPFL); Ph.D. of Science (awarded by IMT/Univ. of Neuchâtel).

Annual budget: ~SFr 18 million; ~25 of 140 faculty/staff/students are university funded, others are externally funded.

Site: two buildings in Neuchâtel

Mission: Advance the science and technology of miniaturization, including micro- and nanotechnologies.

INTRODUCTION

The region around Neuchâtel has a considerable economic base in precision manufacturing, particularly watches and their components. In the 1970s, in response to major job losses in the electronic watch industry, IMT was founded to provide education in those microtechnologies relevant to modern precision manufacturing. In microelectronics, this implies a focus on low-power CMOS circuitry as well as mixed analog/digital circuits.

For example, in addition to the educational programs for university students, summarized above, IMT offers a hands-on MEMS training course for industrial users. IMT also houses a design center for optical MEMS, funded in part through the Europractice III program of the EU, which has a strong focus on technology transfer to the commercial sector.

The DeRooij group is 57 in total, including 16 Ph.D. students and 8 full-time engineering staff; ~10% of all salaries are covered by the university, the rest of the funding is raised externally. The laboratories are very well equipped by European and U.S. standards, and the 8 full-time technical staff keep the labs, particularly the microfabrication facility, running smoothly, allowing the students to focus more exclusively on thesis research than students in such facilities in many other universities. This professional upkeep of the facilities only occasionally limits the creative exploration of new ways to utilize equipment or facilities that might result in downtime.

FABRICATION CAPABILITIES AND FACILITIES

A central technical focus of IMT is the design, fabrication, and applications of sensors and actuators based upon micro- and nanotechnologies. In recent years, focus has shifted to some degree from "traditional" MEMS devices to optical MEMS, microfluidics, and bioMEMS devices, including biosensors. The microfabrication facility, known as the SAMLAB (Sensors, Actuators, and Microsystems Laboratory) is very flexible in the range of materials and processes that are accessible, although the fact that MOS circuitry (ISFETs in particular) is built in the facility means that gold is an unpopular material in parts of the silicon processing line. Tools available in the SAMLAB include deep reactive ion etching, Si-based CMOS circuitry in general, electron-beam direct-write lithography suitable for nanodimensioned features, glass processing, and polymer patterning, particularly PDMS and polyimide. Characterization tools include a TEM and environmental scanning electron microscope (ESEM), which has a pumping system that allows water-containing devices/samples to be viewed (the rapid pumping creates ice in locations with geometries and fluid content such that water cannot be quickly pumped away). In general, the SAMLAB facility is very well equipped for a university fabrication facility, and as noted above, a dedicated technical staff keeps the labs running smoothly.

SENSOR & ACTUATOR DEVICES

Mechanical and Optical MEMS Devices

One example of an IMT MEMS technology transfer success is the "T- Touch" watch manufactured by Tissot, which includes a MEMS pressure sensor with a sealed vacuum absolute reference that enables precision measurement of pressure, allowing the watch to function as an altimeter. The watch also includes a MEMS-based Hall effect magnetic field sensor, developed at EPFL, that provides the input for an integrated electronic compass.

In the optical MEMS area, deep reactive ion etching (DRIE) was harnessed to make channels in a Si substrate that position input/output optical fibers relative to a vertical mirror, fabricated in the same etching process. This technology is the core of a family of (nxm) optical switches as well as variable optical attenuators (VOAs), which are being marketed worldwide by Sercalo, Inc., a start-up company of SAMLAB.

Scanning Near-Field Optical Microscopy (SNOM) Tips

Advances were made in this high-resolution measurement technology through the micromachining of a tip on the surface of a Si wafer covered by a thick layer (10–12 μm) of SiO_2 to act as the optical tip, with addition of the appropriately patterned metal layer. Using this tip, resolution of 30 nm was obtained for an image of the binding of immunoglobulin G (IgG) to an antibody. This tip technology is currently being commercialized. An eventual goal is to use a single tip for both atomic force microscopy and SNOM in a commercial instrument.

BioMEMS Devices and Environmental Monitoring

Dr. Koudelka-Hep described the work she has led with microbioelectrochemical systems for studies of brain tissue slices in order to monitor signal transmission. These devices utilize a grid of microfabricated Pt-black or Ir electrodes to make an array of contacts with a slice of living tissue *in vitro*. The native tissue environment may offer a more unperturbed look at the functioning of this biological system than single cells studied in isolation or with limited connectivity. One issue with keeping such tissue slices viable on a solid electrode array support is the lack of perfusion on the side of the sample that contacts the chip. To address this, a grid of 30 μm diameter holes that extend clear through the 300 μm thickness of the Si wafer was provided by DRIE, allowing solution access to the chip side of the tissue. The brain tissue slices are in the 200– 300 μm thickness range, comparable to the wafer thickness.

It was found that electrodes that protrude into the tissue slice lead to larger signals (and better signal:noise ratios), presumably a result of enhanced contact area and/or contact with a larger number of neural cells. Two methods were used to provide 3-dimensional electrodes. Electroplating was used to build a "hillock" of metal atop a planar electrode; the resulting bumps are some 20–30 μm high (and about twice this large in diameter). To make higher aspect ratio electrodes, anisotropic etching was utilized to realize an array of pyramidal structures, then the entire chip was metallized, then covered with silicon nitride, then patterned to expose just the electrodes, which protruded some 35 μm into the tissue slice. Separate studies have established that such pointed structures do not have significant damaging/disruptive effects.

Work currently underway in collaboration with CSEM (also in Switzerland) seeks to integrate a conventional CCD as the charge measuring system with an aligned array of electrodes to interface with biological samples. The electrical measurement parameters differ significantly from those used with such arrays for optical detection, so development of modified measurement circuitry is necessary.

Similar microelectrode arrays to those utilized for brain tissue characterization have also been developed for environmental monitoring of heavy metals in water and in sediments. The capability to produce Ir electrodes allows the deposition of a mercury thin-film electrode ~100 nm thick (feasible because Hg does not amalgamate Ir as it does Au and many other metals). For water monitoring, an array of 100 5-μm-diameter electrodes was utilized; for sediments, a linear array of extended length electrodes, each a few mm long, was used to penetrate the sediment. Hg is deemed essential for the detection of some metals at the part-per-trillion level, and one such thin-film electrode can be used for as long as 6 days. Heavy metals are detected by first concentrating them using cathodic deposition into the mercury layer, then stripping voltammetry is used for quantification.

Chemical and Glucose Microsensors

Chemical sensors developed at IMT include a pH-sensitive ISFET (which has been commercialized); a thin-film Si-based glucose sensor, utilizing poly(urethane) as the permeable covering membrane; and various gas sensors. A multisensor device, the "Pentrode," is manufactured by IMT and marketed by Thermo/Orion; it includes pH, redox potential, solution conductivity, temperature (using a diode), and a reference electrode. A rapid titration system is being developed in collaboration with Dr. Steve West at Thermo/Orion.

The glucose sensor has deliberately fabricated pinholes in the poly(urethane) coating to control permeability to the analyte. It has been tested in clinical trials for subcutaneous measurement of glucose for implant periods of as long as 3 days.

Microfluidics/Micro total analytical systems (μTAS)

This area of research is led by Dr. Sabeth Verpoorte, who will be departing from IMT to accept a chaired position at Gronigen in The Netherlands.

A particularly active area of research is the manipulation and utilization of bio/chemically modified beads in microfluidic systems. Such beads are a mainstay of modern biochemical analysis, and they are commercially available in a range of well-characterized sizes, shapes, and surface chemistries. The application of the same sorts of biofunctional chemistries to the inner surfaces of microfluidic channels is being examined as well. The combination of electroosmotic (EO) flow and pressure flow—which differ in their geometric pressure/flow profiles on the scale of the tens-of-micron-dimension channels—enables such processes as recirculating flow and trapping of microbeads. Hydrostatic head of 4–20 cm is combined with EO flow driven by a potential of about 200 V. Beads move by electrophoresis (they have non-zero charge) as well as by pressure- and EO-driven flow, while the bulk solution is unaffected by electrophoresis. This offers the chance to flow solution past the beads, or vice versa, to preconcentrate target analytes without the actual physical capture of the beads. A physical approach to capturing beads was also demonstrated, using a narrowly tapered fluidic channel to capture beads via a "keystoning" effect, whereby a 16 μm constriction is sufficient to capture 3 μm diameter beads if they are flowed through at sufficient flux.

Another area of research is solid-phase PCR, conducted in collaboration with GenInEx SA in Switzerland. DNA is captured on a surface and amplified in situ; ultimately, the goal is also to sequence the DNA by the addition of optically specifically labeled bases. Binding is quantitative with reproducibility of < 10%; heaters for the PCR are made from deposited indium-tin oxide (ITO) films. So-called molecular beacons, the fluorescence of which is a sensitive function of temperature, have been used to calibrate the temperature within the microchannel.

An in-channel immunoassay system is under development, using protein A to attach anti-IgG to form a 200 μm-long affinity column. Electrokinetic flow drives the analytical solution past this modified surface, and the captured antigen can be eluted with an acidic solution at pH = 2.

The IMT group has been one of the leaders in the development of noncontact electrical detection for microfluidic system, i.e., the use of impedance measurements wherein the electrodes are separated from the liquid-filled channel by a thinned region of channel wall. A frequency of 60 kHz is typical, the isolating wall is 10–15 μm thick, and a limit of detection of 18 μM has been demonstrated thus far. The anticipated limit of detection is of the order of 1 μM. While this technique is many orders of magnitude less sensitive than the most sensitive optical (fluorescent) approaches, the noncontact and inexpensive nature of this measurement system makes it a logical choice for higher LOD applications, for example in various process control scenarios.

TECHNOLOGY TRANSFER/COMMERCIALIZATION

In addition to the SNOM tips, ISFET, and multisensor described above, a bioreactor that flew on a number of space missions, including the ill-fated Columbia (from which the bioreactor was recovered, partially intact) has been developed. It uses a Si microfabricated fluid pump based on a classic design from Twente University in The Netherlands, in combination with a precision MEMS flow sensor to provide closed-loop flow rate control. The IMT group has worked under contract with Packard Biosciences to develop the technology for a 96-head piezoelectrically actuated microdispenser system, and technology of this sort was the basis for a spin-off company, Seyonic, founded by Bart van der Schoot, then a member of the IMT group. Seyonic is presently collaborating with Zymark, a U.S. laboratory automation/robotics company.

FUNDING SOURCES/ISSUES

Switzerland has an initiative at the national level for funding nanotechnology at a level of SFr 50 million for applied nanotechnology and SFr 24 million for fundamental nanoscience and technology. Of IMT's 140 positions, 25 are university funded; the rest are covered by external funds. The Swiss canton in which IMT is located has just 160,000 inhabitants, but it provides about one-third of the overall support for the university. The building in which we met is owned by a nonprofit real estate corporation in which the canton is a major stockholder.

REFERENCES

A searchable database of the group's publications is available at www-samlab.unine.ch/Publications/Publications.htm.

Daridon, A., V. Fascio, J. Lichtenberg, R. Wuetrrich, H. Langen, E. Verpoorte, and N.F. de Rooij. 2001. Multi-layer microfluidic glass chips for microanalytical applications. *Fresenius Journal Analytical Chemistry* 371:261–269.

Dodge, A., K. Fluri, E. Verpoorte, and N.F. de Rooij. 2001. Electrokinetically driven microfluidic chips with surface-modified chambers for heterogeneous immunoassays. *Analytical Chemistry* 73 (14):3400–3409.

Juncker, D., H. Schmid, A. Bernard, I. Caelen, B. Michel, N.F. de Rooij, and E. Delamarche. 2001. Soft and rigid two-level microfluidic networks for patterning surfaces. *Journal of Micromechanics and Microengineering* 11:532–541.

Lichtenberg, J., E. Verpoorte, and N.F. de Rooij. 2001. Sample preconcentration by field amplification stacking for microchip-based capillary electrophoresis. *Electrophoresis* 22:258–271.

Lichtenberg, J., N.F. de Rooij, and E. Verpoorte. 2002. A microchip electrophoresis system with integrated in-plane electrodes for contactless conductivity detection. *Electrophoresis* 23:3769–3780.

Lichtenberg, J., N.F. de Rooij, E. Verpoorte. 2002. Sample pretreatment on microfabricated devices. *Talanta* 56:233–266.

Linder, V., E. Verpoorte, W. Thormann, N.F. de Rooij, and H. Sigrist. 2001. Surface biopassivation of replicated poly(dimethylsiloxane) microfluidic channels and application to heterogeneous immunoreaction with on-chip fluorescence detection. *Analytical Chemistry* 73 (17):4181–4189.

Linder, V., N.F. de Rooij, E. Verpoorte, H. Sigrist, and W. Thormann. 2002. Application of surface biopassivated disposable poly(dimenthylsiloxane)/glass chips to a heterogeneous competitive human serum immunoglobulin G immunoassy with incorporated internal standard. *Electrophoresis* 23:740–749.

Michel, Ph., G.C. Fiaccabrino, P. van der Wal, N.F. de Rooij, M. Koudelka-Hep, M.L. Tercier-Waeber, and J. Buffle. 2001. Microelectrode arrays based (bio)analytical systems. *Biocybernetics and Biomedical Engineering* 21:5–9.

Reichmuth, P., H. Sigrist, M. Badertscher, W.E. Morf, N.F. de Rooij, and E. Pretsch. 2002. Immobilization of biomolecules on polyurethane membrane surfaces. *Bioconjugate Chem.* 13:90–96.

Roulet, J.-C., R. Völkel, H.P. Herzig, E. Verpoorte, N.F. de Rooij, and R. Dändliker. 2002. Performance of an integrated microoptical system for fluorescence detection in microfluidic systems. *Analytical Chemistry* 74:3400–3407.

Roulet, J.-C., R. Völkel, H.P. Herzig, E. Verpoorte, N.F. de Rooij, and R. Dändliker. 2001. Microlens systems for fluorescence detection in chemical microsystems. *Optical Engineering* 40(5) 814–821.

Roulet, J.-C., R. Völkel, H.P. Herzig, E. Verpoorte, N.F. de Rooij, and R. Dändliker. 2001. Fabrication of multilayer systems combining microfluidic and micro-optical elements for fluorescence detection. *Journal of Microelectromechanical Systems* 10 (4):482–491.

Weiller, B., L. Ceriotti, T. Shibata, D. Rein, M.A. Roberts, J. Lichtenberg, J.B. German, N.F. de Rooij, and E. Verpoorte. 2002. Analysis of lipoproteins by capillary zone electrophoresis in microfluidic devices: Assay development and surfaces roughness measurements. *Analytical Chemistry* 74:1702–1711.

Site: **University of Regensburg**
Institute of Analytical Chemistry, Chemo- and Biosensors
93040 Regensburg, Germany
http://www-analytik.chemie.uni-regensburg.de/

Date Visited: 20 March 2003

WTEC Attendees: J. Shultz (report author), F. Heineken, M. Mrksich

Hosts: Prof. Dr. Otto S. Wolfbeis, Head
Email: otto.wolfbeis@chemie.uni-regensburg.de
Priv. Doz. Dr. Vladimir M. Mirsky, Faculty, Electrical methods
Dr. Axel Dürkop, Faculty, Dye synthesis
Dr. Michael Schäferling, Faculty, Optical array instrumentation
Prof. Dr. Claudia Steinem, Faculty, Bilayers

OVERVIEW

Approximately 50 people are involved in Institute projects. Following are overviews of their work.

Dr. Mirsky trained in Moscow; his main field of interest was bioelectrochemistry, including membrane biophysics and field-charge distribution at surfaces.

Examples

- Planar lipid bilayers
- Self-assembly monolayers on gold surfaces
- Stability – issue desorption of the monomer from the immobilized film, thiols on the
- Particularly the influence of pH, electrical potential
- Attempts to solve stability problems using a different substrate such as palladium, but Pb did not work at temperatures >60°C
- Use an alternative Typical Membrane
- Use bilayer on gold surface
- Measure rate of change of capacitance with time. Have shown that they can make electrodes using gold/thiol films that are responsive to

antigen adsorption to immobilized antibodies and to molecularly imprinted polymers.

- Another method is called Chemioresistors. For example, one can polymerize a conductive path (wire) on a planar substrate, then cover the exposed electrode wires with a monolayer of a thio polymer to form a protective film or "filter." This coating material prevents aqueous components from passing through the film to the electroactive surface. The primary use has been for a mercury detector system.

T. Sagiv

Dr. Sagiv's team is working on nanostructured monolayers. It is possible to use a substrate or its derivative to form a template for subsequent binding of similar compounds. But this approach was not stable, losing about 15% of its function per cycle.

An unexpected discovery was that the oxidation/reduction behavior of Ferri/Ferro cyanide at low concentrations is dependent on the concentration of single-stranded DNA.

Other studies to rapidly develop optimal structures of electrochemical sensors utilized newly developed methods for combinatorial electropolymerization. About 100 different formulations can be evaluated at one time.

New instrumentation has been developed for monitoring fluorescence energy transfer (FRET) by time-resolved detection based on the changes in fluorescence lifetime. With this instrumentation it may be possible to develop inexpensive instruments that can measure multiple analytes simultaneously, e.g., O_2, CO_2, pH, glucose, Ca^{++}, etc., for metabolic modeling.

This laboratory developed a new fluorescent probe for hydrogen peroxide, a Eu^{+3}-tetracycline complex. This complex absorbs at about 400 nm and emits at about 620 nm—an extraordinary Stokes shift of about 200 nm. The complex is stable in aqueous solution and shows fluorescence behavior in aqueous solution. The complex can be used directly in many enzymatic bioassays that result in the formation of H_2O_2, such as the measurement of glucose via its oxidation glucose oxidase. The linear range of the response is 0 to 400 micro molar. Increased sensitivity and reduction of background noise can be obtained by gating excitation and measurement of emission. The new fluorophore also is expected to be used in two dimensional surface array screening procedures. Very inexpensive instrumentation is being developed.

Institute researchers have also dramatically improved the surface plasmon resonance methodology, another traditional analytical technique, by constructing structures that contain a bimetallic silver/gold layer on the surface of the prism with gold as the exposed layer that protects against oxidation.

Some highly creative research on the preparation of lipid bilayers has offered the potential to use these methods to evaluate, screen, and select membrane-borne bilayers for new potential biorecognition elements. The preparation of a stable supporting structure for the bilayer was achieved by anodizing aluminia in aqueous solution to produce a well-ordered grating with pores on the order of 50 nm. The supporting grating can be made porous so that the properties of transport protein suspended within the lipid bilayer can be evaluated for affinity and transport properties. At the present time only membrane fragments (vesicles) have been successfully immobilized in this fashion.

Site: **University of Twente**
MESA+ Institute
P.O. Box 217
Enschede 7500 AE, The Netherlands
http://www.mesaplus.utwente.nl

Date Visited: 18 March 2003

WTEC Attendees: A.J. Ricco (report author), D. Brady

Host: Professor David Reinhoudt, Tel: +31-53-4892714;
Email: D.N.Reinhoudt@ct.utwente.nl

KEY INFORMATION

Date of founding: July 1999.

Faculty, staff, students at MESA+: ~440 total, distributed among 17 academic research groups.

Scientific Director: Prof. David N. Reinhoudt.

Annual budget: ~€33 million.

Site: several buildings in a number of academic departments at the University of Twente.

Mission: Advance science and technology focused on nanoscale dimensions, including lab-on-a-chip, nanophotonics, and nanostructured materials.

INTRODUCTION AND FUNDING SITUATION

MESA+ is the University of Twente's largest research institute. MESA is an acronym for Microelectronics, Sensors, and Actuators. Its roots are in the Electrical Engineering Dept., which joined with the sensor groups from the faculties of Applied Physics and Chemical Technology in 1990 to establish MESA; in 1999, MESA+ was created from a merger of MESA with the Materials Research Center (CMO) of Applied Physics and Chemical Technology. This occurred in significant measure as a response to funding opportunities that were available preferentially to large teams having a highly multidisciplinary composition in combination with an integrated structure.

MESA+ is engaged in five large interdisciplinary projects, known as Strategic Research Orientations (SROs):

1. Microchemical Systems (MiCS), led by Prof. Albert van den Berg
2. Information technology approaching the molecular scale (known as "Nanolink"), led by Prof. Jürgen P. Brugger

3. Terahertz Signal Processing, led by Prof. G.J. Gerritsma
4. Advanced Photonic Structures, led by Dr. L. Korbus Kuipers
5. Materials Science of Interfaces (MASIF), led by Dr. Dave H.A. Blank.

These faculty leaders are relatively young, which has helped to promote collaboration and to integrate rapidly their research efforts and focus areas with the programs of the more senior, well-established faculty members.

Of the current annual budget of €33 million for MESA+, about 55% is provided by the university (funds originating with the Dutch Ministry of Education, Culture, and Science), and the balance comes from external sources, including competitive funds from the Netherlands Organization for Scientific Research, various industrial contracts, and competitive funding from the European Union. In addition, MESA+ will receive an additional €15 million from 2000–2004 to help develop the five SROs. The external funding sources also include Nanonet, a nationwide government-supported €23 million/year program that funds four major projects in nanotechnology, and an oil and gas tax-based national infrastructure program that awards 10% of the overall budget to knowledge-base infrastructure, focusing in such areas as genomics, nanotechnology, and information technology.

The research groups included in MESA+ are Biosensor Technology (Prof. Piet Bergveld), Lightwave Devices (Dr. Paul V. Lambeck), Applied Analysis and Mathematical Physics (Prof. E. Brenny van Groesen), Micromechanical Transducers (Prof. Miko Elwenspoek), Information Storage Technology (Dr. Cock Lodder), Applied Optics (Prof. Niek F. van Hulst), Supramolecular Chemistry and Technology (Prof. David N. Reinhoudt), Biophysical Techniques (Prof. J. Greve), Materials Science and Technology of Polymers (Prof. G. Julius Vancuso), and seven more groups in device engineering and physics. This clustering of relatively diverse research groups is significant for biosensing, as they represent many of the different disciplines that must be harnessed to make headway in this field.

About 25 Ph.D.s are awarded by MESA+ each year, with about 125 enrolled in the Ph.D. program overall.

FABRICATION CAPABILITIES & FACILITIES

Infrastructure includes a 1000 m^2 cleanroom (class 1000 overall, class 10 at workstations) with the requisite equipment for all stages of device design and lithographic fabrication, including CMOS devices, MEMS, and lab-on-a-chip systems, as well as a range of characterization equipment.

The many research groups have their own decentralized laboratories in the range of disciplines represented above.

TECHNOLOGY TRANSFER/COMMERCIALIZATION/COLLABORATION

MESA+ focuses on small / medium-size companies for much of its technology transfer activity. The Twente MicroProducts (TMP) Foundation, founded within the university in 1995, has focused on creating infrastructure for engineering and production. This "technology accelerator" employs about 20 staff and has had a large number of patents in its two years of existence. To some degree, such accelerators, in combination with available government funds for commercial startups, play part of the role of the U.S. system of venture capital and "angel" investors, though on a smaller financial scale. Recently, TMP spun off as a separate company and was recently acquired by the Dutch integrated optics company Kymata.

About 30% of the work carried out in the MESA+ cleanroom facility is in direct support of commercial products, showing another way in which the university helps to support new commercial ventures. This may also be part of the reason that the MESA+ laboratory facilities are run as a business unit, with the academic and industrial users alike paying about €200/m^2 per year for dedicated space and €5000/year overhead on each full-time worker. As a consequence of this policy, 30% of the laboratory space has been returned and reallocated to groups that have both the need and the funds to use the space.

RESEARCH HIGHLIGHTS

Prof. Reinhoudt briefly highlighted research in a number of focus areas germane to biosensing. These included the eventual replacement of stand-alone chemical/biological sensors by integrated lab-on-a-chip systems, many of which will take advantage of nanotechnology; the combination of lab-on-a-chip technologies with mass spectroscopy to tackle the challenging characterization of the proteome; the merger of nano-fabrication methods—which he referred to generally as "nonconventional" fabrication technologies—with more conventional microfabrication methods, as employed in typical MEMS cleanroom facilities; noncovalent synthetic methods for large organic molecules to enable "error free" synthesis of molecules with molecular weights approaching 50,000 and above; and self-assembly as a synthetic method for the creation of nanostructures. He focused on the role that "bottom-up" nanotechnology can play, where there is no physical template for the structure to be constructed, so fabrication relies instead on self-assembly to match the desired master pattern, and the reproduction technology is based ultimately more on chemical synthesis than on physical replication/patterning. As a supporting technology, he described experiments using an atomic force microscope to pull molecular guests out of their hosts, in an experiment whose results are sensitive to both the energetics of the interaction and the rate at which the measurement is obtained.

REFERENCES

A complete list of references is available at: http://smct.ct.utwente.nl/people/davidr/publyst.html.

Antonisse, M.M.G., B.H.M. Snellink-Ruël, A.C. Ion, J.F.J. Engbersen, and D.N. Reinhoudt. 1999. Synthesis of novel uranyl salophene derivatives and evaluation as sensing molecules in chemically modified field effect transistors (CHEMFETs). *J. Chem. Soc., Perkin Trans.* 1211–1218.

Antonisse, M.M.G., B.H.M. Snellink-Ruël, R.J.W. Lugtenberg, J.F.J. Engbersen, A. Van den Berg, and D.N. Reinhoudt. 2000. Membrane characterization of anion-selective CHEMFETs by impedance spectroscopy. *Anal. Chem.* 72:343–348.

Casnati, A., J. De Mendoza, D.N. Reinhoudt, and R. Ungaro. 1999. Determination of calixarene conformations by means of NMR techniques. In *NMR in supramolecular chemistry,* ed. M. Pons, 307–310. Dordrecht: Kluwer Academic Publishers.

Crego Calama, M., P. Timmerman, and D.N. Reinhoudt. 2000. Guest-templated selection and amplification of a receptor by noncovalent combinatorial synthesis. *Angew. Chem. Int. Ed.* 39:755–758.

De Jong, M.R., J.F.J. Engbersen, J. Huskens, and D.N. Reinhoudt. 2000. Cyclodextrin dimers as receptor molecules for steroid sensors. *Chem. Eur. J.* 6:4034–4040.

Flink, S., F.C.J.M. Van Veggel, and D.N. Reinhoudt. 1999a. A self-assembled monolayer of a fluorescent guest for the screening of host molecules. *Chem. Commun.* 2229–2230.

———. 1999b. Recognition of cations by self-assembled monolayers of crown ethers. *J. Phys. Chem. B.* 103: 6515–6520.

———. 2000. Sensor functionalities in self-assembled monolayers. *Adv. Mater.* 12. :1315–1328.

Flink, S., H. Schönherr, G.J. Vancso, F.A.J. Geurts, K.G.C. Van Leerdam, F.C.J.M. Van Veggel, and D.N. Reinhoudt. 2000. Cation sensing by patterned self-assembled monolayers on gold. *J. Chem. Soc., Perkin Trans.* 2:2141–2146.

Friggeri, A., F.C.J.M. Van Veggel, D.N. Reinhoudt. 1999. Recognition of steroids by self-assembled monolayers of calix[4]arene-resorcin[4]arene receptors. *Chem. Eur. J.* 5:3595–3602.

Kanekiyo, Y., K. Inoue, Y. Ono, M. Sano, S. Shinkai, and D.N. Reinhoudt. 1999. Molecular-imprinting of AMP utilising the polyion complex formation process as detected by a QCM system. *J. Chem. Soc., Perkin Trans.* 2:2719–2722.

Kim, B.J., M. Liebau, J. Huskens, D.N. Reinhoudt, and J. Brugger. 2001. A self-assembled monolayer-assisted surface microfabrication and release technique. *Microelectronic Engineering* 755–760.

Liebau, M., J. Huskens, and D.N. Reinhoudt. 2001. Microcontact printing with heavyweight inks. *Adv. Funct. Mater.* 11:147–150.

Lugtenberg, R.J.W., and D.N. Reinhoudt. Selective ion recognition with durable sensors. 1999. In *Supramolecular Technology*, ed. D.N. Reinhoudt, 193–223. London: John Wiley & Sons Ltd.

Prins, L.J., P. Timmerman, and D.N. Reinhoudt. 1999. Non-covalent synthesis of organic nanostructures. In *Current Trends in Organic Synthesis*, ed. C. Scolastico and F. Nicotra, 35–46. New York: Kluwer Academic/Plenum Publishers.

Reinhoudt, D.N., P. Timmerman, F. Cardullo, and M. Crego-Calama. 1999. Synthesis and Ccharacterization of hydrogen-bonded assemblies: Toward the generation of binding site diversity. In *Supramolecular science: Where it is and where it is going*, ed. R. Ungaro and E. Dalcanale, 181–195. Dordrecht: Kluwer Academic Publishers.

Schönherr, H., M.W.J. Beulen, J. Bügler, J. Huskens, F.C.J.M. Van Veggel, D.N. Reinhoudt, and J. Vancso. 2000. Individual supramolecular host-guest interactions studied by dynamic single molecule force spectroscopy. *J. Am. Chem. Soc.* 122:4963–4967.

Timmerman, P., and D.N., Reinhoudt. 1999. A combinatorial approach to synthetic receptors. *Adv. Mater.* 11:71–74.

Wróblewski, W., K. Wojciechowski, A. Dybko, Z. Brzózka, R.J.M. Egberink, B.H.M. Snellink-Ruël, and D.N. Reinhoudt. 2000. Uranyl salophenes as ionophores for phosphate-selective electrodes. *Sensors and Actuators B*. 313–318.

Site: **University of Twente**
Laboratory of Biosensors
Faculty of Electrical Engineering
P.O. Box 217
7500 AE Enschede, The Netherlands
Tel: +31 53 489 2760; Fax: +31 53 489 2287

Date Visited: 18 March 2003

WTEC Attendees: D. Brady (report author), A. Ricco

Hosts: Prof. dr. ir. Albert van den Berg,
A.vandenBerg@el.utwente.nl
Prof. dr. ir. Piet Bergveld, P.Bergveld@el.utwente.nl

BACKGROUND

The Laboratory of Biosensors (BIOS) is a component of the MESA+ Institute at the University of Twente. The laboratory includes over 30 faculty, staff, and students. The laboratory has a long history in integrated electronic chemical sensors and pioneered FET-based sensors. The current emphasis of the laboratory focuses is on the "lab-on-a-chip" vision that integrates microfluidic, CMOS, and biochemical systems on a single chip. The laboratory's particular emphasis on electrically controlled microfluidics and microchemistry has resulted in several significant sensor systems. One of the more successful is a multistage conductivity-based ammonia sensor.

Microchemistry includes a variety of techniques and interfaces. One may use microfluidic systems in combination with mass spectrometry and NMR to produce miniaturized laboratory instruments. Some of the more successful lab-on-a-chip work at Twente exploits inherent characteristics of microfluidic systems. For example, hydrodynamic chromatography exploits flow profiles and molecular distributions in micromachined channels to achieve size-based molecular separation.

BIOS also includes several sensor projects based on polymer hydrogels. Hydrogels respond to changes in pH or temperature. While hydrogels may not be directly applicable to biosensing, BIOS has used mechanical swelling in hydrogels to build pumps, which may be useful in developing nanofluidic systems for biosensors.

Recent work extends the BIOS integrated system vision to "lab-on-a-cell" technology. The lab has developed a vision for biochemical sampling of living cells based on nanofluidic sensor systems. The system would use nanometer-scale probes to sample femtoliters of cellular material, with a goal of sampling molecules associated with realtime metabolism. New technologies, particularly with respect to nanofluidics and cell interfaces, will be developed in pursuit of this vision.

REFERENCES

Bergveld, P. 2003. Thirty years of ISFETOLOGY: What happened in the past 30 years and what may happen in the next 30 years. *Sensors and Actuators B-Chemical* 88(1):1–20.

Chmela, E., R. Tijssen, M. Blom, H. Gardeniers, and A. van den Berg. 2002. A chip system for size separation of macromolecules and particles by hydrodynamic chromatography. *Analytical Chemistry* 74(14):3470–3475.

Guijt, R.M., E. Baltussen, G. van der Steen, R.B.M. Schasfoort, S. Schlautmann, H.A.H. Billiet, J. Frank, G.W.K. van Dedem, and A. van den Berg. 2001. New approaches for fabrication of microfluidic capillary electrophoresis devices with on-chip conductivity detection. *Electrophoresis* 22(2):235–241.

Schasfoort, R.B.M., S. Schlautmann, L. Hendrikse, and A. van den Berg. 1999. Field-effect flow control for microfabricated fluidic networks. *Science* 286(5441):942–945.

Tiggelaar, R.M., T.T. Veenstra, R.G.P. Sanders, J.G.E. Gardeniers, M.C. Elwenspoek, and A. van den Berg. 2002. A light detection cell to be used in a micro analysis system for ammonia. *Talanta* 56(2):331–339.

Site: **The University of Warwick**
Department of Chemistry
Coventry CV4 7AL
United Kingdom

Date Visited: 17 March 2003

WTEC Attendees: C.L. Wilkins (report author), C. Kelley, D. Walt

Hosts: Professor Peter Derrick, Chair, Department of Chemistry; Director, Institute of Mass Spectrometry, Tel: +44 (0)24 7652 3818; Fax: +44 (0)24 7652 3819; Email: P.J.Derrick@warwick.ac.uk
Professor Nick Dale, Department of Biological Sciences

INTRODUCTION

The University of Warwick is a moderate-sized University with about 14,000 students, of whom approximately 5,000 are graduate students. The Department of Chemistry has 25 faculty members and about 100 graduate students. The primary biosensor-related research is focused on instrumental analysis of biological systems. The Department of Biological Sciences has approximately 50 faculty and 140 graduate students.

NEUROCHEMISTRY, SIGNALING, AND ADENOSINE BIOSENSOR

Professor Dale's group uses electrophysiological and molecular techniques to study neural communication in spinal cord, cerebellum, hypothalamus, and hippocampus. The group uses the Xenopus embryo as a model system as well as brain slices and hemisected spinal cord from rats. Specific areas of research as summarized on Professor Dale's webpage (http://www.bio.warwick.ac.uk/research/dale/framepage.htm) are listed below:

- Purinergic signalling and the role of ectonucleotidases in neural circuits
- Microsensors for amperometric measurement of purine release from nervous system
- Developmental regulation of ion channel expression during maturation of spinal cord and brain

- The characterisation and development of synaptic transmission at cerebellar synapses
- Electrical synapses in mammalian central neurones
- Energy homeostasis and hormonal communication to hypothalamic neurones
- Nitric oxide-dependent signalling in sympathetic neurones.
- Chemical and electrical synaptic transmission in sympathetic neurones

During the visit, it was emphasized that one of the leading edge research areas currently under investigation is development of an adenosine biosensor intended to be small, have rapid response, and be minimally invasive. The current design employs a three-enzyme cascade polypyrrole-based strategy. A new electrodeposition method that permits deposition of all three enzymes in less than 1 minute promises to make the new sensor even more practical for physiological studies. Some relevant recent publications from this research group include Brown and Dale 2000; Dale, Pearson, and Frenguelli 2000; Spanswick et al. 2000; and Trudeau et al. 2000.

MASS SPECTROMETRY RESEARCH

Professor Derrick briefly reviewed the state-of-the art mass spectrometry research underway at Warwick. He discussed his interest in atmospheric pressure mass spectrometry, ion funnel technology, and protein-protein interactions (Lafitte et al. 1999; Heck et al. 1998). He noted that one of the issues potentially having a great impact on pursuit of biologically oriented mass spectrometry research is the increasing difficulty in attracting properly qualified individuals to participate in the research. He perceives this as a general problem in the United Kingdom.

REFERENCES

Brown, P.E., and N. Dale. 2000. Adenosine A1 receptors modulate HVA Ca2+ currents and motor pattern generation in the Xenopus embryo. *Journal of Physiology* 525:655–667.

Dale, N., T. Pearson, and B.G. Frenguelli. 2000. Direct measurement of adenosine release during hypoxia in the CA1 region of the rat hippocampal slice. *Journal of Physiology* 526:143–155.

Heck, A.J., J.T.D. Jorgensen, M. O'Sullivan, M. von Raumer, and P.J. Derrick. 1998. Gas-phase non-covalent interactions between vancomycin-group antibiotics and bacterial cell-wall precursor peptides probed by hydrogen/ deuterium exchange. *J. Am. Soc. Mass Spectrom.* 9:1255–1266.

Lafitte, D., A.J.R. Heck, T.J. Hill, K. Jumel, S.E. Harding, and P.J. Derrick. 1999. Evidence of noncovalent dimerisation of calmodulin. *Eur. J. Biochem.* 261:337–344.

Spanswick, D., M.A. Smith, S. Mirshamsi, V.H. Routh, and M.L.J. Ashford. 2000. Insulin activates ATP-sensitive potassium channels in hypothalamic neurones of lean, but not obese rats. *Nature Neuroscience* 3:757–758.

Trudeau, V.L.D. Spanswick, E.J. Fraser, K. Lariviere, D. Crump, S. Chui, M. MacMillan, and R. Schulz. 2000. The role of amino acid neurotransmitters in the regulation of pituitary gonadotropin release in fish. *Biochemistry & Cell Biology* 78:241–259.

APPENDIX C. SITE REPORTS—JAPAN

Site: **Initium, Inc.**
5f, Yamashiro Bldg.
1-15-16 Minami-Aoyama, Minato-ku
Tokyo 107-0062, Japan
http://www.initium2000.com

Date Visited: 28 January 2003

WTEC Attendees: A.J. Ricco (report author), D. Brady, D. Walt, C. Wilkins, H. Ali

Hosts: Ms. Izumi Ishii, President, Tel: +81-3-5772-2145; Fax: +81-3-5772-2141; Email: izumi@initium2000.com
Mr. Joseph Itoh, Assistant Manager, Business Development
Dr. Tomofumi Jitsukawa, Director

INTRODUCTION

Initium was founded in 1999 by Professor Okahata (Professor of Biochemistry, Tokyo Institute of Technology) and Ms. Ishii with the goal of developing and commercializing the instrumentation for simple, accurate measurements of mass changes on the surface of quartz crystal microbalance (QCM) devices, also known as thickness-shear mode resonators. Initium has approximately 10 employees and a capitalization of ¥45 million.

Professor Okahata has a long history of research and innovation in the use of QCMs for bioanalytical measurements, with many publications in the peer-reviewed literature over the past 15+ years.

Initium's principal product is the AFFINIXQ quartz microbalance instrument (single-channel and 4-channel models) for liquid-, thin-film, and gas-phase measurements.

The QCM enables so-called "label-free" measurements of biomolecules, because it responds directly to any mass that is bound to the device surface. The most closely analogous technique in terms of both sensitivity and interfacial attachment schemes for biosensing is Surface Plasmon Resonance (SPR). Like SPR, QCM biosensing requires a method of capturing the target analyte on the device surface, which in many cases means immobilizing on the device an antibody specific to the target species (if such is available), or tagging the analyte in some way that promotes its specific binding (thus, the term label-free must be used advisedly).

DEVICE AND INSTRUMENT DETAILS

The Initium QCM utilizes a 27 MHz AT-cut quartz crystal; it functions in direct contact with aqueous solution. The typical frequencies utilized by other commercially available QCM systems are in the 5–9 MHz range, although academic research has been reported on devices at frequencies up to 100 MHz. Mass increases on the device surface result in directly proportional decreases in the resonant frequency of the device, so the frequency readout can be calibrated in terms of surface mass per unit area. Higher frequencies can be advantageous because mass sensitivity is inversely proportional to frequency, although the noise level must be managed carefully so that noise does not also increase with frequency. Thus, provided thermal stability—often the main source of noise and drift in QCM biosensing applications—is excellent, and the oscillator circuit is well designed, the limit of detection (minimum detectable mass) improves in inverse proportion to frequency.

In the case of the 27 MHz crystal used by Initium, a frequency change of 1 Hz corresponds to a mass change of 600 pg/cm^2 (this corresponds to a surface mass change of about 30 pg for the ~5 mm^2 area of the crystal). The short-term frequency stability is of the order of 0.3 Hz (verified by the WTEC visitors in a working system at the company's laboratories), so allowing for a three-times-noise limit of detection, sub-ng/cm^2 mass changes are readily detected. The frequency stability is a consequence particularly of an active temperature control system that uses a thermoelectric device to keep the temperature of the liquid-filled cell in which the QCM is immersed stable to better than 0.05°C in a laboratory environment. Stability is also supported by well-designed oscillator circuitry (general type of circuit is Colpit), good immunity to electromagnetic noise sources, and an enclosure that protects the operating QCM from environmental perturbations (drafts, etc.). The operating temperature range of the thermoelectric system is 0.1–50°C. Mounting of the crystal, which must expose only one side of the device to solution, is accomplished with a custom holder and adhesive; currently this assembly is performed manually.

The company's first product is a single-QCM instrument with a magnetically stirred test-tube-like sample cell having a volume of several mL and the temperature control and frequency stability characteristics described above. To date, 70 of these instruments have been supplied to customers. The sales price is ¥5 million (about U.S. $42,000). The next-generation instrument will support four QCMs operating in parallel, also at 27 MHz, each with its own liquid cell. Sample cell volume has been reduced to 0.5 mL for this system, and stirring is accomplished by a vibrating "plunger" that dips into each of the cells. Unlike the single-device system, where the mounted crystal is lowered in its fixture into the liquid cell, each of the four devices forms the bottom of its low-volume sample cell for the multisensor instrument. Our Initium hosts said the challenging problem of cross-talk between adjacent devices in this system has been solved, and stability is comparable to the single-device system. The instrument sales price will be about double that of the single-device instrument.

APPLICATION EXAMPLES AND SURFACE CHEMISTRY

One of the keys to success of "label-free" measurement systems is a robust method to attach the analyte-selective moiety to the device surface. For example, Biacore's surface chemistry for its SPR devices, which provides a reproducible mechanism to immobilize capture moieties of many types, is an important part of the success of that commercial system. Like most QCM devices, the crystals used by Initium have gold electrodes. Hence, they are readily functionalized with thiol-based self-assembled monolayers. Three general methods to immobilize capture moieties were described: (1) incubation of the device with a sorptive material (typically for a period of many minutes to an hour or more); (2) deposition of a dilute solution and evaporation to dryness, leaving the nonvolatile materials on the surface (requiring a few minutes); and (3) attachment of a dithiol monolayer via the Au–S interaction. In the last case, the example of dithiopropionic acid to produce a monolayer terminated in carboxylic acid functionalities was shown. The acid groups can attach proteins (e.g., antibodies) to the device surface; this was demonstrated using EDC/NHS coupling over a period of about 2 hours.

The predicted mass sensitivity of the device was confirmed using the binding of DNA oligomers of known mass, and also by the casting of poly(styrene) films of known mass. In this way, the linearity of response for mass loadings ranging from 100 pg to 10 μg was confirmed; these measurements were made both for a crystal operating in air and in liquid phase. Confirming linearity and consistency of the mass loading coefficient for different surface analytes and in both aqueous and dry conditions is important, because under some conditions QCMs can respond to the mechanical (viscoelastic) properties of material films, causing significant deviation from ideal behavior. (Such deviation from ideal mass loading behavior is most often seen for relatively "rubbery" surface layers, an example being polydimethyl siloxane). It was also pointed out that temperature changes can affect the signal in a significant way, due to changes in the coverage (in cases of dynamic equilibrium) as well as the temperature dependence of solution viscosity and density for aqueous phase measurements. For gas-phase measurements, it was pointed out that changes in the relative humidity of the contacting ambient are likely to affect the quantity of water absorbed by the surface film, which in turn will generate a change in sensor signal.

Attachment of single-stranded DNA followed by hybridization of the complementary strand was demonstrated, and the expected 1:1 relationship between frequency changes for strands of equivalent mass was confirmed. An experiment showing the realtime tracking of the process of elongation of DNA was also demonstrated, with an initial observable mass change corresponding to the binding of the polymerase, followed by additional mass change when dNTPs were added and polymerase-catalyzed extension occurred.

Additional applications discussed include the binding of distamycin to a dAT 30-mer oligonucleotide, showing the binding of 3–4 distamycins per dAT-30; specific binding of bZIP by the appropriate double-stranded immobilized DNA; binding of ligase enzyme by DNA; detection of enterotoxin B in milk; binding of red blood cells; and ligand fishing using cells adhered to the device surface.

SUPPORT OF TECHNOLOGY COMMERCIALIZATION; INTELLECTUAL PROPERTY

Initium's laboratory and conference facilities are housed in Keio University's very modern collaborative ventures building, constructed at a cost of roughly U.S. $50 million. Keio, a private university, funded the building construction and charges rent to the companies that are "incubated" in this building.

In years past, many Japanese academics were permitted by their universities to own patents on developments from their laboratories, but the patenting process was not considered a priority by universities. Professor Okahata owned 4 patents on his work on the QCM biosensing system resulting from his work at Tokyo Institute of Technology, and these patents are now the property of Initium. Two additional patents have been filed by Initium since its founding, in protein and cell applications.

Beginning in 1999, the Japanese government began to promote more seriously the filing of patents at universities and, in addition, began to allow faculty at publicly funded universities to participate in private companies and to be compensated financially for this work. This process has been further streamlined, and it is now possible to obtain permission for such activities by filing a single-page document.

FUTURE ACTIVITIES

Initium's future plans include starting sales of the 4-channel unit; developing systems with improved mass senstivity; developing an automated sample-loading system; and exploring applications of its technology to drug discovery. Its management also foresees applications in environmental monitoring and in medical diagnostics.

The Initium QCM system seems to be a very well-engineered laboratory measurement tool, making it possible for researchers to focus their efforts on selective surface chemistry rather than custom building their own hardware for such direct mass measurements.

REFERENCES

Furusawa, H., Y. Kitamura, N. Hagiwara, T. Tsurimoto, and Y. Okahata. 2002. Binding kinetics of the toroid-shaped PCNA to DNA strands on a 27 MHz quartz crystal microbalance. *ChemPhysChem.* 3 (5):446–8.

Matsuno, H., H. Furusawa, Y. Okahata. 2002. Direct monitoring of DNA cleavages catalyzed by an ATP-dependent deoxyribonuclease on a 27 MHz quartz crystal microbalance. *Chem. Comm.* Mar. 7 (5):470–1.

Site:	**Japan Advanced Institute of Science and Technology (JAIST)** **School of Materials Science** **1-1, Asahidai, Tatsunokuchi, Nomi** **Ishikawa, Japan** **http://www.jaist.ac.jp**
Date Visited:	31 January 2003
WTEC Attendees:	J. Brewer (report author), D. Brady, D. Walt, C.Wilkins, J. Schultz, S. Bhatia, S. Green, T. Ricco, H. Ali
Hosts:	Prof. Eiichi Tamiya, Professor, School of Materials Science, Tel: +81-761-51-1660; Fax: +81-761-51- 1665; Email: tamiya@jaist.ac.jp Dr. Quamrul Hasan, Visiting Professor of Mandom Corp., Email: hassan@po.broad2000.co.jp Dr. Yasutaka Morita, Assistant Professor, School of Materials Science, Tel: +81-761-51-1662; Fax: +81-761-51-1665; Email: tamori@jaist.ac.jp Mr. Fabrice Morin, Ph.D. Student Dr. S. Ramachandra Rao, Post Doc

BACKGROUND

The Japan Advanced Institute of Science and Technology (JAIST), a national university, was founded in 1990. It offers masters and doctoral programs. JAIST is comprised of three schools, the School of Knowledge Science, the School of Information Science, and the School of Materials Science. JAIST has 327 faculty members and 911 students, of whom 352 attend the School of Materials Science. JAIST expenditures in FY2001 were ¥6.6 billion. The researchers of the School of Materials Science seek to extend the understanding of the interrelation between structures and the functions from the atomic/molecular level to the aggregated state. The Bioelectronics and Bioengineering Lab is part of the School of Materials Science and conducts research on on-chip biotechnology and bioelectronics; design and creation of molecular recognition; nano-biotechnology based on scanning probe microscopy (SPM); and environmental biotechnology.

R&D ACTIVITIES

On-chip Biotechnology and Bioelectronics

Dr. Tamiya's work on on-chip biotechnology and bioelectronics focuses on pico/nano-chamber arrays, microfluidic biochip/biosensors, and microfluidic array chips for cellular signal analyses. The pico/nano chamber arrays hold 10 molecules each, and there can be as many as 10,000 chambers in an array for polymerase chain reaction (PCR) analysis. The microfluidic biochip/biosensors detect enzymes, immunology, genes, and cells through on-chip cultivation after introduction of an antigen. The microfluidic array chips are used for neuronal cell network, ES cell differentiation, and drug screening. Researchers at this lab are also developing a new type of electrochemical DNA sensor concept that binds the DNA with bisbenzimide and gets DNA amplification after PCR. DNA aggregation initiated by Hoechst 33258 to form a complex is used for detection. They developed a commercial PCR analysis system using a rotating table and high heat. This device can run 30 cycles in 10 minutes.

Design and Creation of Molecular Recognition

Researchers at the Bioelectronics and Bioengineering Lab are studying molecular recognition using combinatorial molecular approaches and by designing very tough enzymes that are thermostable (50% reduction in activity after 3–4 hours at 100°C).

Nanobiotechnology based on SPM

The researchers of the School of Materials Science are developing a new type of DNA sensor using imaging by AFM of a single molecule. They developed a nano-dissection technique to slice chromosomes. They are also working on bioimaging based on scanning near-field optical/atomic force microscopy (SNOAM).

Environmental Biotechnology

The school's work in environmental biotechnology is with a private company, En BioTec Laboratories Co., Ltd. It is developing biosensors for environmental protection and bioremediation. One project found that vitellogenin is an ideal biomarker for detection of the estrogenic effects of endocrine disrupting chemicals.

Constraining Cell Development by Microfluidic Technique: A Cluster Approach to Cell-Based Chips

Mr. Morin presented his work on cell-based microarrays. Issues include cell positioning, immobilization and culture handling, distribution of liquids, and control of surface effects. This group uses a Panasonic Med64 system using P19 cell line and primary cultures. They have characterized the impedance and signal transduction. In their work on neuronal cultures they are interested in the interface between the electronics and neurons. Their aims are long-term survival of the biological material, signal transmission from biological to electronics, and control of the topology. Their work in neuronal signaling found signals from a few hundred microvolts raised to 10 microvolts. They found that neuronal cells can be constrained and partially guided. Cell placement can be improved through use of a Teflon-like coating under the SU8 pattern.

High-Throughput Drug Screening and Multiplexed Immunoassays Using Microarray Chip Technology

Dr. Rao works in high-throughput drug screening; multiplexed immunoassays can screen hundreds of thousands of chemical entities against a biological target in a short timeframe. Through miniaturization, only minute amounts of the component and assay are needed (volume reduction about 2500 times), and hundreds of cells can be analyzed at one time. This group's cell-based assays look at secondary messengers, reporter-gene assay, and cell proliferation. Their micromachine-encoded particle array for multiplexed immunoassay allows the simultaneous transmission of several messages along a single channel.

REFERENCES

Degenaar, P., B. Le Pioufle, L. Griscom, A. Tixier, Y. Akagi, Y. Morita, Y. Murakami, K. Yokoyama, H. Fujita, and E. Tamiya. 2001. A method for micrometer resolution pattening of primary culture neurons for SPM analysis. *J. Biochem.* 130:367–376.

Helianti, I., Y. Morita, A. Yamamura, Y. Murakami, K. Yokoyama, and E. Tamiya. 2001. Characterization of native glutamate dehydrogenase from an aerobic hyperthermophilic archaeon Aeropyrum pernix K1. *Applied Microbiol Biotechnol.* 56:388–394.

Kobayashi, M., T. Mizukami, Y. Morita, Y. Murakami, K. Yokoyama, and E. Tamiya. 2001. Electrochemical gene detection using microelectrode array on a DNA chip. *Electrochemistry* 69 (12).

Morita, Y., Y. Murakami, K. Yokoyama, and E. Tamiya. 2002. Synthesis and analysis of peptide ligand for biosensor application using combinatorial chemistry. Chapter 17 in *Biological Systems Engineering.* Washington, D.C.: American Chemical Society.

Murakami, Y., T. Morita, T. Kanekiyo, and E. Tamiya. 2001. On-chip capillary electrophoresis for alkaline phosphatase testing. *Biosensors and Bioelectronics* 16:1009–1014.

Nagai, H., Y. Murakami, Y. Morita, K. Yokoyama, and E. Tamiya. 2001. Development of a microchannel array for picoliter PCR. *Anal. Chem.* 73:1043–1047.

Tamiya, E., S. Iwabuchi, N. Nagatani, Y. Murakami, T. Sakaguchi, and K. Yokoyama. 1997. Simultaneous topographic and fluorescence imaging of recombinant bacterial cells containing a green fluorescent protein gene detected by a scanning near-field optical/atomic force microscope. *Anal. Chem.* 69:3697–3701.

Yamamura, A., T. Sakaguchi, Y. Murakami, K. Yokoyama, and E. Tamiya. 1999. Purification and characterization of cold-active l-glutamate dehdrogenase independent of NAD(P) and oxygen. *J. Biochem.* 125:760–769.

Yamamura, S., Y. Morita, Q. Hasan, K. Yokoyama, and E. Tamiya. 2002. Keratin degradation: A cooperative action of two enzymes from Stenotrophomonas sp. *Biochemical and Biophysical Research Communications* 294:1138–1143.

Site: **Kyushu University**
Department of Applied Chemistry
6-10-1 Hakozaki-ku, Higashi-ku
Fukuoka 812-8581, Japan
(Review presented at Yaesu Fujiya Hotel, Tokyo)

Date Visited: 28 January 2003

WTEC Attendees: D. Brady (report author), D. Walt, A. J. Ricco, H. Ali

Host: Dr. Shigeori Takenaka, Associate Professor,
Tel/Fax: +81-92-642-3603
Email: staketcm@mbox.nc.kyushu-u.ac.jp

BACKGROUND

Kyushu University is in Fukuoka on Kyushu Island in Southern Japan. Fukuoka is approximately 1200 kilometers from Tokyo. Kyushu University enrolls 16,000 students in 10 faculties. In view of the distance from the university to Tokyo, Professor Takenaka generously agreed to meet the WTEC panel in Tokyo.

Professor Takenaka works in the Applied Chemistry laboratory. The Department of Applied Chemistry is in the School of Engineering.

BIOSENSING ACTIVITY

Professor Takenaka has been developing electrochemical DNA analysis systems for the past decade. He uses ferrocenyl napthalene diimide (FND) as a hybridization indicator, which is a key component of the detection scheme

In electrochemical DNA detection, single-strand bioprobe DNA is immobilized on an electrode. Target complementary molecules hybridize with the probe molecules (Gooding 2002; Wang 2002). In the Kyushu work, FND binds selectively with the resulting double-strand oligonucleotides. FND enables charge transfer; thus the binding process produces an electrochemical signal proportional to the amount of target DNA (Takenaka, Yamashita, et al. 2000; Yamashita, Takenaka, et al. 2000; Sato, Takenaka, et al. 2001; Takenaka 2001; Takenaka, Miyahara, et al. 2001; Yamashita, Takenaka, et al. 2002).

Professor Takenaka's Lab has shown that ferrocene-based ECD detects target sequences at femtomole levels. Development of a blocking layer to inhibit charge transfer without hybridization is key to these results. While details of the blocking layer used by Professor Takenaka were not disclosed, his group demonstrated impressive results for various RNA and gene targets and particularly impressive results for detection of mutation polymorphisms.

TUM-gene, Inc. (http://www.tum-gene.com/), a company started in 1999 with support from MITI to develop ECA systems, is developing a chip and reader system based on Professor Takenaka's work.

SUPPORT

Primary support for Professor Takenaka's program comes from the New Energy and Industrial Technology Department Organization and from the Ministry of Education, Culture, Sports Science, and Technology. This support comes from programs without specific biosensor focus.

ASSESSMENT

The Kyushu University electrochemical analysis work is world-class. It is highly cited and, in achieving femtomole sensitivity, has taken a well known technique to a level with broad utility.

REFERENCES

Gooding, J.J. 2002. Electrochemical DNA hyhridization biosensors. *Electroanalysis* 14 (17):1149–1156.

Sato, S., S. Fujii, et al. 2001. Ferrocenyl naphthalene diimide can bind to DNA-RNA hetero duplex: Potential use in an electrochemical detection of mRNA expression. *Journal of Organometallic Chemistry* 637:476–483.

Takenaka Lab. 2003. http://www.takenaka.cstm.kyushu-u.ac.jp/index-e.html.

Takenaka, S. 2001. Highly sensitive probe for gene analysis by electrochemical approach. *Bulletin of the Chemical Society of Japan* 74:217–224.

Takenaka, S., H. Miyahara, et al. 2001. Base mutation analysis by a ferrocenyl naphthalene diimide derivative. *Nucleosides Nucleotides & Nucleic Acids* 20 (4–7):1429–1432.

Takenaka, S., K. Yamashita, et al. 2000. DNA sensing on a DNA probe-modified electrode using ferrocenylnaphthalene diimide as the electrochemically active ligand. *Analytical Chemistry* 72 (6):1334–1341.

TUM-gene, Inc., website: http://www.tum-gene.com/.

Wang, J. 2002. Electrochemical nucleic acid biosensors. *Analytica Chimica Acta.* 469 (1):63–71.

Yamashita, K., M. Takagi, et al. 2000. Electrochemical detection of base pair mutation. *Chemistry Letters* (9):1038–1039.

Yamashita, K., M. Takagi, et al. 2002. Electrochemical detection of nucleic base mismatches with ferrocenyl naphthalene diimide. *Analytical Biochemistry* 306 (2):188–196.

Site:	**Matsushita Electric Industrial Co., Inc.** **(National/Panasonic)** **(a) Living Environment Development Center** **(b) Advanced Technology Research Laboratories: Humanware** **3-4 Hikaridai, Seika-cho, Soraku-gun** **Kyoto 619-0237, Japan**
Date Visited:	28 January 2003
WTEC Attendees:	S. Bhatia, D. Brady, S. Green, A.J. Ricco, J. Schultz, D. Walt, C. Wilkins, H. Ali
Hosts:	Dr. Hirokazu Sugihara, General Manager, Bioelectronics and Molecular Electronics Group, Tel: +81-774-98-2566; Fax: +81-774-98-2566; Email: sugihara.hirokazu@jp.panasonic.com Dr. Fumiaki Emoto, Manager, Living Environment Development Center Dr. Nobuhiko Ozaki, Nanotechnology Research Laboratory, Advanced Technology Research Laboratories Dr. Ichiro Yamashita, Ph.D., Advanced Research System Theme Leader, Advanced Technology Research Laboratories Dr. Hiroaki Oka, Manager, Bioelectronics and Molecular Electronics Group Dr. Kentaro Onizuka Mr. Hidenobu Yaku Ms. Maki Katagiri Mr. Tetsuo Yukimasa

BACKGROUND

Matsushita is organized in four major locations in Japan. The Tokyo location is the center for the organization's marketing and sales efforts. Product development is located in the Kyoto and Osaka areas. Matsuyama is the location for production.

(A) LIVING ENVIRONMENT DEVELOPMENT CENTER

(C.Wilkins, report author)

In the formal presentation, the overall evolution of the company's highly successful glucose sensor was reviewed. It was explained that the present glucose sensor and glucose monitor products are the result of over thirty years of research and development. A summary of the major milestones in that effort appears below.

Glucose Sensor Development History

1964 Fundamental Research
1977 Government-supported Fuel Cell Research
1980 Glucose Sensor Development
1983 Development of the Disposal Glucose Sensor
1991 Blood Glucose Monitor (marketed by Akray)
1992 Bayer Markets Blood Glucose Monitor (worldwide)
1996 Introduction of Blood Lactate Monitor (worldwide, Akray)

Marketing is handled by the Matsushita Health Products group, located in Tokyo. It was explained that the key to the product's success was the successful solution of problems related to the reproducibility and reliability of the analytical results. As a consequence, the end users can simply purchase small test strips that can be inserted into a monitor device. According to the presentation, the monitor unit itself is given to the user and the profits for the device derive from sale of the test strips at a per unit cost of about 60 cents.

SUMMARY

Currently, the Matsushita blood glucose monitor represents about one third of the worldwide market and is representative of a highly successful commercial biosensor, with hundreds of millions of test strips sold annually. No new research was a part of this presentation. Rather, it was a case study of the way biosensing research can ultimately be converted to a practical and highly profitable biosensor product.

(B) ADVANCED TECHNOLOGY RESEARCH LABORATORIES

(A.J. Ricco, report author)

Introduction

The Advanced Technology Research Laboratories (ATRL) of Matsushita is distributed over four sites: Keihanna, near Nara/Kyoto; Osaka; Shinagawa in the Tokyo area; and Kawasaki. The WTEC group visited the Kyoto/Nara site, which is in the Kansai "Science City" area known as Keihanna, where in addition to Matsushita's ATRL, there are laboratories of Shimadzu, Kyocera, and LTT Communications, among others.

ATRL was established in November of 1993 and opened in April of 1994. The building has about 22,500 m^2 (225,000 ft^2) of floor space on 4 floors and was constructed at a cost of U.S. $117 million. Of the 250 researchers and staff working in the building, about 120 are assigned to ATRL. Including the 3 other sites around Japan, ATRL has a staff of 248.

Matsushita Electric Company's activities are divided among three main divisions: Multimedia and Software; Semiconductors; and Device, Environmental, and Manufacturing. As well as at ATRL, Matsushita R&D activities are carried out in the corporate research labs, divisional labs, and the engineering departments of the various product divisions.

ATRL has four mission areas:

1. creating new business via new technologies (focused on advanced projects and future studies)
2. studying and developing emerging technologies (including analytical and biological technologies)
3. advanced platform technologies, including speech, imaging, and networking security
4. advanced core technologies, including SiGe devices.

The ATRL has three focus areas: Mobile Networking (~50 staff); Humanware, focusing on the human/machine interface (~50 staff); and Nanotechnology Research, which includes Molecular Electronics and Bioelectronics (~100 staff). In support of these 200 technical staff, there are an additional 50 administrative and support staff.

Humanware has five centers: speech and language; vision and image processing; cognitive processes and computing; humanware technology; and the humanoid group (i.e., robots with some human characteristics). The Nanotechnology Research Lab also has five centers: millimeter wave devices; brainwave devices; photonic devices; nanostructure devices; and bioelectronics and molecular electronics, the group of which our host is the manager.

R&D ACTIVITIES

Cellular Ion Flux Measurement Device

A device called the Electrophysiological Biosensor was described by Mr. Nobuhiko Ozaki. The goal of this device is to speed the high-throughput screening (HTS) drug discovery process by addressing measurement of cell toxicity of candidate pharmaceutical compounds or predecessors, a process that presently is relatively costly and low in throughput. The device bears some resemblance to a patch-clamp system, but differs in that no breach of the cell membrane is made. Rather, the cell is trapped in a 5 μm-diameter through-hole in a microfabricated chip, permitting a different bio/chemical environment to be introduced on either side of the cell, leading in some cases to different ion fluxes through the ion channels on opposite sides of the cell. The result of this difference in fluxes is voltage transients measured between electrodes in contact with the physically separated electrolyte solutions on either side of the cell and chip. In the HTS/toxicity application, one side of the cell is exposed to buffer and the other side to the same solution with the addition of the pharmaceutical candidate, thereby probing the action of this compound on the cell's ion channels. The measurement made by this approach was referred to as "total ion recording."

Each chip has a total of 16 electrodes, and each electrode has a total of 100 through-holes in contact with it. Each hole holds one cell. (Rather large cells must be used in this case for the 5 μm holes, but scaling down to smaller holes should be possible). Mollusk neuronal cells, in particular *Lymnaea* ganglions, were utilized due to their size and ease of manipulation. The response represents a composite of roughly 100 cells. Therefore, the activity of an ion channel cannot be attributed to any particular cell, and with this approach there is no way to know if only some of the 100 cells are causing the bulk of the measured activity, or if most of them are responding similarly. Sealing of each cell to the edges of the hole is critical in order to observe a signal due to differential ion flux, and achieving a good seal is one of the most challenging aspects of this measurement approach.

In preliminary experiments, cells were stimulated by application of a 10 μM solution of glutamate, with very noticeable difference in the distribution and magnitude of voltage pulses associated with the activity of ion channels.

Nanoprocessing

Dr. Ichiro Yamashita described Matsushita's activities in the area of "nanoprocessing." He focused particularly on the concept of biomineralization, the central precept of which is to utilize biological systems—proteins—to build nanostructured inorganic materials. Some of the company's activities in this area are being conducted in collaboration with NAIST (in Nara, Japan), RIKEN (numerous locations around Japan), and Montana State University in the United States.

Dr. Yamashita explained the concept of a floating-gate memory element that relies upon a single-electron transistor, wherein a change of one electron on the transistor's gate changes its transconductance by a significant amount. The hope is that such devices could enhance significantly the density of solid-state memory elements by making each device both smaller and much lower in its power dissipation than present field effect transistors that are the heart of memory chips. These devices have been much researched since the early 1990s, and a recent review of the basic concepts of this device was made by Kastner of MIT (Kastner

2000). The near-term target is 100-electron devices; the longer term target is 1-electron devices.

The idea is for proteins to be utilized to fabricate the inorganic "nanodots" that are the key to the single-electron transistors, with a target density of 10^{12} dots/cm^2. The protein under examination for this role is ferritin, a spherical protein supramolecule with a diameter of 12 nm, composed of 24 hetero subunits. Ferritin has a 6 nm cavity that accommodates about 4,000 iron atoms in its native form (Iwahori et al. 2001). Yamashita and colleagues employed a recombinant L-ferritin, composed of only L-type subunits from horse liver and its mutants (Takeda et al. 1993). They were able to transfer a two-dimensional crystal of iron oxide-loaded L-ferritin molecules that self-assembled at an air/water interface onto a hydrophobic Si surface (Furuno, Sasabe and Ulmer 1989). A well-ordered array of L-ferritin on the Si surface was observed by high-resolution scanning electron microscopy. L-ferritin shells were eliminated by heat-treatment under nitrogen or UV-ozone treatment, leaving an array of nanometer-size iron oxide dots with little deformation. Fourier-transform IR spectroscopy and X-ray photoelectron spectroscopy were consistent with the notion that both treatments remove the protein shell completely. Some amount of negative charge was trapped on the surface of each particle by this process, helping to prevent aggregation of the resultant nanoparticles. The well-ordered array of nanodots produced in this way was deemed suitable for use as a key component of quantum electronics.

It was also reported that the use of apoferritin results in a good-quality colloid of iron oxide, with a narrow and very monodisperse particle size distribution centered at 7 nm. These particles were formed into ordered monolayers on a liquid surface using a Langmuir trough system.

REFERENCES

Furuno T., H. Sasabe, and K.M. Ulmer. 1989. *Thin Solid Films* 180:23–30.

Kastner, M.A. 2000. *Ann. Phys. (Leipzig)* 9:885–894.

Iwahori K., R. Tsukamoto, M. Okuda, H. Yoshimura, and I. Yamashita. 2001. Nanometer-size structures fabricated by bio-nano-process. Paper presented at the Ninth Foresight Conference on Molecular Nanotechnology, Nov. 9–12, Santa Clara, CA. Abstract available online at http://www.foresight.org/Conferences/MNT9/Abstracts/Kenji/. Palo Alto, CA: Foresight Institute.

Takeda S., M. Ohta, S. Ebina, and K. Nagayama. 1993. *Biochim. Biophys. Acta.* 1174:218–220.

Site: **The National Institute of Advanced Industrial Science and Technology (AIST)**
Special Division of Human Life Technology (HLT)
Osaka National Research Institute
AIST Kansai (Osaka) Center
1-8-31 Midorigaoka, Ikeda
Osaka 5638577, Japan

Date Visited: 30 January 2003

WTEC Attendees: S. Green (report author), S. Bhatia, D. Brady, A. Ricco, J. Shultz, D. Walt, C. Wilkins, H. Ali

Hosts: Dr. Yoshiro Ohmiya, Cell Dynamics Research Group
Tel: +81-72-751-7997; Fax: +81-72-751-9628;
Email: y-ohmiya@aist.go.jp
Dr. Mitsuo Ataka, Protein Fine Structure Research Group
Tel: +81-72-751-9526; Fax: +81-72-751-9628

OVERVIEW

The Special Division for Human Life Technology (HLT) is a cross-disciplinary research center at AIST Kansai that relies on collaboration between academia, industry, and the Japanese government to further research in biomedical and materials sciences. Within HLT there are seven distinct research groups focusing on biomolecular dynamics, protein fine structure, cell dynamics, neuronics, mesophase technology, dynamic materials, and green biotechnology.

R & D ACTIVITIES

Biomolecular Dynamics Research Group

The WTEC group briefly spoke to Yoshiro Tatsu. This group is developing protein cages that inactivate specific functional groups on the main chain or side chain of proteins. Recent work has shown progress in this area and success in caging small proteins.

Protein Fine Structure Research Group

This group, led by Dr. Mitsuo Ataka, has an interest in characterization of highly thermostable archaebacterial enzymes. Recent work by national laboratories has provided complete genomic sequences for a number of species, including the anaerobe *P. horikoshi* and the aerobe *A. pernix*. Using these species as models, the Ataka group has studied a number of enzymes, including Glycerol 1-phosphate dehydrogenase, thioredoxine, peptidases, and Cysteine Synthetase. Glycerol 1-phosphate dehydrogenases construct lipids with ether bonds, which are significantly more stable than typical lipids and lend stability to protein membranes at high temperatures. Genomic analyses indicate the presence of three thioredoxines in *A. pernix* and none in *P. horikoshi*. Pepsidases in archaebacteria are significantly more stable at high temperatures than those in higher species. Cysteine synthetases are present in many of these species, and their enzymatic activities may prove useful for industrial applications.

Cell Dynamics Research Group

This group is led by Yoshiro Ohmiya, and its work is centered around analyses of bioluminescent and fluorescent proteins. The group's goal is to successfully identify and characterize reporter genes for further analysis of cellular processes. The laboratory has isolated approximately five distinct luciferin molecules in addition to working with GFP and CFP. The group has isolated a luciferase from *Vargula hilgendorfi* that is secreted rather than remaining in cytosol, as in firefly species (Tanahashi et al. 2001). The laboratory is currently refining a perfusion system that allows continuous monitoring of cells with *Vargula* luciferin under the control of a Growth Hormone (GH) promoter (Tanahashi et al. 2001). An additional luciferase molecule is pH sensitive, though the group has yet to include this in a functional assay. In the future, the Ohmiya group seeks to incorporate multiple bioluminescent agents into each functional assay, including a newly characterized luciferin from the railroad worm *Phrixothrix*, which emits red luminescence (Viviani et al. 1999).

SUMMARY

This group of laboratories has an academic focus, and, while their research is quite interesting, it is not directly related to the focus of this study. The newly characterized luciferins studied in the Cell Dynamics Research Group may find some application in sensing systems, as they might not exhibit the same level of toxicity as GFP.

REFERENCES

AIST. N.d. National Institute of Advanced Industrial Science and Technology. Brochure.

———. 2002. An independent administrative institution under Ministry of Economy, Trade, and Industry. Handout.

Hirano, T., I. Mizoguchi, M. Yamaguchi, F.Q. Chen, M. Ohashi, Y. Ohmiya, and F.I. Tsuji. 1994. Revision of the structure of the light emitter in aequorin bioluminescence. *J. Chem. Soc. Chem. Commun.* 165–167.

Hirano, T., R. Negishi, M. Yamaguchi, F.Q. Chen, Y. Ohmiya, F.I. Tsuji, and M. Ohashi. 1995. Chemi- and bio-luminescent properties of coelenterazine analogues possessing an adamantyl group. *J. Chem. Soc. Chem. Commun.* 1335–1336.

Tanahashi, Y., Y. Ohmiya, S. Honma, Y. Katsuno, H. Ohta, H. Nakamura, K. Honma. 2001. Continuous measurement of targeted promoter activated by a secreted bioluminescence reporter, *Vargula hilgendorfii* luciferase. *Anal. Biochem.* 289:260–266.

Viviani, V.R., E.J.H. Bechara, and Y. Ohmiya. 1999. Cloning, sequence analysis, and expression of active *Phrixothrix* railroad-worms luciferases: Relationship between bioluminescence spectra and primary structure. *Biochemistry* 38 (26):8271–8279.

Site: **National Institute of Advanced Industrial Science and Technology (AIST)**
Research Center of Advanced Bionics
AIST Tsukuba Central, 4
1-1-1 Higashi, Tsukuba
Ibaraki 305-8562, Japan

Date Visited: 27 January 2003

WTEC Attendees: D. Brady (report author), C. Wilkins, D. Walt, H. Ali

Hosts: Dr. Norihiko Minoura, Deputy Director
Phone: +81-29-861-2987;
Fax: +81-29-8553833;
Email: n.minoura@aist.go.jp
Dr. Kenji Yokoyama, Deputy Director,
Email: ke-yokoyama@aist.go.jp

BACKGROUND

The National Institute of Advanced Industrial Science and Technology was established 1 April 2001. AIST is a unit of Japan's Ministry of Economy, Trade, and Industry (METI, formerly known as the Ministry of International Trade and Industry or MITI). AIST comprises 15 research institutes under the former Agency of Industrial Science and Technology in MITI and the Weights and Measures Training Institute. AIST is Japan's largest public research organization and has a staff of 3,200, located at 9 research bases throughout Japan. AIST promotes "industry-academia-government collaborative research." In addition to life science and biotechnology, AIST includes programs in information technology, nanotechnology, environmental science and energy technology, and earth and marine science. AIST is charged with assisting in the development of new industries, facilitating technology transfer, and application of basic science advances to societal problems. AIST's mission is analogous to that of the U.S. National Institutes of Standards and Technology (NIST) with respect to standards development. In other areas, AIST is analogous to the U.S. Geological Survey; in addition, it includes industrial development and research functions with no direct U.S. analog.

WTEC visited the Research Center of Advanced Bionics and the Division of Biological Resources and Functions (see following site report), both in Tsukuba, the location of one of the largest of the AIST centers. The Research Center of Advanced Bionics is a "research initiative" in the AIST structure. According to the AIST website www.aist.go.jp/aist_e/basic_policy/main.html, "the purpose of a Research Initiative is to promote specific research projects with flexibility, for a specified period of time, especially those with a high possibility of cross-field application or with relevance to immediate administrative needs." As indicated in Figure C.1 below, the Research Center seeks to apply fundamental technologies in molecular recognition and transduction to specific applications in toxin and protein analysis.

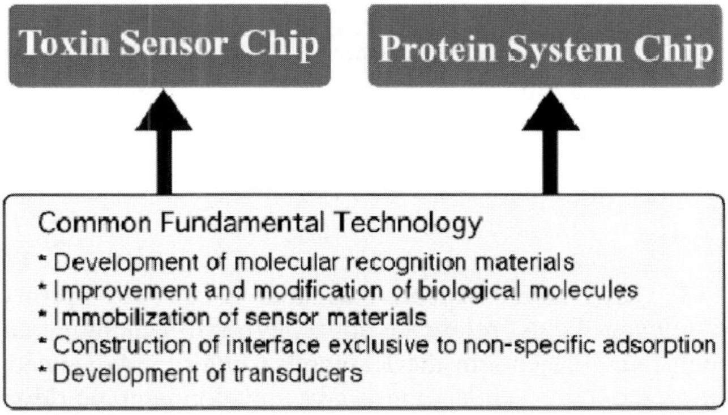

Fig. C.1. AIST Research Center of Advanced Bionics: Applying fundamental technologies in molecular recognition and transduction to specific applications in toxin and protein analysis.

BIOSENSING RESEARCH ACTIVITIES

The Research Center of Advanced Bionics is developing systems for rapid detection and functional analysis of biological targets, such as toxins, DNA, and proteins. The center develops fundamental enabling technologies, such as molecular recognition materials, transducers and sensor materials. Specific current projects include an ultrasensitive toxin sensor chip and a protein system chip.

The toxin sensor relies on quartz crystal microbalance technology (Uzawa et al. 2002). Dr. Minoura presented an overview of this project. The sensing process is based on target-specific molecular probes immobilized on a gold surface. When the target species binds to the probe, material accumulates and changes the oscillation frequency of the microbalance. Dr. Minoura described the use of sugar molecules expressed on kidney cell surfaces as probes for verotoxin secreted from pathogenic bacteria O-157. The pentameric B-subunits of verotoxin recognize sugars on the cell membrane, so a natural sugar mimic was designed and synthesized. The sugar is then attached to the sensor substrate using Langmuir-Blodgett deposition. A similar approach might be applied to *C. botulinum* toxin detection.

Rapid turnaround point-of-care biosensing is a primary focus of the center's toxin sensor project. Current techniques require days to transmit samples and return results. Point-of-care QCM systems could return results in minutes.

A group is exploring the protein system chip with collaborators at the Tokyo University of Technology and Keio University. The goal is to recognize 300–1000 proteins using a capillary isoelectric focusing chip (CIEF). The chip is a plasma polymerized film in a glass capillary system. CIEF (Rodriguez-Diaz, Wehr, and Zhu 1997; Shen et al. 2000; Shimura 2002) is one of several technologies under development for multidimensional protein analysis (Liu, Lin, and Yates 2002). The AIST effort in this biosensor was less than one year old at the time of the WTEC visit.

ASSESSMENT

The AIST QCM toxin sensor was one of the few projects with substantial relevance and reference to bioterrorism encountered on the panel's Japan tour. Projects in the Research Center of Advanced Bionics focused on technologies under aggressive development worldwide. The center seems to be particularly successful in developing molecular recognition for specific toxins in QCM systems.

REFERENCES

AIST. 2003. National Institute for Advanced Industrial Science and Technology website, http://www.aist.go.jp/.

Liu, H.B., D.Y. Lin, and J.R. Yates III. 2002. Multidimensional separations for protein/peptide analysis in the post-genomic era. *Biotechniques* 32 (4): 898–911.

Rodriguez-Diaz, R., T. Wehr, and M. Zhu. 1997. Capillary isoelectric focusing. *Electrophoresis* 18 (12–13):2134–2144.

Shen, Y.F., S.J. Berger, G.A. Anderson, and R.D. Smith. 2000. High-efficiency capillary isoelectric focusing of peptides. *Analytical Chemistry* 72 (9):2154–2159.

Shimura, K. 2002. Recent advances in capillary isoelectric focusing: 1997–2001. *Electrophoresis* 23 (22–23):3847–3857.

Uzawa, H., S. Kamiya, N. Minouta, H. Dohi, Y. Nishida, K. Taguchi, S. Yokoyama, H. Mori, T. Shimizu, and K. Kobayashi. 2002. A quartz crystal microbalance method for rapid detection and differentiation of Shiga toxins by applying a monoalkyl globobioside as the toxin ligand. *Biomacromolecules* 3 (3):411–414.

Site: **National Institute of Advanced Industrial Science and Technology (AIST)**
Biosensing Technology Research Group
Division of Biological Resources and Functions
AIST Tsukuba Central
1-1-1 Higashi, Tsukuba
Ibaraki 305-8566, Japan

Date Visited: 27 January 2003

WTEC Attendees: C. Wilkins (report author), D. Brady, D. Walt, H. Ali

Hosts: Dr. Fumio Mizutani, Leader, Biosensing Technology Research Group
Tel: +81-298-61-6166,67; Fax: +81-298-61-6177;
Email: mizutani.fumio@aist.go.jp
Dr. Soichi Yabuki
Dr. Yukari Sato

BACKGROUND

Please refer to previous AIST site report (Research Center for Advanced Bioelectronics) for background on AIST and its Tsukuba center. The biosensing technology research group of AIST's Division of Biological Resources and Functions in Tsukuba emphasizes fundamental research in the biosensor field. Examples of this group's work are represented in the references list (below) for this site report.

MATERIAL SCIENCE IN BIOSENSING DEVICES

The guiding interests of AIST's biosensing technology research group are focused on the fabrication of biosensing devices and systems with high performance characteristics and the preparation and characterization of electrodes modified with ultrathin layers of biofunctional molecules. Research presentations emphasized progress toward these goals.

Some of the current research involves development and preparation of two-dimensional cross-linked polysiloxane Languir-Blodgett films for the purpose of blocking electroactive interferences such as ascorbic acid, cystein, etc. These monolayers were effective for eliminating interferences with a glucose sensor, based on the use of glucose oxidase chemistry, which was used to validate the technology (Hirata et al. 2001; Kato et al. 2002). In a recent report, these workers established that a trienzyme/polu (dimethylsiloxane) bilayers-based sensor could be used effectively for the determination of acetic acid in ethanol-containing food samples (Mizutani et al. 2001). This appears to be the first example of such a biosensor. The electrode was sufficiently stable that it could be used for more than one month.

PREPARATION AND CHARACTERIZATION OF ELECTRODES WITH ULTRATHIN BIOFUNCTIONAL MOLECULES

Here, the group's research involves use of fabrication of electrodes based upon thiol self-assembled monolayers on gold. Specifically, ferrocene functionalized alkanethiol monolayers were used to permit luminol-based chemiluminescence sensors to be developed. This approach was also used to develop a sensor utilizing an ordered monolayer of 4-mercaptopyrindine on gold, which was used as an electron transfer mediator for cytochrome c, which was used as an O^{2-} sensor. In another example of the use of thiol-based chemistry, chemisorption/reductive desorption of thiolcholine on the electrode service was demonstrated to permit the sensitive determination of the activity of acetylcholinesterase (Matsuura et al. 2002). Thus, the group is actively involved in development of practical and sensitive biosensors with a wide variety of useful applications in health-related applications.

REFERENCES

Hirata, Y., M.-L. Laukkanen, K. Keinänen, H. Shigematsu, M. Aizawa, and F. Mizutani. 2001. Microscopic characterization of Langmuir-Blodgett films incorporating biosynthetically lipid-tagged antibody. *Sensors and Actuators B.* 76:181–186.

Kato, D., M. Masaike, M. Majima, Y. Hirata, F. Mizutani, M. Sakata, C. Hirayama, and M. Kunitake. 2002. Permselective monolayer membrane based on two-dimensional cross-linked polysiloxane LB films for hydrogen peroxide detecting glucose sensors. *Chem. Commun.* 2616–2617.

Matsuura, H., Y. Sato, T. Sawaguchi, and F. Mizutani. 2002. Highly sensitive determination of acetylcholinesterase activity based on the chemisorption/reductive desorption/process of thiol compound on a silver electrode. *Chem. Lett.* 618–619.

Mizutani, F., T. Sawaguchi, Y. Sato, S. Yabuki, and S. Iijima. 2001. Amperometric determination of acetic acid with a trienzyme/poly(dimethylsiloxane)-bilayer-based sensor. *Anal. Chem.* 73:5738–5742.

Site: **National Rehabilitation Center for Persons with Disabilities**
Department of Rehabilitation Engineering, Research Institute
Bioengineering Division
4-1 Namiki, Tokorozawa City
Saitama Pref. 359-8555, Japan

Date Visited: 28 January 2003

WTEC Attendees: S. Green (report author), J. Brewer, S. Bhatia, and J. Schultz.

Host: Dr. Shigeru Toyama, Tel: +81-42-995-3100

INTRODUCTION

The National Rehabilitation Center for Persons with Disabilities is the largest rehabilitation facility in Japan, with a maximum capacity of 200 inpatients, and it is the most comprehensive rehabilitation center in Japan, treating any disability that impairs quality of life. The most common disabilities treated include physical, auditory, and visual disabilities, but there is also treatment for other ailments, including internal disabilities. Other national rehabilitation centers in Japan are generally less comprehensive and focus on only a few special research areas. The National Rehabilitation Center is funded through the Ministry of Health, Labor, and Welfare; its funding is considered more stable than that of other government-funded facilities, which are under pressure to privatize due to difficulties in the Japanese economy.

The complete facility has four main divisions, including a hospital, training center, college, and the research institute. The hospital is specialized to provide care for disabled persons, while the training center provides physical and vocational training to patients. The college trains rehabilitation professionals in a wide variety of areas, including speech pathology, orthotics, and living skills education. Scientists in the research institute work to advance scientific understanding of medical problems and seek to develop concrete solutions for medical problems.

The research institute has approximately twenty permanent researchers and twenty visiting staff divided among six departments. Research in the institute ranges from assessment of physical and mental problems to engineering prosthetics and assistive devices. Dr. Toyama's laboratory is within the Department of Rehabilitation Engineering and, along with a robotics laboratory, comprises the Division of Bioengineering.

RESEARCH AND DEVELOPMENT ACTIVITIES

Dr. Toyama specializes in electrochemistry and has experience in materials science. Dr. Toyama reported work on four major projects: (1) a miniaturized Au-Pt electrode with enhanced surface area, (2) immunodetection using surface plasmon resonance (SPR), (3) blood osmotic pressure sensors, and (4) a voice module for a commercially available blood glucose sensor.

One of Dr. Toyama's goals is to construct glucose sensors with enhanced surface area and increased signal strength. In order to construct a sensor in this fashion, Dr. Toyama uses electrochemical deposition to produce a fine mixture of Au and Pt on a 100 μm Pt wire. Dr. Toyama theorizes that Au within the sensor immobilizes proteins at the material surface, while Pt acts as a transducer for hydrogen peroxide. Chronocoulometric testing indicates that the electrode surface area is approximately 36 times larger than a comparable electrode, and SEM analyses of the electrode surface confirm its porous structure and expanded surface area (Toyama et al. 2001). Testing suggests an Au-Pt electrode is capable of receiving as much as ten times more current than a comparably sized platinum electrode. This system is designed for use in solution, and practical applications for this electrode might be in blood-drop sensors. Au-Pt electrodes have not been used in vivo within the Toyama laboratory.

Recent work in the laboratory produced a prototype voice module for a commercially available blood glucose sensor developed by the Matsushita Corporation. This voice module is designed to aid patients with diabetic retinopathy to accurately gauge their blood glucose levels, and Dr. Toyama reports that since its development there have been 6,000 units produced. This work marks the laboratory's first exposure to product commercialization.

The Toyama laboratory has significant interest in developing an optical sensor based on Surface Plasmon Resonance (SPR). The laboratory constructed an SPR sensor with multilayer film refractive surface and quartz prism. Dr. Toyama would also like to develop a disposable SPR sensor using a glass plate, which would be used in a clinical setting.

Attempts to incorporate an Au-Pt electrode into an SPR sensor failed, as the high surface area of the electrode interfered with propagation of the plasmon wave. Experimentation with this system yielded irregular absorption spectra, with peaks too broad for analysis. Dr. Toyama was able to model this behavior precisely and at the time of the WTEC visit was working on optimization of other aspects of the SPR sensor.

A side project involved the use of a Tris (1,10 – phenanthroline) ruthenium (II) complex in a screen for DNA binding compounds. Ruthenium is naturally luminescent when unbound, but its luminescence is quenched when the compound intercalates with DNA. Dr. Toyama, in conjunction with another laboratory, constructed a system in which this property is used to identify DNA binding proteins, since the presence of competing binding proteins increases average luminescence.

A final project worked on in this laboratory is a dehydration sensor. The goal of this project is to develop a functional sensor that could be used in disabled individuals who are unable to determine or communicate their level of thirst. The operating principal of this sensor is the measurement of ion levels as an estimate of total blood ion pressure. Dr. Toyama plans to have a sensor with four individual membranes that measure blood levels of sodium, potassium, glucose, and urea. A completed sensor might find uses in other practical applications, including athletic training and care for the elderly.

FACILITIES

In addition to having three separate SPR sensing systems, two obtained through commercial distributors and one built in the laboratory, the Toyama laboratory has access to AFM, SEM with digital processing, several spectrometers, and FTIR. In addition, the laboratory is in close proximity to a high quality genomics laboratory and tissue culture facility managed by Dr. Seishi Kato, formerly of the Sagami Chemical Research Center.

SUMMARY AND ASSESSMENT

In addition to his knowledge of electrochemistry and materials science, Dr. Toyama completes many technical operations himself, as he had no technical support staff until April 2003. Luckily, he can operate AFM, SEM, and is able to build software programs and circuits. This laboratory is working on a variety of sensor designs, but it appears that this laboratory may benefit from greater communication with the clinical staff, which is a great strength in this institution. The dehydration sensor they are working on would be a useful device and could have applications in a number of areas, including athletics, military training, rehabilitation training, childcare, and clinical operations.

REFERENCES

Ikariyama, Y., and S. Yamauchi. 1988. Surface control of platinized platinum as a transducer matrix for micro-enzyme electrodes. *Journal of Electroanalytic Chemistry* 251:267–274.

Ikariyama, Y., and S. Yamauchi. 1989. Electrochemmical fabrication of amperometric microenzyme sensor. *Journal of the Electrochemical Society*. 136:702–706.

Kato, S., S. Sekine, S. Oh, N. Kim, Y. Umezawa, N. Abe, M. Yokoyama-Kobayashi, and T. Aoki. 1994. Construction of a human full-length cDNA bank. *Gene* 150:243–250.

Kuwabara, T., T. Noda, H. Ohtake, T. Ohtake, S. Toyama, and Y. Ikariyama. 2003. Classification of DNA-binding mode of antitumor and antiviral agents by the electrochemiluminescence of ruthenium complex. *Analytical Biochemistry*. 314:30–37.

2000. National Rehabilitation Center for the Disabled. 2000. (Brochure).

Takei, O., S. Toyama, M. Someya, T. Kurokawa, R. Usami, K. Horikosi, and Y. Ikariyama. 2001. Glucose Ssensor bBased on Au-Pt bBlack Eelectrode: Preparation of functionally different sites on electrode surface. *Electrochemistry* 69 (12): 956–958.

Toyama, S., A. Shoji, Y. Toshida, S. Yamauchi, and Y. Ikariyama. 1998. Surface design of SPR-based immunosensor for the effective binding of antigen or antibody in the evanescent field using mixed (?) polymer matrix. *Sensors and Actuators B* 52:65–71.

Toyama, S., M. Someya, O. Takei, T. Ohtake, R. Usami, K. Horokishi, and Y. Ikariyama. 2001. Fabrication and characterization of gold-platinum black electrode. *Chemistry Letters* (2):160–161.

Toyama, S., N. Doumae, A. Shoji, and Y. Ikariyama. 2000. Design and fabrication of a waveguide-coupled prism device for surface plasmon resonance sensor. *Sensors and Actuators B* 65:32–34.

Toyama, S., O. Takei, M. Tsuge, R. Usami, K. Horikoshi, and S. Kato. 2002. Surface plasmon resonance of electrochemically deposited Au-black. *Electrochemistry Communications* 4:540–544.

Yamauchi, S., Y. Ikariyama, and M. Yaoita. 1989. Enzyme Eembodied Eelectrode: A new amperometric biosensing device. *Chemical Sensor Technology* 2:205–223.

Site:	**RIKEN (Wako Main Campus)** **Discovery Research Institute, Bioengineering Laboratory** **2-1, Hirosawa, Wako,** **Saitama 351-0198, Japan**
Date Visited:	29 January 2003
WTEC Attendees:	J. Schultz (report author), D. Brady, D. Walt, H. Ali
Hosts:	Dr. M. Maeda, Tel: +81-48-467-9311; Fax: +81-48-462-4658; Email: mizuo@riken.go.jp

BACKGROUND:

RIKEN (The Institute of Physical and Chemical Research) is a public corporation supported by the government. The objectives of RIKEN are to conduct comprehensive research in science and technology (excluding only humanities and social sciences) and to disseminate the results of its scientific research and technological developments. RIKEN was founded in 1917 as a private research foundation (Taisho 6). In 1958 it was reorganized as a public corporation under the jurisdiction of the Science and Technology Agency (now the Ministry of Education, Culture, Sports, Science and Technology), and has since engaged in wide-ranging research activities that span basic to applied science. Recently, RIKEN has added five more research projects, mainly in the field of bio-science.

The Wako Main Campus includes 41 Laboratories, as well as the Frontier Research System (FRS) and the Brain Science Institute. The WTEC team visited both the Bioengineering Laboratory at the Discovery Research Institute and the Local Spatio-Temporal Functions Laboratory of the Frontier Research Program (see following site report).

TOPIC I. SUPRAMOLECULAR DESIGN OF NANOPARTICLES FOR BIOLOGICAL APPLICATIONS.

Dr. Maeda and colleagues have developed a colloidal-aggregation technique for detecting DNA sequences. They have applied the method to detect point-mutation assay in gene diagnosis.

The colloidal particle is prepared from a graft copolymer of poly (N-iopropylacrylamide). This copolymer has a phase change in aqueous solution and becomes insoluble at higher temperatures. A sketch of the colloidal system is shown in Figure C.2 below.

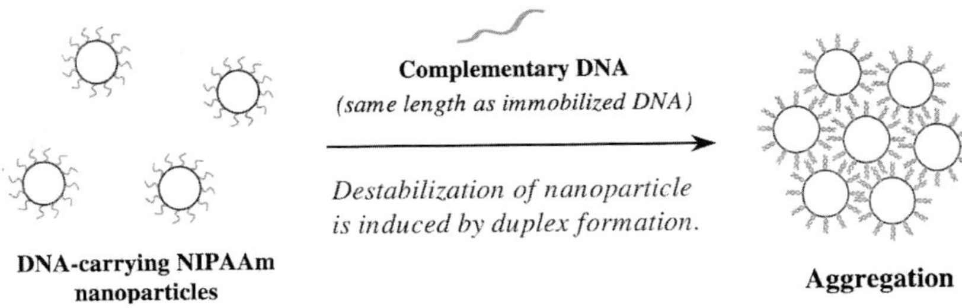

Fig. C.2. A sketch of a colloidal-aggregation technique for detecting DNA sequences. (Courtesy RIKEN)

The particle size is on the order of 50 nm. The colloidal particles of various compositions had between 30 and 300 polymer macromolecules per particle. There is a high concentration of DNA in the particle, with an apparent surface concentration of one DNA per 24 nm^2.

This group showed a novel aggregation behavior; in particular, at certain critical salt concentrations, the stability of the colloidal particle is destabilized only by the hybridization of the surface DNA with complementary DNA in solution.

The detection strategy is based on the stability decrease of colloidal particles accompanied by duplex formation of the shell DNA with complementary DNA in solution. This system is very sensitive to exact matches in DNA sequence between the immobilized DNA in the colloid and DNA fragments in the suspending fluid. The phenomenon appears to depend on a delicate balance between electrostatic forces between particles that promote dispersion and phase transitions of the polymer that favor aggregation. By appropriately adjusting the ionic strength of the solution, the molecular weight, amount of DNA on the backbone, and the temperature of the solution, single-base discrepancies can be easily determined. A rather simple turbidimetric assay has been devised that can provide results in a matter of minutes as compared to hours for other assay methods.

The example in Figure C.3 shows that aggregation occurs only when the DNA-carrying colloid is exposed to a solution that contains DNA strands that provide complete complimentarity (System I). Other mixtures of DNA, even with an one-base mismatch, do not exhibit aggregations (Systems II, III, IV, V).

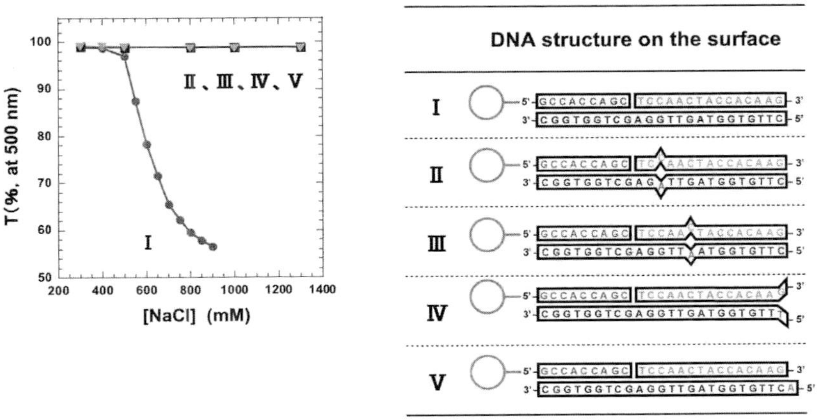

Fig. C.3. Results of a detection strategy based on the stability decrease of colloidal particles accompanied by duplex formation of the shell DNA with complementary DNA in solution. (Courtesy RIKEN)

The assay system can be used to detect single nucleotide mutations. To quote Dr. Maeda,

> A single nucleotide mutation on certain genes can cause inheritable disorders and cancers. Consequently, the development of a simple and practical detection method for single nucleotide difference has been one of the most important subjects in analytical biochemistry.
>
> In this study, we applied a turbidity change of colloidal particle dispersion by salting out to single nucleotide difference assay. The colloidal nanoparticle comprises a hydrophobic core of poly(N-isopropylacrylamide) (polyNIPAAm) and a hydrophilic shell of oligonucleotide. The particle was constructed by self-assembly of amphiphilic random copolymers of DNA derivative and NIPAAM (DNA-polyNIPAAm conjugate). When the conjugate solution was incubated above the phase transition temperature of polyNIPAAM, the colloidal nanoparticles was spontaneously formed and kept dispersed. The nanoparticles aggregated rapidly when the complementary DNA was added into the dispersion. In contrast, they kept dispersed in the presence of point-mutated DNA. These distinct phenomena should be applied for the point-mutation assay in gene diagnosis.

TOPIC II. NEW METHOD FOR SEPARATION/DETECTION OF DNA MUTATIONS BY CAPILLARY ELECTROPHORESIS

Another application of oligo-DNA-polyacrylamide copolymers is the separation of DNA oligomers in aqueous solutions (Figure C.4). The method has an advantage over the standard slab gel electrophoresis technology in that the analysis time is on the order of minutes. The principle of the method is to first load the copolymer into a capillary (pseudoimmobilized affinity ligand) and then place an aliquot of the sample at one end of the capillary. An electric field is then applied. Because of the weak interactions between DNA strands in the sample and the gel immobilized DNA strands, there is distribution in the mobility of the various DNA species in the sample. This system contrasts to a typical affinity chromatography approach that is based on a high affinity of the analyte with the immobilized bioreceptor. Affinity chromatography often requires the change of solvents to remove the bound analyte from the column. In the new method, a separate washing step is not required.

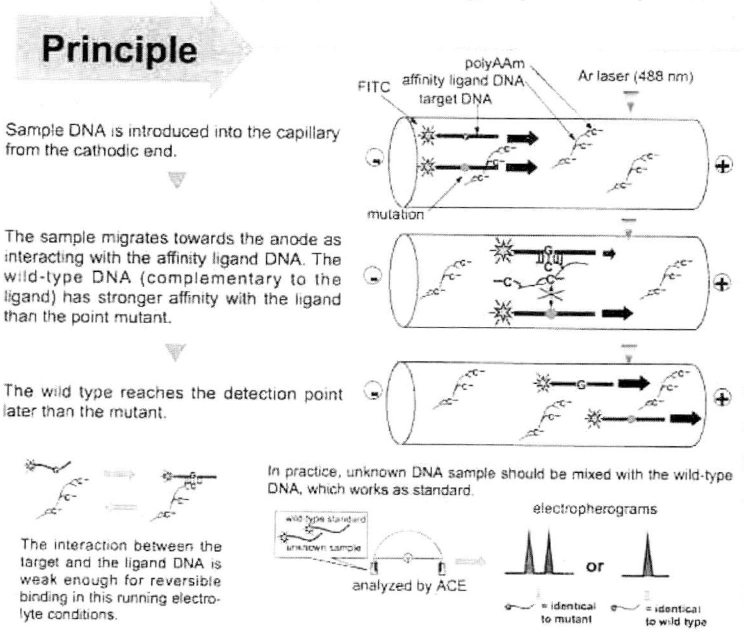

Fig. C.4. Examples of the capillary electrophoresis technique showing the detection of a K-ras point mutation in codon 12 in less than ten minutes. (Courtesy RIKEN)

REFERENCES

Anada, T., T. Arisawa, T. Ozaki, T. Takarada, Y. Katayama, and M. Maeda. 2002. The separation of oligodeoxynucleotides having a single-base difference by affinity capillary electrophoresis using oligodeoxynucleotide-polyacrylamide conjugate. *Electrophoresis* 23:2267–2273.

Katayama, Y., T. Arisawa, T. Ozaki, and M. Maeda. 2000. An affinity capillary electrophoresis for the separation of sequence isomers of oligonucleotide. *Chem. Lett.* 2:106–107

Mori, T., and M. Maeda. 2002. Stability change of DNA-carrying colloidal particle induced by hybridization with target DNA. *Polymer Journal* 34:624–628.

Ozaki, Y., Y. Katayama, T. Ihara, and M. Maeda. 1999. An affinity capillary electrophoresis for the detection of gene mutation using immobilized oligonucleotides-polyacrylamide conjugate. *Anal. Sci.* 15:389–392.

Site:	**RIKEN (Wako Main Campus)** **Frontier Research Program** **Local Spatio-Temporal Functions Laboratory** **2-1, Hirosawa, Wako** **Saitama 351-0198, Japan**
Date Visited:	29 January 2003
WTEC Attendees:	D. Brady (report author), J. Schultz, D. Walt, H. Ali
Host:	Dr. Masahiko Hara, Laboratory Head, Tel: +81-48-467-9600, Email: masahara@postman.riken.go.jp

BACKGROUND

The following information is adapted from the RIKEN brochure.

The Frontier Research Program (FRP) was started in 1986. At that time the research system consisting of fixed-term contract researchers was very rare in Japan even for RIKEN. Typically, researchers were guaranteed lifetime employment.

This experimental organization was successfully operated consecutively by the first three directors-general Prof. Kubo, Prof. Masao Ito, and Prof. Yoshitaka Nagai. The organization has introduced dynamism into Japanese research system and achieved remarkable research results.

RIKEN Brain Science Institute developed out of FRP in 1997. Also, numerous short-term research organizations were recently established both in and out of RIKEN, following FRP's example. From this point, one can easily understand why FRP is so highly regarded.

The name changed to the Frontier Research System (FRS) in October 1999 and it continues to occupy a unique place among the many fixed-term projects within RIKEN. Because of its diverse nature, FRS has a generic name, which does not reflect a single research area.

FRS is distinctive because it possesses the freedom and flexibility to bring together high caliber scientists from different disciplines to work together on cutting edge research projects. High expectations are placed on the FRS to continue to develop and incubate novel interdisciplinary research areas.

FRS's aims are to create new fields in science/technology, contribute towards benefiting society, and impact industry and the economy.

RIKEN's Frontier Research System, unlike other projects, has a flexible organization structure, which attracts high-caliber domestic and foreign scientists from a wide range of fields. FRS promotes "frontier research" in cutting-edge, basic scientific fields previously not accessed.

The FRS Spatio-Temporal Functions materials research program, which began in October 1999, combines nanoscience, supramolecular chemistry and nonlinear physics in exploring the emergence of space-time patterns, self assembly and functional applications

The spatio-temporal functions materials research program includes the

- Local spatio-temporal functions laboratory
- The dissipative hierarchy structures laboratory
- The exciton engineering laboratory
- The topochemical design laboratory

The program includes approximately 40 researchers and excellent facilities for surface science and chemistry digital systems, and space-time neural pattern analysis. The program emphasizes international approaches and includes collaborators in the United States and Germany. The lab includes well-developed facilities for international video-conferencing with these collaborators.

According to Toyoki Kunitake, Group Director

> Incorporation of such temporal elements into artificial materials will open up unprecedented possibilities for materials research. Non-equilibrium aspects, non-linear properties, hierarchical structures, and fractal structures are typical examples of the temporal element. Our Spatio-Temporal Function Materials Research sets the development of such nano-precision materials as major targets. Creation and detailed analysis of molecular organizations that arise from hierarchical structures, and spontaneous pattern generation should be important themes of our project. Materials design based on dissipative structures, and new opto-electronic functions due to manipulation of exciton, and topochemical design of molecular organization with nanometer-precision are other major targets.

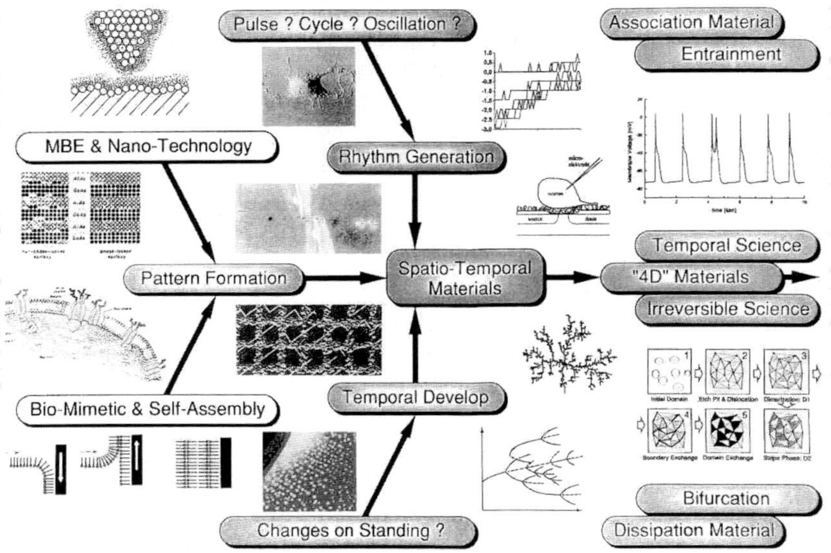

Fig. C.5. Spatio-Temporal Function Materials Research.

LOCAL SPATIO-TEMPORAL FUNCTIONS LABORATORY

In current materials research, limitations exist on conventional methodology, which has centered on the spatial control of atomic and molecular assemblies. Meanwhile, all creatures in nature show advanced functionality together with spatial order and additionally temporal factors under non-equilibrium states. RIKEN aimes to extract fundamental factors of such spatio-temporal functions and create new material research fields that differ from present approaches, mainly aiming at the following

1. Correlation between structure/functionality based on local probe studies from molecular proximity to mesoscopic scale
2. Molecular Architecture from the view point of hierarchy where functions are produced
3. Molecular communication for function and information transfer

Research Focus

The Local Spatio-Temporal Functions Laboratory focuses on fundamental factors in spatial order and the dynamics of stochastic systems. The group explores systems ranging from nanoscale organic and biological materials up to living neurons. Manipulation of spontaneous activities of embryonic cardiomyocytes has been a particular success (Micheletto, Denyer, et al. 1999).

The Local Spatio-Temporal Functions Laboratory has worked on immobilization and molecular recognition technologies of relevance to biosensing (Nakamura, Mitsui, et al. 2000; Kimura-Suda, Nakamura, et al. 2001; Nakamura, Mitsui, et al. 2001). It is currently focusing more on information transfer and information dynamics in biological systems. These studies raise issues of potential interest to advanced biosensors. While most sensor efforts focus narrowly on local measures, the "local" aspect of the RIKEN effort actually refers to relationships between local probes and global dynamics and space-time structure.

The group is investigating how biomaterials handle information transfers; for example by imaging wave dynamics in chemical and biological systems on patterned surfaces.

ASSESSMENT

Discussions with Dr. Hara ranged over wide scientific issues ranging from stochastic resonance as a processing and sensing tool, through multiscale sampling, biocomputation, mathematical physiology, bifurcation in dynamics systems, harmony in cell population dynamics, and artificial intelligence.

As an example of biocomputation and spatio-temporal dynamics, the Local Spatio-Temporal Functions Laboratory participated in a recent study of maze navigation by slime mold. (Nakagaki, Yamada, et al. 2000). While such studies do not focus on near-term biosensor development, the core issues the lab considers are of large potential significance. On the other hand, it is not clear that chaos theory or nonlinear dynamic studies can be more effective than direct engineering in biosensor development.

REFERENCES

Kimura-Suda, H., F. Nakamura, et al. 2001. Orientation of DNA thin films fabricated on substrates. *Molecular Crystals and Liquid Crystals* 370:367–370.

Micheletto, R., M. Denyer, et al. 1999. In vitro monitoring of live cardiomyocytes dynamics by a scanning near field optical microscope setup. *Optical Review* 6 (3):268–271.

Nakagaki, T., H. Yamada, et al. 2000. Maze-solving by an amoeboid organism. *Nature* 407 (6803):470–470.

Nakamura, F., K. Mitsui, et al. 2001. Adsorption behavior of DNA onto self-assembled monolayer containing intercalator. *Molecular Crystals and Liquid Crystals* 370:359–362.

Nakamura, F., K. Mitsui, et al. 2000. Immobilization of DNA on self-assembled monolayer. *Molecular Crystals and Liquid Crystals* 349:219–222.

Site: **Tokyo Institute of Technology**
Graduate School of Bioscience and Biotechnology
4259, Nagatsuta, Midori-ku
Yokohama 226-8501, Japan

Date Visited: 29 January 2003

WTEC Attendees: S.N. Bhatia (report author), S. Green, A. Ricco, C. Wilkins

Hosts: Dr. Masuo Aizawa, President, Tokyo Institute of Technology, Tel: +81-3-5734-2001; Fax: +81-3-5734-3446; Email: titech.pres@jim.titech.ac.jp
Dr. Eiry Kobatake, Associate Professor, Department of Biological Information,
Tel: +81-45-924-5760; Fax: +81-45-924-5779;
Email: ekobatak@bio.titech.ac.jp

BACKGROUND

The Tokyo Institute of Technology (Tokyo Tech) was established in 1881. There are three campuses: Ookayama, Suzukakedai, and Tamachi. It is organized as a research university with 5 graduate schools, 3 undergraduate schools, 4 attached research facilities, and 15 joint-use research centers. The total number of students is approximately 10,000, about 800 of them foreign. In 2002, the university began transmitting lectures to the Asian Institute of Technology in Thailand. The institute has planned further university reform in the near future with the plan of becoming an independent administrative organization shortly. Approximately 70% of the undergraduates proceed to graduate school. Annual research grants are 4 billion yen for 900 research grants.

OVERVIEW

Dr. Aizawa presented an overview of his ongoing research interests. Dr. Kobatake has taken on the role of principal investigator in the past two years, since Dr. Aizawa has assumed the role of president. The laboratory of the Department of Biological Information contains approximately 25 students, of which 4 are doctoral students, 2 are post-doctoral fellows.

The group is focused on combining a number of technologies to improve biosensor technology, including: genetic engineering, protein engineering, microfabrication technology, measurement technology, information technology, systems technology, and nanotechnology. Their efforts are divided into four areas:

1. **ultimate biosensing**, including in vivo monitoring, single molecule detection, biosensing of massive information, and biosensing under extreme environments
2. **massive information screening**, including genome analyses, proteome analyses, and high throughput analyses
3. **biological effects evaluation**, including animal test alternatives, pharmacological evaluation for drug discovery, environmental monitoring, safety assessment, health care, and point-of-care
4. **intelligent sensing**, including artificial nose, artificial tongue, health care, and intelligent matrices.

The WTEC panelists were introduced to three main aspects in detail: (1) single molecule studies, (2) cell-based sensing, and (3) protein engineering.

Single molecule studies in biological information systems

Projects in this area primarily utilize AFM imaging to visualize DNA interacting with proteins. One project (BBRC (291): 361) demonstrated the ability to track individual bending of DNA molecules upon binding of a transcription factor, IHF. Similarly, gene mapping could be achieved using an AFM approach. Specifically, a single-stranded DNA probe was combined with RecA protein to form a nucleoprotein filament. This structure will localize to a complementary sequence on a double-stranded DNA target and create a visible complex at the region of interest (*Anal. Chem.* (72):1288, 2000). Finally, cell lysates can be screened for transcription factors using a similar strategy. Biotin was deposited on a mica surface together with streptavidin-derivatized doubled-stranded DNA probes. NF-kappaB binding to the DNA probe can be detected by AFM and the distance from the streptavidin terminus can be utilized to map the transcription-factor binding domain (*Anal. Biochem.* (309):241, 2002).

Cell-based sensing for assessing biological effects

Projects in this area are focused on detecting cellular responses to exogenous compounds and sensing cellular responses to external stimuli such as electric fields. The cell-based systems of interest are endothelial cell-based systems for vasodilation, macrophage-like cell-based systems for immunomodulatory drugs, and neuron-based systems for assessing narcotics. A nitric oxide-based sensor was utilized to detect NO as a marker of both constitutive and inducible nitric oxide synthetase production. The sensor consisted of a gold electrode coated with a perm-selective membrane of 2:1 poly-L-lysine and ply (4-styrenesulfonate). This layer can be fabricated with controlled porosity and hydrophilicity and promotes cell adhesion. The nominal molecular weight cutoff is 100 Da. Cells of interest are plated on the membrane and a platinum counter electrode and Ag/AgCl reference electrode is utilized. Endothelial cells were interrogated with model vasodilatory compounds (acetylcholine, NOC 7 a NO donor, and L-NMMA an NOS inhibitor) resulting in activation of constitutively expressed eNOS and a rapidly detectable NO signal. Similarly, a macrophage-like cell line, RAW 264.7, was plated on the sensor and interrogated with lipopolysaccharide and interferon-gamma. These immunomodulatory compounds act at the transcriptional level to increase the synthesis of inducible NOS and a detectable NO response was observed after several hours.

The effects of electrical stimulation on mammalian cells was also investigated. Cells were grown on an ITO/glass electrode at relatively low density (~104 cells/cm^2) and exposed to AC potentials of approximately 300 mV for 1 hour. Their data indicate that astroglial cells and a number of other cell types respond to low frequency stimulation (a few 100 Hz). The time course of response indicates a rapid (~1 hour) c-fos mRNA induction, followed by elevated NGF and c-jun mRNA levels (~3 h) (*Nat. Biotech.* (15): 964, 1997). Accordingly, PC12 cells were shown to switch from a proliferative state to a differentiated state (as seen under the influence of NGF) in a calcium-channel specific response and protein kinase C dependent response (*J. Biotech.* (63):55, 1998). The promoter of HSP70 has been identified as a proported electrical-response element. Using a luciferase reporter strategy, genetically-engineered 3T3 fibroblasts were generated that were responsive to electrical stimulation (*J. Biotech.* (79):53, 2000).

PC12 (neuronal-like cell line) were also grown on 64-electrode ITO/glass arrays that were commercially purchased. Cells are plated in monolayer and monitored for patterns of electrical activity.

Protein engineering

Projects in this area include design of thermostable proteins and multifunctional proteins for biosensing. The approach is to use molecular biology to form fusion proteins with modular functions. Building blocks include elastin-based hexapeptide (APGVGV)12, a hydrophobic thermo stable peptide, an antibody-binding domain (B domain of protein A), cell adhesive peptides (RGD), and luciferase. Proteins are expressed in recombinant E. coli. This approach allowed development of an auto-clavable cell-adhesive coating (elastin-like peptide/RGD), a solid-state immunoassay (protein A/luciferase), an extracellular ATP-production assay (protein A/luciferase), and an immunoliposome sensor of approximately 200 nm in diameter (fusion of E. coli phosphatidylcholine and protein A). (*Anal. Biochem.* (282):65, 2000; *Am. J. Phys.* (276):C267, 1999; *Anal. Chem.* (69):1295, 1997; *Bioconjugate Chem.* (11):789, 2000).

TECHNOLOGY TRANSFER

The institute jointly administers the Frontier Collaboration Research Center (FCRS), the Venture Business Laboratory (VBL), the Incubation Center, and the Technology Licensing Office, in order to create a system that utilizes the technology developed by Tokyo Tech and promotes interaction between academia and industry.

Assessment: Novel NO sensor, modern molecular biology, with relatively primitive cell biology.

REFERENCES

Aizawa, M., S. Koyama, K. Kimura, T. Haruyama, Y. Yanagida, E. Kobatake. 1999. Electrically Stimulated Modulation of Cellular Function in Proliferation, Differentiation, and Gene Expression. *Electrochemistry* 67 (2), :118–125.

Kimura, K., Y. Yanagida, T. Haruyama, E. Kobatake, M. Aizawa. 1998. Gene Expression In The Electrically Stimulated Differentiation Of PC12 Cells. *Journal of Biotechnology* 63, 55–65.

Kobatake, E., H. Sasakura, T. Haruyama, M. Laukkanen, M. Aizawa. 1997. A Fluoroimmunoassay Based on Immunoliposomes Containing Genetically Engineered Lipid-Tagged Antibody. *Biomacromolecules* 69, 1295–1298.

Kobatake, E., K. Onoda, Y. Yanagida, M. Aizawa. 2000. Design and Gene Engineering Synthesis of an Extremely Thermostable Protein with Biological Activity. *Biomacromolecules* 1, 382–386.

Koyama, S., T. Haruyama, E. Kobatake, M. Aizawa. 1997. Electrically Induced NGF Production by Astroglial Cells. *NatureBbiotechnology* 15:164–166.

Seong, G.H., E. Kobatake, K. Miura, A. Nakazawa, M. Aizawa. 2002. Direct Atomic Force Microscopy Visualization of Integration Host Factor-Induced DNA Bending Structure of the Promoter Regulatory Region on the Pseudomonas TOL Plasmid. *Biochemical and Biophysical Research Communications* 291:361–366.

Seong, G.H., T. Niimi, Y. Yanagida, E. Kobatake, M. Aizawa. 2000. Single-Molecular AFM Probing of Specific DNA Sequencing Using RecA-Promoted Homologous Pairing and Strand Exchange. *Analytical Chemistry* 72 (6):1288–1293.

Seong, G.H., Y. Yanagida, M. Aizawa, E. Kobatake. 2002. Atomic force microscopy identification of transcription factor NF B bound to streptavidin-pin-holding DNA probe. *Analytical Biochemistry* 309:241–247.

Sugihara, T., G.H. Seong, E. Kobatake, M. Aizawa. 2000. Genetically Synthesized Antibody-Binding Protein Self-Assembled on Hydrophobic Matrix. *Bioconjugate* 11:789–794.

Yanagida, Y., A. Mizuno, T. Motegi, E. Kobatake, M. Aizawa. 2000. Electrically stimulated induction of hsp70 gene expression in mouse astroglia and fibroblast cells. *Journal of Biotechnology* 79:53–61.

Zhang, X., E. Kobatake, K. Kobayashi, Y. Yanagida, M. Aizawa. 2000. Genetically Fused Protein A-Luciferase for Immunological Blotting Analyses. *Analytical Biochemistry* 282:65–69.

Site: **Tokyo University of Agriculture and Technology**
Department of Biotechnology
2-24-16, Naka-cho, Koganei
Tokyo 184-8588, Japan

Date Visited: 27 January 2003

WTEC Attendees: S.N. Bhatia (report author), S. Green, A. Ricco, J. Schultz

Hosts: Dr. Koji Sode, Chair- Department of Biotechnology,
Tel/Fax: +81-42-388-7027;
Email: sode@cc.tuat.ac.jp
Dr. Tadashi Matsunaga, Dean of Technology,
Tel: +81-42-388-7020;
Fax: +81-42-385-7713;
Email: tmatsuna@cc.tuat.ac.jp;
http://www.tuat.ac.jp/~matsunaga/

BACKGROUND

The Tokyo University of Agriculture and Technology (TUAT) was established in 1874. There are two main campuses: the technology campus is in Koganei and the agriculture campus is in Fuchu, approximately 30 minutes away. The university is populated by approximately 6000 students and 400 faculty. The Faculty of Agriculture are focused on problems of food production and resource management, and on environmental and health issues. The Faculty of Technology conduct research in molecular biology, soft materials, fiber information and communication systems, and nanotechnology, used in support of technologies such as micromechanics. In 2004 the university will be reclassified from a national university to an independent administration, as is the trend throughout much of Japan.

GROUP OVERVIEW

Dr. Matsunaga presented an overview as the Dean of Technology. The division consists of approximately 23 faculty, 100 undergraduates, 60 Master's students, and 24 doctoral students per year. A typical laboratory contains 10 undergraduates, 20 Master's students, 5 doctoral students, 2 post-doctoral fellows, 2 research associates, and the principal investigator. Funds are raised from both government and industry, with approximately 70% of the budget coming from the national government. The annual budget is approximately $150 M.

Sode Laboratory

Dr. Sode's group uses molecular bioengineering to design novel molecules for use in biosensors applied to diabetes. The general approach is to improve molecules by: (1) mutating naturally occurring products, (2) mutating known proteins mined from existing databases, or (3) mimicking enzymes by molecular imprinting into polymers.

Novel enzymes for continuous glucose monitoring

One enzyme of interest was PQQDH, an oxygen-independent enzyme that can be used as an alternative to glucose oxidase for glucose sensing, alleviating the potential oxygen-dependence of continuous glucose monitoring in diabetics. The native enzyme PQQDH has higher catalytic efficiency than glucose oxidase but suffers from reduced specificity (i.e., it interacts with galactose, maltose, and other sugars), the need for a specific oxidizing agent, and relative thermal instability. The specificity of the enzyme was increased by modification of the active site via random mutagenesis and deliberate protein engineering. The thermal stability of the enzyme was improved by addition of a disulfide bond between the dimer subunits. The resultant enzyme showed reduced activity against galactose and enzymatic activity up to 70°C. This molecule requires a rather specific electron acceptor in order to be utilized in an amperometric sensing scheme, in analogy to the role of oxygen with glucose oxidases. For in vitro applications, this was achieved by addition of relatively toxic electron acceptors such as ferrocene or methoxy-PMS. Therefore, another area of interest is in utilizing a naturally occurring electron acceptor such as cytochrome b562 to interface PQQDH with the electrode surface. Organization of these molecules in the appropriate configuration would be achieved using DNA oligonucleotides (see section C).

Novel reagents for real-time detection of Hemoglobin A1C (HbA1C)

In diabetics, glycation of hemoglobin at the N-terminal valine produces a fructose-valine product (HbA1C) in erythrocytes that serves as a time-average of blood glucose levels over approximately 120 days. The detection of HbA1C levels is achieved clinically by HPLC; however, real-time assays are not currently available. By screening of microorganisms, Dr. Sode's laboratory has determined that the enzyme fructosyl amine oxidase could be utilized to electrochemically detect the cleavage of the fructose-valine bond. The enzyme was recombinantly expressed, immobilized in a polyvinyl alcohol matrix, and used in conjunction with carbon/platinum electrodes to detect the production of hydrogen peroxide. The reaction was specific to fructose-valine when compared to fructose-lysine, which occurs on glycated albumin. Efforts to transfer catalytic activity of this enzyme to a polymer were explored using molecular imprinting. The resultant "imprinted" polymer demonstrated a 1.8-fold increase in reaction rate over non-imprinted controls.

DNA nanowiring

This project aimed at utilizing DNA and sequence-specific DNA binding proteins as building blocks to arrange enzymes and electron acceptors in a high-density three-dimensional array (with substantially higher volume density of enzyme than typical surface attachment) proximal to an electrode surface. Sequence-specific DNA binding proteins were expressed as fusion proteins of interest, in a similar manner to "his" (histidine) tags, and used to attach and orient proteins via DNA linkers on a surface.

DNA detection using enzymatic reporter.

This work was presented by Dr. Ikebukeno, an associate professor in Dr. Sode's group.

This research focused on the use of enzymatic, rather than fluorescent, labeling of DNA oligonucletides to facilitate electrochemical detection of DNA. Enzymatic amplification was accomplished using the thermally stable mutant of PQQDH described above. Methoxy-PMS was used as an electron acceptor. Using this technique, the sensitivity of DNA detection improved from 10^{-6} to 10^{-10} M. A similar approach was utilized to detect single nucleotide polymorphisms in the PPAR gamma gene that confer resistance to type II diabetes. In this case, a DNA polymerase was utilized that only acted on the SNP of interest. Signal detection was achieved by using a biotinylated oligonucleotide in conjunction with an avidin-PQQDH fusion protein.

Another area of interest was the detection of double-stranded DNA resulting from PCR reactions. One approach utilized a sandwich of probes and PQQDH for detection. Alternatively, salmonella-derived DNA protein was utilized to recognize double-stranded DNA. In this scheme, a fluorescent DNA intercalating dye was used to detect the amount of bound DNA. In the future, this work will be extended to a family of double-stranded binding proteins, the zinc finger transcription factor family.

Matsunaga Lab

Dr. Tanaka, a research associate in Dr. Matsunaga's group presented this overview. Dr. Matsunaga's group develops biologically produced magnetic (nano) particles, BMPs, and lab-on-a-chip devices for analytical applications.

Magnetic nanoparticles

The research in this area is centered around the microbe *magnetospirillium magneticum*, which has been shown to produce uniform, nano-sized (50–100 nm), lipid-covered ferromagnetic particles in its cytosol. These magnetic particles are thought to influence migration and/or orientation of these microbes in the earth's geomagnetic field, perhaps helping them to find the mud substrate in which they thrive in aquatic environments. Four types of crystal morphologies, each with good aqueous dispersity, have been previously reported. The particles are thought to form due to a GTPase-mediated vesicle formation process and the localization of an iron transporter in the vesicle bilayer. Dr. Matsunaga is interested in a number of aspects of studying and using these BMPs. He has an interest in understanding the genetic determinants of various crystal morphologies, and has therefore completed the shotgun sequencing of the organism's genome. In addition, his group has developed a genetic system to express fusion proteins in the vesicle bilayer that confer a protein coating on the nanoparticles. They have explored expression of luciferase, protein A (for antibody labeling), the estrogen receptor, and even complex g-protein coupled receptors (dopamine 1) that seem to maintain a normal conformation. These magnetic nanobeads have been incorporated into an immunoassay and a drug-binding assay that simplify wash steps, requiring only the application of a small permanent magnet to the side of a pipet tip to hold the functionalized beads during washing. This process is currently being automated by the group. These biologically derived particles have some advantages over synthetic particles: a relatively uniform size, response to weaker magnetic fields compared to the intense fields needed with paramagnetic beads, good dispersity due to their small size and negative surface charge, and a lipid bilayer coating. Detection of bound analytes was accomplished using the GDH scheme described above.

Lab-on-a-chip

This research is focused on the use of isoelectric point differences to separate free antigen from antigen-antibody complexes. The approach is to utilize a microfluidic network, perfused with different pH solutions to separate the two species according to their isoelectric points. In model experiments, ferrocene was bound to the antibody and histamine to the antigen, to facilitate electrochemical detection. Chip fabrication was performed by precision machining of PMMA with a CAD interface and 100 μm resolution for rapid protoyping, or PDMS casting onto etched silicon masters for high-resolution work.

TECHNOLOGY TRANSFER

In 1983, the Ministry of Education enacted a cooperative research system between universities and communities. The objective of the system was to promote diversification in research and education at universities by offering space for private researchers to conduct cooperative work with university researchers. In 1988, TUAT was authorized to establish a cooperative research center. The building was completed in 1989 and then expanded in 1996 to 2,000 sq. meters. Since 2001, a liaison-coordinator system has been established so that the "seeds" provided by the university can be better coordinated with the needs of the private sector. By 2002, 116 cooperative research projects had been carried out. The current space is subdivided into two areas. The "venture business" space is intended for academics to develop early-stage technologies for commercialization whereas the "collaborative space" is intended for later-stage transitional work and typically incorporates an industrial partner. Approximately 20 companies are currently involved in the collaborative space.

ASSESSMENT

TUAT exhibits current, solid application of molecular biology to widely-recognized existing biomedical problems. It does not exhibit much engineering work, and its facilities are not fully modernized.

REFERENCES

Arakaki, A., J. Webb, T. Matsunaga. 2003. A Novel Protein Tightly Bound to Bacterial Magnetic Particles in *Magnetospirillum magneticum* Strain AMB-1. *The Journal of Biological Chemistry* 278.

Ikebukuro, K., Y. Kohiki, K. Sode. 2002. Amperometric DNA sensor using the pyrroquinoline quinone glucose dehydrogenase-avidin conjugate. *Biosensors and Bioelectronics* 17:1075–1080.

Lim, T., N. Nakamura, T. Matsunaga. 2000. Use of Anion Exchange Resin-Packed Capillary Column for Rapid Detection of Anti-Double-Stranded DNA Antibody in systemic Lupus Erythematosus Serum. *Biotechnology and Bioengineering* 68 (5):571–575.

Lim, T., S. Imai, T. Matsunaga. 2002. Miniaturized Amperometric Flow Immunoassay System Using a Glass Fiber Membrane Modified with Anion. *Biotechnology and Bioengineering* 77 (7):758–763.

Lim, T., Y. Komoda, N. Nakamura, T. Matsunaga. 1999. Automated Detection of Anti-Double-Stranded DNA Antibody in Systemic Lupus Erythematosus Serum by Flow Immunoassay. *Analytical Chemistry* 71 (7):1298–1302.

Ogawa, K., D. Stöllner, F. Scheller, A. Warsinke, F. Ishimura, W. Tsugaa, S. Ferri, K. Sode. 2002. Development of a Flow-Injection Analysis (FIA) Enzyme Sensor for Fructosyl Amine Monitoring. *Anal Bioanal Chem.* 373:211–214.

Okamura, Y., H. Takeyama, T. Matsunaga. 2001. A Magnetosome-specific GTPase from the Magnetic Bacterium *Magnetospirillum magneticum* AMB-1. *The Journal of Biological Chemistry* 276 (51):48183–48188.

Okuda, J., J. Wakai, K. Sode. 2002. The Application of Cytochromes As the Interface Molecule to Facilitate the Electron Transfer for PQQ Glucose Dehydrogenase Employing Mediator Type Glucose Sensor. *Analytical Letters.* 35 (9): 1465–1478.

Sakaguchi, T., J.G. Burgess, T. Matsunaga. .1993. Magnetite formation by a sulphate-reducing bacterium. *Nature.* 365 (6441):47–49.

Sode, K., S. Igarashi, A. Morimoto, H. Yoshida. 2002. Construction of Engineered Water-soluble PQQ Glucose Dehydrogenase with Improved Substrate Specificity. *Biocatalysis and Biotransformation.* 20 (6):405–412.

Takeuchi, A., K. Sode. 2000. A Salmonella Detection System Using an Engineered DNA Binding Protein That Specifically Captured a DNA Sequence. *Analytical Chemistry.* 72 (13):2809–2813.

Tanaka, T., T. Matsunaga. 2000. Fully Automated Chemiluminescence Immunoassay of Insulin Using Antibody-Protein A-Bacterial Magnetic Particle Complexes. *Analytical Chemistry* 72 (15):3518–3522.

Yamazaki, T., S. Ohta, Y. Yanai, K. Sode. 2003. Molecular Imprinting Catalyst Based Artificial Enzyme Sensor for Fructosylamines. *Analytical Letters* 36 (1):73–87.

Site: **Tokyo University of Technology**
Katayanagi Institute
1401-1 Katakura, Hachioji
Tokyo 192-0982, Japan

Date Visited: 27 January 2003

WTEC Attendees: D. Walt (report author), J. Schultz, D. Brady, C. Wilkins, S.N. Bhatia, A.J. Ricco, S. Green, H. Ali

Hosts: Professor Dr. Isao Karube, Director of the Board, Katayanagi Institute,
Tel: +81-426-37-2111; Fax: +81-426-37-3134;
Email: karube@bs.teu.ac.jp
Mr. Koh Katayanagi, Chairman of the Board
Dr. Hideo Aiso, President
Mr. Shigeru Chiba, Vice Chairman
Dr Kazuyohi Yano, Associate Professor
Dr. Hideaki Nakamura, Lecturer,
Tel: +81 (0) 426-37- 2149;
Email: nakamura@bs.teu.ac.jp

BACKGROUND

The Katayanagi Institute had its start in 1947 in downtown Tokyo. It started as a technical college for television technicians and then expanded its activities into computers and environmental technology in the mid-1960s. In 1977, it expanded to the broadcast arts. Tokyo University of Technology and its technical college were established in 1986 in the very large campus of 360,000 m^2, along with 26 other academic institutions, in the Hachioji area, where there was ample land. The university was founded with a School of Engineering and, in 1998, a School of Media Science was established based in the institution's strength in broadcast arts. In the same period, the university began a serious effort to facilitate university/industry collaboration. This effort has blossomed, and in April 2003 a School of Bionics will be completed. An impressive new U.S. $250M building with 15,000 m^2 of space was due to open to house industrial/academic research projects along with the traditional academic research and academic facilities. For example, the Research Center of Advanced Bionics (RCAB), with Prof. I. Karube as director, started a collaboration with the National Institute of Advanced Industrial Science and Technology (AIST) beginning 2003 in a laboratory in the new building. This collaboration is the first one established with a private university in Japan. The new structure of the university contains the School of Media Science, and the School of Engineering (previously housing departments of electronics, mechanical control, information science and communications), which has been split into the School of Bionics and the School of Computer Science. Fifty-plus new faculty and research staff were hired for the two new schools. In two years, the university intended to begin a graduate program in Bionics coupled with Technical Management. Graduate students in the program will work with industry through internships. The university has 4000 undergraduates and 1000 graduate students.

Organization and Facility

The School of Bionics was launched in April 2003. The goal is to have a faculty with the largest group of the highest quality in Japan. Enrollment is 421 students with approximately 12 students per faculty member.

The new building at the Katayanagi Institute has 150,000 m^2 of space. The facility is extremely impressive with open flexible laboratory space, faculty and administrative offices, student recreational space, and ample space for conversation and casual encounters. The building is replete with extensive artwork and provides an ideal intellectual environment. Five floors in the building are designated for specific collaborative projects between industry and academics. The 6th floor is a bionanotechnology center that will serve as a central facility and resource with both nano fabrication and nanostructure determination capabilities. This central facility will contain a $20M fab facility. On the 7th floor is designated as a protein systems chip facility, to be funded exclusively with government funds. The 8th, 9th, and 10th floors are designated for house laboratories for medical engineering (microarrays for disease), coenzyme-Q biotechnology, and environmental engineering, respectively. Half of the funding for these three laboratories is to from industry with the other half coming from the Ministry of Economy, Trade and Industry (METI).

The new building and the financial resources available are substantial and make the institution a highly attractive place to pursue advanced studies.

BIOSENSOR R&D ACTIVITIES

Research Overview

Professor Isao Karube, formerly of Tokyo University, moved here in April 2002 along with his research staff of approximately 60 people. The exclusive effort at Tokyo University of Technology in biosensor studies is within Professor Karube's purview. Karube has been working in the field of biosensors for 30 years. He is internationally recognized as a major figure in the field of biosensors and has written important fundamental texts on the subject and has published extensively. Karube has transferred much of his laboratory's research to industry and has a string of commercial products emanating form a large patent portfolio. The entire biosensors effort at Tokyo University of Technology is extensive, and the WTEC visiting team was unable to hear about all the research projects being pursued.

The Karube laboratory has developed microelectrode biosensors for hydrogen peroxide and glucose, both of which have been transferred to the NEC Corporation. Micromachined microelectrodes for oxygen measurement have been transferred and commercialized by Fujitsu Corporation. His current major research efforts are in immunochips for infectious disease diagnosis, glucose biosensors, SNP chips for medical diagnosis, and environmental biosensors. The infectious disease immunochip is based on a bead aggregation assay. Two companies have expressed interest in licensing the technology.

The glucose chip is designed to measure hypoglycemia in pediatric patients and it is anticipated that the commercialization launch date was to be August 2003.

One of the major projects advocated by Karube starting nearly 10 years prior to the WTEC visit, is the Bio-toilet, in which biosensors are placed in the toilet bowl to perform an analysis of the sample. The ultimate goal of the Bio-toilet is to perform a complete urinalysis and to screen for analytes that signal the presence of disease. The Karube group has successfully developed sensors for urinalysis of glucose and urea that will be deployed soon (August 2003) in the first generation Bio-toilet.

MAJOR RESEARCH PROJECTS

Two major research efforts at Tokyo University of Technology are environmental biosensors and DNA/protein chips. These efforts are well funded and are aimed at addressing specific industrial/medical applications in Japan.

Environmental Biosensors

In the environmental biosensors program, sensors are being pursued for biological oxygen demand (BOD) for bode portable and bench top units, phosphate ions, a photoluminescence plankton sensor, detergents, Chemical oxygen demand (COD), and cyanide ion. The major application for these sensors is for industrial effluent, household water monitoring, and seawater quality, or for natural water quality. The major effort is to develop robust systems, including all the necessary supporting instrumentation, to address these applications.

DNA and Protein Chips

Another major effort is the development of DNA SNP chips. These chips are being fabricated with a printing technique in which probe sequences are printed on membranes. The DNA chip area is being pursued for SNP detection applications. The protein chip work is aimed at designing antibody arrays for detecting proteins in a multiplexed fashion. There is novel work being performed in the protein chip area in which polymers and enzymes are copolymerized using plasma polymerization. This approach enables a convenient method for immobilizing proteins onto polymers while retaining the protein activity (binding, catalysis).

Media Lab

A major strength at Tokyo University of Technology is in the area of broadcast/media studies. The university boasts a state-of-the-art soundstage and production studio with full animation and sound production capabilities. It is also clear that there is substantial expertise on the computational side, with extensive computer processing capabilities to perform real-time complex calculations (e.g., a video camera array is being used to capture coordinates from an individual performer and convert them directly into animated characters).

ASSESSMENT

It was clear that the Katanayagi researchers take a systems approach to biosensor development and create both the sensors as well as the supporting instrumentation to enable the sensors to be used. There is a clear mandate to facilitate the transfer of technology developed within the university setting to the private sector and a welcoming collaborative environment is provided, which enables academic/industrial collaborations and partnership.

REFERENCES

A list of nearly 700 references from the Karube laboratory was provided.

Asai, R., C. Nakamura, K. Ikebukuro, I. Karube and J. Miyake. 2002. Detection technique of a T-PCR-based amplified single-stranded DNA and its application to biosensor for detection of mRNA for cyanobacteria, Anabaena variabilis. *Biotechnology Letters* 24:1677–1682.

Banzai, T., G. Hershkovits, D. J. Katcogg, N. Hanagata, Z. Dubinsky, and I. Karube. 2002. Identification and characterization of mRNA transcripts differentially expressed in response to high salinity by means of differential display in the mangrove, Bruguiera gymnorrhiza. *Plant Science* 162:499–505.

Han, T.S., S. Sasaki, K. Yano, K. Ikebukuro, A. Kitayama, T. Nagamune, I. Karube. 2002. Flow injection microbial trichloroethylene sensor. *Talanta* 57:271–276.

Nakamura, H., I. Karube. Current Research Activities in Biosensors. *Analytical and Bioanalytical Chemistry* (in press; 400 references are summarized).

Nakamura, H., Y. Murakami, K. Yokoyama, E. Tamiya, M. Suda, S. Uchiyama, I. Karube. 2001. A compactly integrated flow cell with chemiluminescent FIA system for determining lactate. *Analytical Chemistry* 73 (2):373–378.

Oyama, M., T. Ikeda, T. K. Lim, K. Ikebukuro, Y. Masuda, I. Karube. 2001. Detection of toxic chemicals with high sensitivity by measuring the quantity of induced P450 mRNAs based on surface plasmon resonance. *Biotechnology and Bioengineering* 71 (3):217–222.

Shimomura, M., Y. Nomura, K. H. Lee, K. Ikebukuro, I. Karube. 2001. Dioxin detection based on immunoassay using a polyclonal antibody against actachlorinated dibenzo-*p*-dioxin (OCDD). *Analyst* 126:1207–1209.

Shimomura, M., Y. Nomura, W. Zhang, M. Sakino, K. H. Lee, K. Ikebukuro, I. Karube. 2001. Simple and rapid detection method using surface plasmon resonance for dioxins, polychlorinated biphenylx and atrazine. *Analytica Chimica Acta.* 434:223–230.

Site: **University of Tokyo**
Department of Applied Chemistry, School of Engineering
7-3-1 Hongo, Bunkyo-Ku
Tokyo 113-0033, Japan

Date visited: 29 January 2003

WTEC Attendees: J. Schultz (report author), S. Bhatia, D. Brady, S. Green, A. Ricco, C. Wilkins, D. Walt; H. Ali

Hosts: Prof. T. Kitamori, Tel: +81-3-5841-7231;
Fax: +81-3-5841-6039;
Email: kitamori@icl.t.u-tokyo.ac.jp

BACKGROUND

The University of Tokyo was established in 1874 as the first national university in Japan. It offers courses in essentially all academic disciplines at both undergraduate and graduate levels and provides research facilities for these disciplines. The university has a faculty of approximately 2,800 professors, associate professors, and lecturers, and a total student enrollment of about 28,000. There are about 2,050 international students, and about 1,400 foreign scholars come to the university each year for short or extended visits. The university is known for the excellence of its faculty and students; many of its graduates are and have always been leaders in the government, in business, and in the academic world.

The university organization consists of the College of Arts and Sciences, nine faculties, and fourteen graduate schools. The nine faculties are Law, Medicine, Engineering, Letters, Science, Agriculture, Economics, Education, and Pharmaceutical Sciences. The traditional eleven graduate schools are Law and Politics, Medicine, Engineering, Humanities and Sociology, Science, Agricultural and Life Sciences, Economics, Arts and Sciences, Education, Pharmaceutical Sciences, and Mathematical Sciences.

The university recently saw the establishment of three new advanced graduate schools, Frontier Sciences, Interdisciplinary Information Studies, and Information Science and Technology. The University also operates the eleven institutes covering a broad area of knowledge.

There are also many research facilities connected to various faculties of the University. All the institutes and research facilities work closely with their related faculties and graduate schools. Many of the faculty members associated with these institutes engage in graduate school teaching and supervise graduate students working towards advanced degrees.

The University of Tokyo is composed of three campuses: Hongo, Komaba, and Kashiwa. In addition, University of Tokyo facilities are situated in other parts of both Tokyo and the nation. The main campus of the university is located in Hongo, Bunkyo-ku, Tokyo; it occupies about fifty-six hectares of the former Kaga Yashiki, the Tokyo estate of a major feudal lord. Parts of the 17th century landscaping of the original estate have been preserved and provide greenery and open space, much needed in an otherwise crowded campus. The celebrated Akamon, or Red Gate, which graces the campus, was a special gate on the Kaga estate and dates back to 1827. It has been designated as an "important cultural property" by the Japanese government. Most of the faculties, graduate schools and research institutes of the university are located on the Hongo Campus.

The main campus, visited by the WTEC group, has the Schools of Agriculture, Science, Pharmaceutical Science, and Engineering. Each school has its own chemistry department.

OVERVIEW

The laboratory of the School of Engineering's Department of Applied Chemistry has 80–100 people, the clean room area is about 8,000 ft^2, and the laboratory area is about 6,000 ft^2.

R&D ACTIVITIES

This laboratory has three major projects: bioanalysis, MEMS, and chip-based chemical processing.

Integrated microchemical systems on a monochip—basic size 3 cm x 7 cm—allows processing of small quantities (nano to pico liters); rapid processing 100 times faster than laboratory scale operations; and high level of process integration up, to 10 process steps per square centimeter. The devices can be used for diagnosis, environmental analysis, combinatorial chemistry, and artificial organs.

The approach is to use a glass substrate that allows the use of organic solvents in the preparation systems. Also, by placing one chip on top of another, 3-dimensional constructs can be fabricated. Communication between levels is accomplished by 300-micron holes. Layers are fused together by heating, no adhesives being utilized.

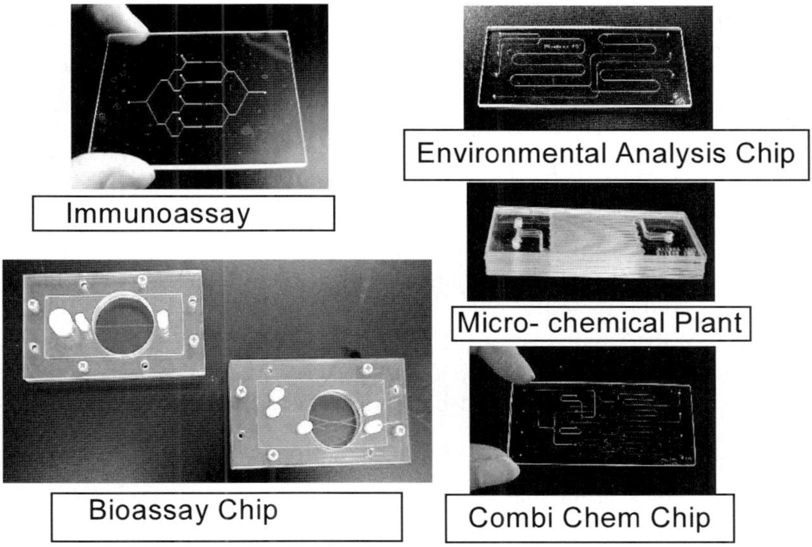

Fig. C.6. Integrated microchemical systems on a monochip.

The flow control for mixing and residence time distribution is based entirely on maintaining laminar flow throughout the system, for the Reynold's number range 0.1–1.0. Typical channel width is 50–100 micron. Flow is accomplished by syringe pumps; electro osmotic flow is not used because of the desire to use organic solvents. Mixing in the system is entirely by diffusion and takes place very rapidly (on the order of minutes) when two streams are brought together because of the small dimensions of the channels. A stable interface is maintained when aqueous and organic streams are contacted, thus allowing the extraction of materials from one phase to another. The stable interface is achieved by chemically modifying the surfaces of opposing walls of the channel to produce hydrophilic and hydrophobic regions. Up to 5 different flow streams can be achieved in a single channel. In addition, technology has been developed for counter flow of organic and aqueous streams in the same channel.

For most of the applications, concentration measurements are accomplished by a thermal lens microscope technique that has been perfected to measure concentrations in the zepto-mole range, or about 50–100 molecules. The thermal lens device (invented by Prof. Kitamori and manufactured by IMT Co. Ltd) interrogates a volume of about 1 micron (Hibara et al. 2002b). Research is under way to miniaturize this desk size instrument to a device the size of a deck of cards. Alternative detection methods, such as laser induced fluorescence, electrochemical techniques, and chemiluminescence, have also been integrated into the chip format.

CAPABILITIES

The system has been designed so that it can be used for chemical processes. Nearly all unit operations—reactors, mixing, extraction, heating, cooling—can be implemented allowing micro factories to be assembled. One to several tens of different flows can be handled simultaneously, and steps can be completed in about 1 minute. Highly controllable microfluidic control valves that are operated by pressure signals allow the implementation of complex processing operations. For expensive reagents, very small quantities, on the order of 200 nanoliters, can be processed efficiently.

Because of the small size of the units, processing times can be reduced from hours to seconds. Also, complex chemical processes can be accomplished, as the system can be engineered to have about 10 process steps per square centimeter.

Because of the high surface to volume ratio, heat and mass fluxes can be hundreds of times faster than in ordinary equipment. Heat exchange rates can be as fast as 1200°C/second.

APPLICATIONS

These systems have been applied to diagnostic, environmental, combinatorial, and polymer chemistry processes. Processing times can be as short as 1 minute.

Some analytical chips have been made for the following species: Ni, Fe, Co, Na, K, ascorbic acid, catecholamines, CEA (cancer marker), estradiol, IgA, and gamma Inteferon.

A prototype analyzer has been constructed that is the size of a 8 x 8 x 10-inch cube. Also, as an example, the researchers in this lab made a 32-channel immunoassay on a microchip. Experiments to test inter-channel reproducibility show variations on the order of about 10%. The range of concentrations that can be measured is about 1000 fold. The sensor card containing the reagents and sample is placed into a reader device (much like a credit card reader), and the result of the analysis is provided on the order of minutes. This rapid response is achieved because the average diffusion distance for antibody-antigen interactions is on the order of 10 microns, whereas in the typical microtiter plate the diffusion distance is about 1000 microns.

As a system for chemical synthesis, devices have been constructed to carry out the following processes: diazo-coupling, asymmetric amide synthesis, combinatorial chemistry, picoline amide hydrolysis and, interfacial polymerization.

Commercial application of the technology has been demonstrated with a process to produce uniform gel particles (about 80-micron diameter) for high pressure chromatography. A manufacturing unit to produce 30 tons per year of this polymer has been constructed. The size of this "plant" is about 2 meter cubed.

Other applications include: micro-chemical plants for diazo-coupling/phase transfer reactions, asymmetric amide compounds, picoline amide/hydrolysis, and nylon/interfacial polymerization.

Because of the highly integrated capability of the system, synthesis of complicated asymmetric molecules can be achieved. By appropriately configuring the flow system, even cell culture bioreactors can be fabricated. Some of the biological processes that have been investigated with this system include: neuron plasticity (mouse hippocampus cells), apoptosis (neuroblastoma-glioma cells), and immunostimulation (mouse macrophage cells).

As the microchip technology is perfected for commercial production, Professor Kitamori envisions that these devices will be coupled with telemetry methods to transfer information to remote sites. Two such concepts are shown in Figure C.7.

Fig. C.7. Two concepts of integrated chip sensor systems coupled with telemetry methods for remote sensing. *Left*, a portable ultrasensitive diagnostic device, and, *right*, an ultratrace pollutant monitoring device.

SOURCES OF SUPPORT

Support for this laboratory comes from a number of government agencies. The Ministry of Agriculture funds research on cell screening, cell breeding, and bio-reactors. The Ministry of Economy, Trade and Industry (METI) supports basic research on analytical systems, synthesis systems, and drug screening systems. The Ministry of Education, Culture, Sports, Science and Technology (MEXT) supports efforts in the basic sciences related to principles of operation, alternative methodologies, and innovative concepts.

Kanagawa Academy of Science and Technology, a foundation of the local government, supported the laboratory's activity of the earliest five years, which were the most important period for establishing its basic methodologies.

Professor Kitamori has been designated for a National Research and Development Project on Advanced Chemical Instrumentation to develop technologies for high-throughput screening devices, disposable diagnostic devices, and micro-chemical plants. Funding for the effort is about $5–6 million per year.

REFERENCES

Hibara, A., M. Tokeshi, K. Uchiyama, H. Hisamoto, T. Kitamori. 2001. Integrated multilayer flow system on a microchip. *Anal Sci.* 17 (1):89–93.

Hibara, A., M. Nonaka, H. Hisamoto, K. Uchiyama, Y. Kikutani, M. Tokeshi, T. Kitamori. 2002a. Stabilization of liquid interface and control of two-phase confluence and separation in glass microchips by utilizing octadecylsilane modification of microchannels. *Anal Chem.* 74 (7):1724–8.

Hibara, A., T. Saito, H.B. Kim, M. Tokeshi, T. Ooi, M. Nakao, T. Kitamori. 2002b. Nanochannels on a fused-silica microchip and liquid properties investigation by time-resolved fluorescence measurements. *Anal Chem.* 74 (24):6170–6. PMID: 12510735

Hisamoto, H., Y. Shimizu, K. Uchiyama, M. Tokeshi, Y. Kikutani, A. Hibara, T. Kitamori. 2003. Chemicofunctional membrane for integrated chemical processes on a microchip. Anal Chem. 75 (2):350–4

Odake, T., K. Tsunoda, T. Kitamori, T. Sawada. 2001. Highly sensitive and direct detection of DNA fragments using a laser-induced capillary vibration effect. *Anal Sci.* 17 (1):95–8.

Sato, K., A. Hibara, M. Tokeshi, H. Hisamoto, T. Kitamori. 2003. Related Articles, Links Integration of chemical and biochemical analysis systems into a glass microchip. [Is this the title of a paper?]Anal Sci. 19 (1):15–22.

Surmeian, M., M.N. Slyadnev, H. Hisamoto, A. Hibara, K. Uchiyama, T. Kitamori. 2002. Three-layer flow membrane system on a microchip for investigation of molecular transport. *Anal Chem.* 74 (9):2014–20.

Tamaki, E., K. Sato, M. Tokeshi, K. Sato, M. Aihara, T. Kitamori. 2002. Single-cell analysis by a scanning thermal lens microscope with a microchip: direct monitoring of cytochrome c distribution during apoptosis process. *Anal Chem.* 74 (7):1560–4.

Tokeshi, M., T. Minagawa, K. Uchiyama, A. Hibara, K. Sato, H. Hisamoto, T. Kitamori. 2002. Continuous-flow chemical processing on a microchip by combining microunit operations and a multiphase flow network. *Anal Chem.* 74 (7):1565–71.

Site: **University of Tokyo**
Department of Chemistry
7-3-1 Hongo, Bunkyo-Ku
Tokyo 113-0033, Japan

Date Visited: 29 January 2003

WTEC Attendees: D. Walt (report author), J. Schultz, S. Bhatia, D. Brady, S. Green, A. Ricco, C. Wilkins, H. Ali

Hosts: Prof. Yoshio Umezawa, Tel: +81-3-5841-4351;
Fax: +81-3-5841-8349;
Email: umezawa@chem.s.u.-tokyo.ac.jp
H. Aoki, M. Sato, T. Ozawa, T. Nishino

OVERVIEW

The Umezawa lab has 30 researchers and operates on a budget of approximately U.S. $1.4M, with U.S. $1M every year coming from the U.S. government for high-throughput screening of endocrine disrupters.

RESEARCH FOCUS

The Umezawa laboratory's major effort is characterized as "Imaging of molecular events in single living cells." The goal of the work is to detect the presence of particular biological species and to localize them in particular cellular sub-regions. The objective of the work is to develop methods for probing chemical processes in living cells and their applications for assay and screening of chemicals that promote/disrupt cellular signaling. The key cellular signaling steps include second messengers such as cGMP, DAG, and PIP3, protein phosphorylation, protein conformational change, protein-protein interaction, and protein localization. The laboratory uses molecular biological techniques and employs fluorescence microscopy as the method of analysis. Endocrine disrupters and drug interactions are being investigated.

The research strategy is to develop new protein constructs that are expressed in cells and to observe their localization and activation using confocal microscopy. The Umezawa group images molecular events in single living cells. An example of the method is exemplified by the group's studies in protein phosphorylation.

In the general protein construct, four protein modules are spliced to one another: cyan fluorescent protein (CFP) and yellow fluorescent protein (YFP) are coupled with an intervening sequence containing a recognition part and a recognized part. The recognized part is designed such that when it undergoes modification, the recognition part binds to it, and the entire construct undergoes a conformational change, leading to proximity of the CFP and YFP moieties and resulting in fluorescence resonance energy transfer (FRET), thereby enhancing the fluorescence of the YFP.

An example is shown in Figure 2.5 in Chapter 2. In this example, designed to measure phosphorylation, a tyrosine residue serves as the recognition part. When the tyrosine is phosphorylated, the recognition part, a phosphotyrosine binding protein, binds and causes the CFP and YFP to approach one another and increase the FRET.

In a more sophisticated construct, an optional localization moiety is attached to the YFP end of the protein so that the protein is targeted to a particular cellular structure. An example given was a membrane localization peptide that causes the construct to bind only to membranes, so that no cytosol fluorescence occurs. Other localization moieties include signaling peptides that transport the construct into specific organelles such as the Golgi apparatus or mitochondria.

The biological studies are designed to observe the effects on fluorescence when the modified cells are stimulated with various agonists and antagonists and to look at up- and down-regulation of various cellular activities.

The laboratory is interested in studying the fundamental aspects of cell signaling pathways.

Another approach is to use a split GFP method. In this approach, akin to the yeast two-hybrid method, when two proteins bind with each protein containing half the GFP sequence (C and N terminal ends), they are spliced and GFP is reconstituted. The system is unique in that it employs the intein system, which excises intein to splice the two GFP halves.

In an elegant recent variation of the approach (Ozawa et al. 2003) the researchers developed a method that allows rapid identification of novel proteins compartmentalized in mitochondria by screening large-scale cDNA libraries. The principle is based on reconstitution of split-enhanced green fluorescent protein (EGFP) by protein splicing of DnaE derived from Synechocystis. The cDNA libraries are expressed in mammalian cells following infection with retrovirus. If a test protein contains a functional mitochondrial targeting signal (MTS), it translocates into the mitochondrial matrix, where EGFP formation occurs by protein splicing. The cells harboring this reconstituted EGFP are screened rapidly by fluorescence-activated cell sorting. From the screened cells, the cDNAs are isolated and identified. The analysis of 258 cDNAs revealed various MTSs. New transcripts corresponding to mitochondrial proteins were identified. This method provides a broadly applicable means for mapping proteins distributed within intracellular organelle in different tissues or disease states.

Other work being conducted in the Umezawa laboratory involves using electrochemical detection to detect DNA hybridization to a surface. The researchers are able to differentiate double-stranded hybridized DNA from its single-stranded form by observing the difference in current due to transfer to an electro active Ruthenium species in solution. A single-stranded peptide nucleic acid (PNA) is used as the probe sequence immobilized on the electrode surface.

Finally, the laboratory is using chemically modified STM tips (molecular tips) for scanning tunneling microscopy (STM) to obtain atomic level resolution using chemically selective imaging.

ASSESSMENT

The Umezawa group is using state-of-the-art molecular biological techniques to develop completely novel protein constructs for cellular imaging. The work is extremely creative and the methods should have broad application to many biologically relevant molecules. The methods should be important for elucidating fundamental cell signaling pathways, as well as applications to high-throughput cell screening and new optical sensing modalities.

REFERENCES

Nishino, T., T. Ito, and Y. Umezawa. 2002. Carbon nanotube scanning tunneling microscopy tips for chemically selective imaging. *Anal. Chem.* 74 (16):4275–4278.

Ozawa, T., Y. Sako, M. Sato, T. Kitamura, and Y. Umezawa. 2003. A genetic approach to identifying mitochondrial proteins. *Nature Biotech.* 21:287–293.

Ozawa, T., and Y. Umezawa. 2001. Detection of protein-protein interactions in vivo based on protein splicing. *Current Opinion in Chemical Biology; Section Analytical Techniques* 5:578–583.

Sato, M., T. Ozawa, K. Inukai, T. Asano, and Y. Umezawa. 2002. Fluorescent indicators for imaging protein phosphorylation in single living cells. *Nature Biotech.* 20:287–294.

Sato, M., Y. Ueda, T. Takagi, and Y. Umezawa. 2003. Production of $PtdInsP_3$ at endomembranes is triggered by receptor endocytosis, *Nature Cell Biology* 5 (11):1016–1022.

Sugawara, M., A. Hirano, B. Bühlmann, and Y. Umezawa. 2002. Design and application of ion-channel sensors based on biological and artificial receptors. *Bull. Chem. Soc. Jpn (Accounts)* 75 (2):187–201.

Umezawa, Y., T. Ozawa, and M. Sato. 2002a. Assay and screening methods for chemicals that disrupt cellular signaling pathways. Risk assessment for potential endocrine disruptors. *Environmental Sciences* 9 (1):23–35.

Umezawa, Y., T. Ozawa, and M. Sato. 2002b. Methods of analysis for chemicals that promote/disrupt cellular signaling. *Anal. Sci.* 18:503–516.

Umezawa, Y., T. Ozawa, and M. Sato. 2002c. Probing chemical processes in living cells: Application for assay and screening of chemicals that disrupt cellular signaling pathways. *Bull. Chem. Soc. Jpn. (Accounts)* 75 (7):1423–1433.

Prof. Umezawa also has patents on the synthesis of the constructs.

Site: **University of Tokyo**
School of Pharmaceutical Sciences
7-3-1 Hongo, Bunkyo-Ku
Tokyo 113-0033, Japan

Date Visited: 29 January 2003

WTEC Attendees: D. Walt (report author), J. Schultz; D. Brady; C. Wilkins; H. Ali

Hosts: Dr. Kazuya Kikuchi, Associate Professor, Graduate School of Pharmaceutical Sciences,
Tel: +81-3-5841-4853: Fax: +81-3-5841-4855;
Email: kkikuchi@mol.f.u-tokyo.ac.jp

OVERVIEW

The Kikuchi laboratory is primarily an organic chemistry laboratory. The group is employing photoinduced electron transfer (PET) to design new sensing materials. The primary goal is to develop soluble indicators for various solution species. One of the goals of the work is to visualize intracellular dynamic events. There are significant limitations with existing Zn^{2+} indicators, including their excitation wavelength in the UV range and their pH sensitivity.

A Zn^{2+} fluorescent sensor molecule has been developed based on aminofluoresceins that coordinate to Zn^{2+} and relieve the PET effect. The resulting indicators have nanomolar sensitivity and are not pH sensitive. The indicators are modified with an ester that is hydrolyzed once it permeates the plasma membrane and becomes trapped within the cell. Zn^{2+} release has been measured within hippocampal cells . Inhibition of NMDA receptors can be observed. Ratiometric Zn^{2+} indicators have also been developed. The final application for Zn^{2+} sensing is for in vivo diagnostics. A Gd complex with a Zn^{2+} binding site has been prepared that has been used for in vitro studies of Zn^{2+} binding using NMR. The eventual goal is to use this system for in vivo cellular imaging.

A second project is ratiometric imaging of hydrolytic enzymes.

Another project discussed involves a FRET cleavage reaction in which donor and acceptor fluorophores are connected by a reactive chain. For example, a phosphodiester linker can be used to assay for the activity of phophoesterases. When cleavage occurs, fluorescence of the donor is recovered. A more direct assay is to attach a masking group that eliminates fluorescence from a chromophore. When the masking group is cleaved, the fluorescence is recovered. For example, a coumarin can be modified with a phosphate group, and when the phosphate is cleaved by a phosphatase, a fluorescent signal returns.

The last project discussed was CALI (chromophore assisted laser inactivation). In this approach, an antibody is conjugated to a chromophore. Light is used to activate the chromophore, which destroys the target species, such as tissue or a specific molecule. The Kikuchi laboratory is studying the IP3 pathway.

ASSESSMENT

The Kikuchi laboratory is performing strategic molecular design coupled with first-rate organic synthesis to study important biological problems. Dr. Kikuchi has strong ties to U.S. researchers and publishes extensively in first-tier international journals.

REFERENCES

Hanaoka, K., K. Kikuchi, Y. Urano, M. Narazaki, T. Yokawa, S. Sakamoto, K. Yamaguchi, and T. Nagano. 2002. Design and synthesis of a novel magnetic resonance imaging contrast agent for selective sensing of zinc ion. *Chemistry and Biology* 9:1027–1032.

Hirano, T., K. Kikuchi, Y. Urano, T. Higuchi, and T. Nagano. 2000. Highly zinc-selective fluorescent sensor molecules suitable for biological applications. *J. Am. Chem. Soc.* 122:12399–12400.

Hirano, T., K. Kikuchi, Y. Urano, T. Higuchi, and T. Nagano. 2000. Novel zinc fluorescent probes excitable with visible light for biological applications. *Communications, Angew. Chem. Int. Ed.* 39 (6):1052–1054.

Inoue, T., K. Kikuchi, K. Hirose, M. Iino, and T. Nagano. 2001. Small molecule-based laser inactivation of inositol 1,4,5-trisphosphate receptor. *Chemistry and Biology* 8:9–15.

Kojima, H., Y. Urano, K. Kikuchi, T. Higuchi, Y. Hirata, and T. Nagano. 1999. Fluorescent indicators for imaging nitric oxide production. *Communications, Angew. Chem. Int. Ed.* 38 (21):3209–3212.

Maruyama, S., K. Kikuchi, T. Hirano, Y. Urano, and T. Nagano. 2002. A novel, cell-permeable, fluorescent probe for ratiometric imaging of zinc ion. *JACS Communications, J. Am. Chem. Soc.* 124: 0650–10651.

Mizukami, S., T. Nagano, Y. Urano, A. Odani, and K. Kikuchi. 2002. A fluorescent anion sensor that works in neutral aqueous solution for bioanalytical application. *JACS Articles, J. Am. Chem. Soc.* 124:3920–3925.

Takakusa, H., K. Kikuchi, Y. Urano, S. Sakamoto, K. Yamaguchi, and T. Nagano. 2002. Design and synthesis of an enzyme-cleavable sensor molecule for phosphodiesterase activity based on flourescence resonance energy transfer. *JACS Articles, J. Am. Chem. Soc.* 124 (8):1653–1657.

APPENDIX D. NIH GRANTS RELATED TO BIOSENSING, CALENDAR YEAR 2002

Grant Number	Principal Investigator	Project Title	Institution
R43AA014115-01	Subramanian, Kumar	MEMS Based Continuous Alcohol Monitoring Device	Phoenix Biosystems
R43AA014116-01	Mo, Jianwei	Minimally Invasive Microsensor for Blood Alcohol Assay	Kumetrix, Inc.
R43AA014118-01	Tempelman, Linda A.	Wireless, Low-Maintenance Transdermal Alcohol Sensor	Giner, Inc.
P30AI028691-14	Kolodner, Richard D.	Core–Molecular Biology Facility	Dana-Farber Cancer Inst.
R01AI047427-02	Tender, Leonard M.	Biosensor for Investigating a Developing Immune Response	U.S. Naval Research Laboratory
R37AI014910-23	Huber, Brigitte T.	B Lymphocytes – Differentiation and Triggering	Tufts University Boston
R41AI052747-01	Saldivar, Enrique N.	Prototype Fabrication of a Cell Migration Sensor	Rainmaker Technology
R43AI049606-02	Israel, Barbara A.	Biophotonics for Detection of West Nile Virus	Platypus Technologies, LLC
R43AI050304-02	Clarke, Jean M.	Novel Biosensor for Detecting Antibiotic Resistance	Nomadics, Inc.
R43AI051772-01	Wavering, Thomas A.	Micromachined Biosensor for Mycobacterial Pathogens	Luna Innovations, Inc.
R43AI052533-01	Niu, Chunming	Biomolecule-Gated Nanowire FET Sensors	Nanosys, Inc.
R43AI052980-01	Wiesmann, William P.	Bead Based Immuno-PCR for Biowarfare Agent Detection	Biostar, Inc.
R43AI053003-01	Spangler, Charles W.	Multifunctional Dendritic Tethers for Biosensor Devices	Mpa Technologies, Inc.
R43AI053032-01	Mosher, Curtis L.	AFM Sensors to Detect Biological Warfare Agents	Bioforce Nanosciences, Inc.

Grant Number	Principal Investigator	Project Title	Institution
R44AI043806-04	Sand, Theodore T.	Biosensor Detection of Water-Borne Cryptosporidium	Disan, Inc.
R44AI046866-02	Montagna, Richard A.	Microchip-Based Field Assay to Detect Dengue Virus	Innovative Biotechnologies International
U01AI053857-01	Kornguth, Steven E	Simultaneous Detection on Multiple Pathogenicity Islands	University of Texas Austin
Z01AI000528-15	Venkatesan, Sundararajan	Structure-Function Studies of Chemokine Receptors and Mo	
R01AR041729-07	Hamilton, Susan L.	Structural Analysis of the Ca++ Release Channel	Baylor College of Medicine
R01AR041802-09	Hamilton, Susan L.	Modulation of Sarcoplasmic Reticulum Calcium Release	Baylor College of Medicine
R01AR048544-01	Fertala, Andrzej	Site Specific Interactions and Collagen Self Assembly	Thomas Jefferson University
R01AT000212-02	Sloane, Philip D.	High Intensity Light Therapy in Alzheimer's Disease	University of North Carolina Chapel Hill
Z01BO003013-01	Kozlowski, Steven	Biosensor Sensitivity	
P01CA049210-13	Jankowiak, Ryszard	Advanced Biomonitoring Techniques for Carcinogenesis	University of Nebraska Medical Center
P01CA078039-05	Taylor, D. L.	High Content Screening of Anticancer Lead Compounds	University of Pittsburgh at Pittsburgh
P01CA091597-07A2	Clarkson, Robert B.	Carbon Based Sensors for in Vivo EPR Oximetry	Dartmouth College
P01CA091597-07A2	Gallez, Bernard	Coating of Paramagnetic Oxygen Sensitive Compounds	Dartmouth College
P30CA010815-34S2	Speicher, David W.	Core – Protein Microchemistry/Mass Spectrometry Facility	Wistar Institute
P30CA010815-34S3	Speicher, David W.	Core – Protein Microchemistry/Mass Spectrometry Facility	Wistar Institute
P30CA016520-27	Chaiken, Irwin M.	Core – Biosensor/Interaction Analysis Facility	University of Pennsylvania

Grant Number	Principal Investigator	Project Title	Institution
P30CA042014-15	Myszka, David G.	Core–Protein Interaction	University of Utah
R03CA089705-02	Luck, Linda A.	Estrogenic Substance Detection By a Modified Nanobalance	Clarkson University
R21CA092581-01A1	Tan, Weihong	Molecular Beacon Aptamer for Diagnostic Cancer Imaging	University of Florida
R21CA097945-01	Kelley, Shana O.	Detection of H. Pylori Using Electrical DNA Sensing	Boston College
R33CA083229-04	Meyer, Tobias	Cell Arrays for Screening Signal Transduction Processes	Stanford University
R43CA092796-02	Shen, Shanxiang	Microcantilever Array Device for Protein Profiling	Protiveris, Inc.
R43CA094430-01	Tang, Cha-Mei	Sensitive, Integrating Multi-Waveguide Biosensor	Creatv Microtech, Inc.
R43CA097569-01	Szmacinski, Henry K.	Metallic Nanosensor Matrix With Enhanced Fluorescence	Microcosm, Inc.
R44CA082079-03A1	Nelson, Randall W.	Biosensor-Chip Mass Spectrometry	Intrinsic Bioprobes, Inc.
U19CA052995-14	Lazo, John S.	Core–Cell Fluorescent Biosensor	University of Arizona
R44CI000069-03	Tabb, Joel S.	Rapid Diagnostic Biosensor for Foodborne Pathogens	Agave Biosystems
R21DA014944-02	Gerhardt, Greg A.	Neurochem Chip: Study of Neurotransmitter Release	University of Kentucky
F32DC005580-01A1	Ault, Addison D.	Modeling Olfaction in Yeast	Princeton University
R01DC004712-02	Lewis, Nathan S.	Biomedical Application of an Electronic Nose	California Institute of Technology
R01DC006201-01	Wall, Conrad	Motion Sensor Array for Vestibular-Deficient Individuals	Massachusetts Eye and Ear Infirmary
R44DC004261-03	Hatt, Brian W.	Multielectrode Arrays for Olfactory Investigations	Bionic Technologies, Inc.
R44DC004261-04	Hatt, Brian W.	Multielectrode Arrays for Olfactory Invest.	Cyberkinetics, Inc.
U01DE014950-01	Walt, David R.	Microsensor Arrays for Saliva Diagnostics	Tufts University Medford

Grant Number	Principal Investigator	Project Title	Institution
U01DE015017-01	Anslyn, Eric V.	Saliva Analysis With an Array Sensor	University of Texas Austin
U01DE015018-01	Wong, David T.	UCLA Collaborative Oral Fluid Diagnostic Research Center	University of California Los Angeles
K25DK002925-01A1	Olesberg, Jonathon T.	On-Line, Near-Infrared Urea Sensor for Hemodialysis	University of Iowa
P01DK043881-09	Evan, Andrew P.	Effect of Shock Wave Lithotripsy on Renal Function/Structure in the Pig	Indiana Univ.–Purdue Univ. at Indianapolis
R01DK046960-09	Kennedy, Robert T.	Design and Use of Methods for Peptide Secretion Studies	University of Florida
R01DK046960-10	Kennedy, Robert T.	Design and Use of Methods for Peptide Secretion Studies	University of Michigan at Ann Arbor
R01DK054932-04	Reichert, William M.	Biosensor Biocompatibility	Duke University
R01DK054932-04S1	Reichert, William M.	Biosensor Biocompatibility	Duke University
R01DK057210-03	Rebrin, Kerstin	Support System for Subcutaneous Insulin Delivery By Pump	Medtronic Minimed
R01DK057284-03	Birder, Lori A.	Role of Nitric Oxide in Interstitial Cystitis	University of Pittsburgh at Pittsburgh
R01DK057583-02	Bradbury, Neil A.	Mechanisms of CFTR Internalization	University of Pittsburgh at Pittsburgh
R01DK058839-02	Mahvi, David M.	Hepatic Rf Ablation: Development of Effective Devices	University of Wisconsin Madison
R01DK059063-01A1	Ward, W. K.	Assessment of a Chronic Subcutaneous Glucose Sensor	Legacy Health System
R01DK060369-02	Farber, Steven A.	in Vivo Biosensor Screen for Mutants in Lipid Metabolism	Thomas Jefferson University
R01DK060369-02S1	Farber, Steven A.	in Vivo Biosensor Screen for Mutants in Lipid Metabolism	Thomas Jefferson University

Grant Number	Principal Investigator	Project Title	Institution
R01DK060369-01S2	Farber, Steven A.	in Vivo Biosensor Screen for Mutants in Lipid Metabolism	Thomas Jefferson University
R01DK060770-01	Bjorkman, Pamela J.	HFE/Transferrin Receptor/Transferrin Interactions	California Institute of Technology
R01DK063493-01	Philipson, Louis H.	Imaging Beta Cell Function With Biosensors	University of Chicago
R01DK064567-01	Steil, Garry M.	Long Term Glucose Sensing & Physiologic Insulin Delivery	Medtronic Minimed
R01DK064569-01	Arnold, Mark A.	Continuous Near Infrared Glucose Sensor	University of Iowa
R03DK062152-01	Boozer, Carol N.	Free-Living Physical Activity and Energy Expenditure	St. Luke's-Roosevelt Inst. for Health Sciences
R15DK061316-01	Hu, Jun	New Method for Creating Polymer Encapsulated Nanosensors	University of Akron
R43DK059690-01A1	Ghanem, Abdel-Halim	Improved Non-Invasive Blood Glucose Monitoring Device	Aciont, Inc.
R43DK060308-01A1	Ghanem, Abdel-Halim	Non-Invasive Blood Phenylalanine Monitor	Aciont, Inc.
R43DK061117-01	Fernandez, Salvador M.	Surface Plasmon Resonance Protein Array Phenotyping	Ciencia, Inc.
R44DK056544-03	Tierney, Michael J.	High Performance Biosensor Electrode Materials	Cygnus, Inc.
R44DK057347-02	Wolf, David E.	Optimization of Kinetics in a Novel Glucose Sensor	Sensor Technologies, Inc.
R01EB000127-01	Churchill, Bernard M.	Uropathogen Detection Using DNA Biosensors	University of California Los Angeles
R01EB000205-01	Daugherty, Patrick S.	Combinatorial Optimization of Protein Biosensors	University of California Santa Barbara
R01EB000433-01A1	McKnight, Timothy E.	Nano Arrays for Real-Time Probing Within Living Cells	UT–Battelle, LLC–Oak Ridge National Lab
R01EB000657-01	Doktycz, Mitchel J.	Nanosensing and Actuation Using Cell Mimetics	UT–Battelle, LLC–Oak Ridge National Lab

Grant Number	Principal Investigator	Project Title	Institution
R01EB000660-01	Szivek, John A.	Sensate Scaffolds for Orthopaedic Tissue Repair	University of Arizona
R01EB000675-01	Stojanovic, Milan N.	DNA-Based Arrays of Cross-Reactive Molecular Sensors	Columbia University Health Sciences
R01EB000682-01	Lakowicz, Joseph R.	Metal-Enhanced Fluorescence Sensing	University of Maryland Baltimore Prof. School
R01EB000708-01	Shcoenfisch, Mark H	Nitric Oxide-Releasing Glucose Biosensors	University of North Carolina Chapel Hill
R01EB000720-01	Shih, Wan Y.	Quantitative Array Piezoelectric Microcantilever Sensors	Drexel University
R01EB000726-01	Dandy, David S.	Multianalyte Physiological Optical Waveguide Sensing	Colorado State University
R01EB000734-01	Ward, W. K.	Passivating Proteins in Implantable Glucose Sensors	Emanuel Hospital and Health Center
R01EB000739-01	McShane, Michael J.	Fluorescent Glucose Sensors From Polyion Microshells	Louisiana Technological University
R01EB000741-01	Auner, Gregory W.	Novel Acoustic Sensor Arrays for Biomedical Applications	Wayne State University
R01EB000763-01	Esenaliev, Rinat O.	Novel Sensor for Measurement of Blood Oxygenation	University of Texas Medical Br Galveston
R01EB000782-11A1	Ruzicka, Jaromir	Flow Injection Cytometry in Analytical Biology	University of Washington
R01EB000783-05	Meyerhoff, Mark E.	Biocompatible Chemical Sensors Via Nitric Oxide Release	University of Michigan at Ann Arbor
R01EB000784-21	Meyerhoff, Mark E.	Polymer Membrane Ion/Polyion Sensors: New Frontiers	University of Michigan at Ann Arbor
R01EB000823-01	Frangos, John A.	Applications With Industrial Partners	La Jolla Bioengineering Institute
R21EB000481-01	Blair, Steven M.	Exploration of Nanoparticle Optical Bisensor Arrays	University of Utah

Grant Number	Principal Investigator	Project Title	Institution
R21EB000672-01	Zeng, Xiangqun	Engineered Self-Assembling FVS for Piezoimmunosensors	Oakland University
R21EB000735-01	Przybycien, Todd M.	A MEMS Membrane-Based Gravimetric Biosensor	Carnegie-Mellon University
R21EB000767-01	Colton, Jonathan S.	Plastic Microcantilevers for Biodetection	Georgia Inst. of Technology
R21EB000778-01	Bashir, Rashid	Micromechanical Sensors for Virus Detection	Purdue University West Lafayette
R21EB000807-01	Swanson, Basil I.	Optical Biosensor for the Early Detection of Cancer	University of California–Los Alamos National Lab
R21EB000982-01	Bashir, Rashid	Rapid Determination of Viability of Anthrax Spores	Purdue University West Lafayette
P42ES004699-16	Kennedy, Ian M.	Rapid Miniaturized Sensors for the Detection of Environmental Toxins	University of California Davis
P42ES004699-16S1	Kennedy, Ian M.	Rapid Miniaturized Sensors for the Detection of Environmental Toxins	University of California Davis
P42ES007380-06	Daunert, Sylvia	Sensing Superfund Chemicals With Recombinant Systems	University of Kentucky
R43ES010920-02	Gatewood, Joe M.	Nanotechnology-Based DNA Sequencing Instrumentation	Seirad, Inc.
R43ES011229-01A1	Ehret, Anne	Low Cost, Single Use Sensor for Human Exposure to VOCS	Chemmotif, Inc.
R43ES011469-01	Yang, Cathy Z.	An in Vitro Robotic Assay for Estrogenic Activity	Certichem, Inc.
R43ES011684-01	Cantor, Hal C.	Personal Monitor to Detect Exposure to Toxic Agents	Advanced Sensor Technologies, Inc.
R43ES011702-01	Sarangapani, Shantha	A Novel Sensor for Total Mercury in Fish Tissue	Innovative Chemical /Environmental Tech
R43ES011882-01	Larkin, Patrick M.	Arrays to Measure Endocrine Disruption in Fish	Aquagene, Inc.

Grant Number	Principal Investigator	Project Title	Institution
R44ES010076-03	Erb, Judith L.	Biosensor Studies of Estrogenic Compounds With Her-A & B	IA, Inc.
R03EY014177-01	Von Wiegand, Thomas E.	Haptic Display of Space Through Portable Nav Aids	Sensimetrics Corporation
F31GM066386-01	Tubbs, Julie L.	Minority Predoctoral Fellowship Program	Scripps Research Institute
F32GM020510-03	Franz, Kathrine J.	Synthesis of Peptide-Based Luminescent Lanthanide Probes	Massachusetts Institute of Technology
F32GM020878-02	Kam, Lance C.	Cell Adhesion on Protein-Micropatterned Lipid Bilayers	Stanford University
F32GM066501-01	Clark, Matthew A.	Synthesis of a Fluorescent Sensor for Nitric Oxide	Massachusetts Institute of Technology
P01GM056550-06		Core C: Proteins and Interactions	University of Pennsylvania
P01GM066521-01	Hill, Christopher P.	Structural Biology of HIV Budding	University of Utah
R01GM035556-18	Goldstein, Byron B	Receptor Aggregation and Its Effects	University of California-Los Alamos National Lab
R01GM042618-11	Remington, Stephen J.	Biosensors and Dynamics of Green Fluorescent Protein	University of Oregon
R01GM043768-13	Gilmore, James R.	Assembly and Transfer of N-Linked Oligosaccharides	Univ. of Massachusetts Medical School Worcester
R01GM044842-11	Weber, Stephen G.	Sensitive and Selective Detection of Peptides	University of Pittsburgh at Pittsburgh
R01GM047372-06	Cramer, Steven M.	Low Molecular Weight Displacers for Protein Purification	Rensselaer Polytechnic Institute
R01GM047645-09	Satterlee, James D.	Structure and Dynamics of Heme Protein Active Sites	Washington State University
R01GM048400-07	Shea, Kenneth J.	Template Polymerization	University of California Irvine

Grant Number	Principal Investigator	Project Title	Institution
R01GM059716-03	Bakker, Eric	Improving the Detection Limit of Potentiometric Sensors	Auburn University at Auburn
R01GM060562-03	Nie, Shuming	Luminescent Quantum Dots as Biological Labels	Emory University
R01GM061077-03	Barton, Jacqueline K.	Electrochemical DNA-Based Sensors	California Institute of Technology
R01GM061358-02	Colton, Richard J.	Single Cell Detection and Analysis	U.S. Naval Research Laboratory
R01GM061789-02	Ellington, Andrew D	Ribozymes for Peptide- and Protein- Sensing Chip Arrays	University of Texas Austin
R01GM062836-02	Laue, Thomas M.	Analytical Ultracentrifugation for Complex Systems	University of New Hampshire
R01GM062958-02	Plaxco, Kevin W.	Bio-Optical Composites for Rapid Analyte Detection	University of California Santa Barbara
R01GM062998-02	Rotello, Vincent M.	Biomolecular Recognition Using Nanoparticle Receptors	University of Massachusetts Amherst
R01GM063702-01A1	Meyer, Tobias	Chemotactic Signal Transduction	Stanford University
R01GM065507-01	Robinson, Anne S.	Sensing and Analyzing Stress During Protein Expression	University of Delaware
R01GM066137-01	Tan, Weihong	Real-Time and Quantitative Determination of Genes	University of Florida
R01GM067244-01	Milewski, Paul	Transport and Heterogeneity in Surface Volume Reactions	University of Wisconsin Madison
R15GM057855-02	Heagy, Michael D.	Fluorescent Chemosensors for Carbohydrates	New Mexico Institute of Mining & Technology
R15GM065840-01	Rucker, Joseph B.	Biochemical Studies of Retroviral Receptor Pseudotypes	Villanova University
R25GM056931-05	Caple, G.	Conducting Polymers in Biomedicine	Northern Arizona Univ.
R43GM064898-01A1	Izenson, Michael G.	A Practical, Low-Cost Xenon Anesthesia Circuit	Creare, Inc.

Grant Number	Principal Investigator	Project Title	Institution
R43GM064924-01	Doranz, Benjamin J.	Viracore Pseudotype Production Optimization	Integral Molecular
R43GM064979-01	Oldenburg, Steven J.	Bioassays Capable of Detecting Single Molecules	Seashell Technology, LLC
R43GM065676-01	Vulfson, Evgeny N.	Novel Lithographic Protein-Bioreceptor Immobilization	Avatar Biotechnologies, LLC
R44GM056598-03	Campbell, Ellen R.	Production of Recombinant Nitrate Reductase in Pichia	Nitrate Elimination Company, Inc.
R44GM058342-03	Guire, Patrick E.	Photoreactive Self-Assembled Monolayers	Surmodics, Inc.
R44GM060884-02	Smith, Richard H.	Integrated Fiber Optic Sensor for DNA Hybridization	IA, Inc.
R44GM062100-02	Powell, Richard D.	Gold Quenched Molecular Beacons	Nanoprobes, Inc.
S06GM008047-29	Tachikawa, Hiroyasu	Conducting Polymer Enzyme Based Biosensor for No	Jackson State University
S06GM008101-31	Zhou, Feimeng	Characterization and Quantification of Immobilized DNA	California State University Los Angeles
S06GM008101-31S2	Zhou, Feimeng	Characterization and Quantification of Immobilized DNA	California State University Los Angeles
S06GM008102-30S1	Guadalupe, Ana R.	Reactivity & Energetics of Dehydrogenase Enzymes & Redox Mediators	University of Puerto Rico Rio Piedras
S06GM008102-31S2	Guadalupe, Ana R.	Reactivity & Energetics of Dehydrogenase Enzymes & Redox Mediators	University of Puerto Rico Rio Piedras
S06GM008102-31	Guadalupe, Ana R.	Reactivity & Energetics of Dehydrogenase Enzymes & Redox Mediators	University of Puerto Rico Rio Piedras
S06GM008102-31S1	Guadalupe, Ana R.	Reactivity & Energetics of Dehydrogenase Enzymes & Redox Mediators	University of Puerto Rico Rio Piedras
S06GM008102-30S1	Quinones, Edwin	Reactions Catalyzed By Enzymes Entrapped in Gel Glasses	University of Puerto Rico Rio Piedras
S06GM008102-31S2	Quinones, Edwin	Reactions Catalyzed By Enzymes Entrapped in Gel Glasses	University of Puerto Rico Rio Piedras

Grant Number	Principal Investigator	Project Title	Institution
S06GM008102-31S1	Quinones, Edwin	Reactions Catalyzed By Enzymes Entrapped in Gel Glasses	University of Puerto Rico Rio Piedras
S06GM008102-31	Quinones, Edwin	Reactions Catalyzed By Enzymes Entrapped in Gel Glasses	University of Puerto Rico Rio Piedras
S06GM008194-22S2	Gorski, Waldemar	Enzyme Electrodes Based on Chitosan Scaffoldings	University of Texas San Antonio
S06GM008194-23	Gorski, Waldemar	Enzyme Electrodes Based on Chitosan Scaffoldings	University of Texas San Antonio
S06GM008205-17	Tao, Nongjian	Electron Transfer/Biosensors–Immobilized Proteins	Florida International University
S06GM008205-17S3	Tao, Nongjian	Electron Transfer/Biosensors–Immobilized Proteins	Florida International University
S06GM008205-17S1	Tao, Nongjian	Electron Transfer/Biosensors–Immobilized Proteins	Florida International University
S06GM008247-15	Khan, Ishrat M.	Development of Methods for Synthesis of Polymers–Generation of Biomaterials	Clark Atlanta University
S06GM060654-03	Brazill, Derrick T.	Signal Transduction of Cell Density Sensing in Dictyostelium Discoideum	Hunter College
S06GM060654-03S1	Brazill, Derrick T.	Signal Transduction of Cell Density Sensing in Dictyostelium Discoideum	Hunter College
S06GM060654-03S3	Brazill, Derrick T.	Signal Transduction of Cell Density Sensing in Dictyostelium Discoideum	Hunter College
U54GM062114-02S1	Mayer, Tobias	Evanescent Wave Microscopy and Plasma Membrane Signals	University of Texas SW Medical Center Dallas
U54GM062114-02S2	Mayer, Tobias	Evanescent Wave Microscopy and Plasma Membrane Signals	University of Texas SW Medical Center Dallas
U54GM062114-03S1	Mayer, Tobias	Evanescent Wave Microscopy and Plasma Membrane Signals	University of Texas SW Medical Center Dallas
U54GM062114-03S2	Mayer, Tobias	Evanescent Wave Microscopy and Plasma Membrane Signals	University of Texas SW Medical Center Dallas

Grant Number	Principal Investigator	Project Title	Institution
U54GM062114-03	Mayer, Tobias	Evanescent Wave Microscopy and Plasma Membrane Signals	University of Texas SW Medical Center Dallas
R01HD039099-03	Loeb, Gerald E.	Injectable Sensors for Control of FES	Univ. Southern California
R43HD041853-01	Schmidt, Robert N.	Ultrathin Shear Force Sensor for Direction and Magnitude	Cleveland Medical Devices, Inc.
U10HD041906-02	Tamborlane, William V.	Yale Center in the Children's Glucose Sensor Network	Yale University
U10HD041908-02	Buckingham, Bruce A.	Near-Continuous Glucose Monitoring in Pediatrics	Stanford University
U10HD041919-02	Chase, Peter H.	Glucose Sensors in Children with Type I Diabetes	University of Colorado Health Sciences Center
F31HG002520-01	Gore, Mitchell R.	Minority Predoctoral Fellowship Program	University of North Carolina Chapel Hill
F32HG002463-02	Ginger, David S.	Nanoscale Devices on DNA Functionalized Semiconductors	Northwestern University
P01HG001984-03S1	Mastrangelo, Carlos H.	Advanced Fabrication and Sensor Development	University of Michigan at Ann Arbor
P01HG001984-03S1	Burke, David T.	Core–Fabrication, Assembly, and Testing Support	University of Michigan at Ann Arbor
P01HG001984-03S1	Burke, David T.	Micromechanical Integrated DNA Analysis Technology	University of Michigan at Ann Arbor
P01HL006296-42	Stull, James T.	Myosin Phosphorylation in Skeletal Muscle	University of Texas SW Medical Center Dallas
R01HL026043-22	Stull, James T.	Myosin Light Chain Kinase Function in Smooth Muscle	University of Texas SW Medical Center Dallas
R01HL032132-14	Herron, James N.	Multi-Analyte Waveguide Immunosensing	University of Utah
R01HL050676-08	Peterson, Cynthia B.	Macromolecular Interactions of Human Vitronectin	University of Tennessee Knoxville

Grant Number	Principal Investigator	Project Title	Institution
R01HL056143-06	Webster, John G.	Electrode Design for Cardiac Tachyarrythmia Rf Ablation	University of Wisconsin Madison
R01HL064038-03	Gilles-Gonzalez, Marie A.	Mutagenesis of Fixl, an O_2 Sensing PAS Domain Protein	Ohio State University
R01HL066147-01A1	Burstyn, Judith N.	COoa: A Hemoprotein CO Sensor	University of Wisconsin Madison
R01HL066315-03	Schumacker, Paul T.	O_2 Sensing By Mitochondria During Intermittent Hypoxia	University of Chicago
R43HL070360-01	Rotman, Boris	Real-Time Detection of Bacteria in Platelet Concentrates	BCR Diagnostics
R43HL070399-01	Orser, Cindy S.	A Catalytic Conformational Prion Sensor	Arete Associates
R43HL070463-01	Elson, Edward C.	Biosensor for Contaminant-Free Platelets	Opto-Gene, Inc.
R44HL062038-03	Sawatari, Takeo	Optical Pressure Sensor Built in Angioplasty Guidewire	Sentec Corporation
R44HL062777-02	Walker, James K.	Novel Mass Production Method for Optical Bio-Sensors	Nanoptics, Inc.
R01MH066199-01	Verselis, Vytautas	Biophysics of CNS Connexins	Yeshiva University
R01MH067531-01	Gallant, Jack L.	Neural and Metabolic Activity in Vision and Attention	University of California Berkeley
R21MH062444-02	Thakor, Nitish V.	Integrated Electrochemical Microsensor Array	Johns Hopkins University
R21MH063123-02	Michael, Adrian C.	Innovations for In Vivo Neurochemical Analysis	University of Pittsburgh at Pittsburgh
F32NS010998-03	Martin, Heidi B.	Diamond Microelectrodes for Neurotransmitter Detection	University of North Carolina Chapel Hill
P01NS030606-10	Eisenberg, Roselyn J.	Herpes Simplex Virus Entry Into Cells of Neural Origin	University of Pennsylvania
P50NS038367-04	Dunn, Bruce S.	Core–Neuroengineering	University of California Los Angeles

Grant Number	Principal Investigator	Project Title	Institution
P50NS038367-04S2	Dunn, Bruce S.	Core–Neuroengineering	University of California Los Angeles
P50NS038367-04S1	Dunn, Bruce S.	Core–Neuroengineering	University of California Los Angeles
R01NS028389-09	Cooper, Dermot M.	Intracellular Calcium Control of Camp Synthesis	University of Colorado Health Sciences Center
R01NS029549-10	Peckham, P. Hunter	Multichannel Implantable System for Neural Control	Case Western Reserve University
R01NS040547-03	Triolo, Ronald J.	Automatic Control of Standing Balance With FNS	Case Western Reserve University
R01NS040628-03	Woodward, Donald J.	Multichannel Sensors for Neurosciences	Wake Forest University Health Sciences
R43NS042953-01	Kanaan, Abed K.	Long-Term, Implantable, Intra-Cranial Pressure Sensor	Foster-Miller, Inc.
R44NS037608-03	Johnson, David A.	Wireless Biosensor Array for In-Vivo Monitoring	Pinnacle Technology, Inc.
R44NS039714-02	Cogan, Stuart F.	Model for Electrode Testing and Reduced Animal Use	EIC Laboratories, Inc.
U54NS045309-01	Breaker, Ronald	Engineering RNA Switches that Respond to Dopamine/Analog	University of Rochester
R43OH007673-01	Masterman, Michael F.	Bioelectronic Telemetry System for Firefighter Safety	Extreme Endeavors and Consulting
R44OH004174-02	Deininger, Debra J.	On-Board Diagnostic Sensor for Respirator Breakthrough	Nanomaterials Research, LLC
M01RR000052-41	Saudek, Michael	Clinical Research Toward Closed Loop Insulin Delivery	Johns Hopkins University
M01RR000069-40	Townsend, Susan	Comparison of Two Monitors in Detection of Infantile Apnea	University of Colorado Health Sciences Center
M01RR000633-30	Raskin, Philip	Clinical Evaluation of Glucose Microelectrodes	University of Texas SW Medical Center Dallas

Grant Number	Principal Investigator	Project Title	Institution
M01RR000997-27	Eaton, R. P.	Development and Evaluation of a Noninvasive Glucose Sensor	University of New Mexico Albuquerque
M01RR001032-27	Laham, Roger J.	Biosense DMR Safety & Feasibility Investigational Trial	Beth Israel Deaconess Medical Center
R01RR016230-01	Westbrook, Edwin M.	A Micromachined Silicon Crystallographic X-Ray Detector	Molecular Biology Consortium
R01RR016334-01	Westbrook, Edwin M.	A CCD Crystallographic X-Ray Detector With Lens Optics	Molecular Biology Consortium
R21RR017329-01	Andrade, Joseph D.	Multi-Analyte Micro-Devices for Biomedical Applications	University of Utah
R21RR017414-01	Weiss, Shimon	High Performance Photon-Counting Imager	University of California Los Angeles
R21RR017420-01	Larson, Dale N.	Development of an Unlabeled Macromolecule Detector	Harvard University (Medical School)
R43RR016832-01	Doranz, Benjamin J.	Viracore Biosensor Optimization	Integral Molecular
R44RR014385-02A2	Asanov, Alexander N.	Fluorescence System for Sensing Biospecific Interactions	Bioelectrospec, Inc.
S10RR015885-01A1	Kane, William H.	Biacore 3000 Biosensor	Duke University
S10RR016787-01	Myszka, David G.	Biacore S51 Optical Biosensor	University of Utah

APPENDIX E. NSF-SPONSORED PROJECTS RELATED TO BIOSENSING, CY2002

Investigator CPIC=Co-Principal Investigator Current	**Project Title**
Doug Schulz	SBIR Phase I: CdSe Nanoparticle/Metal-Organic Inks for Printable Electronics
Jiri Janata	The New Challenges of Chemical and Biological Sensing (Workshop: January 9-10, 2002)
Todd A. McAdams	SBIR Phase II: Clinical-Scale Suspension Bioreactor for Primary Hematopoietic Culture
Daniel T. Chiu	CAREER: Elucidation of Surface Affects on Biochemical Reactions
Wei Chen	SBIR Phase II: Nanoparticle Photostimulated Luminescence Based Optical Storage
Jean'ne M. Shreeve	Idaho EPSCoR Research Infrastructure Improvement Grant
F. Peter Schloerb Rafael Millan-Gabet (CPIC)	Research on Infrared Imaging with IOTA
W. Ronald Fawcett	Molecular Level Effects for Simple Electrode Reactions at Well Defined Polarizable Electrodes
Joseph M. DeSimone	US-Turkey Cooperative Research: Processing for Sub-Micron Imaging in Supercritical CO2: An Integrated Approach to the Deposition and Development of Photoresists
Hector D. Abruna	US-Spain Cooperative Research: Designed Interfacial Assembly of Redox Active Enzymes and Recognition Layers for Biosensor Applications
Steven M. Blair	CAREER: Integrated-Optic Nanoparticle Biosensor Arrays
Eniko T. Enikov	CAREER: Optically Transparent Gripper for Microassembly
Richard L. Collins Mark Conde (CPIC)	CEDAR: Ground-Based Optical Imaging of Sporadic Sodium Clouds Near the Summer Mesopause, Using Resonantly Scattered Sunlight
Audra M. Bullock	CAREER: Improvement and Integration of Laser-based Sensors for Advanced Situational Awareness: A Combined Research and Education Program for Students in Lasers and Optics
Terry E. Whitledge	SGER: Installation of Biochemical/Optical Sensors on Shelf-Basin Interactions (SBI) Moorings

Investigator CPIC=Co-Principal Investigator Current	**Project Title**
Brian R. Crane	CAREER: Correlating Metalloenzyme Structure with Reactivity By Tunneling Electrons in Crystals
Luis Echegoyen	Fullerenes: Reactivity, Supramolecular Interactions, and Devices
Christopher K. Mathews	Protein-Protein Interactions in DNA Precursor Biosynthesis
Mary Jo Ondrechen	THEMATICS: Development and Application of a New Computational Tool for Functional Genomics
Jiri Janata Miroslawa Josowicz (CPIC)	Design of Advanced Sensing Materials
Omowunmi A. Sadik Walker Land (CPIC)	SGER: Molecular Design of Intelligent Sensors for Selected Chemical Warfare Agents using Support Vector Machines
George W. Luther	Deployable In Situ Electrochemical Analyzer (ISEA) for Remote and Automatic Analysis of O_2, H_2S and Sulfur Species in Hydrothermal Vent Environments
Robert M. Corn	Fabrication of Biopolymer Microarrays for SPR Imaging Measurements
Janice E. Reutt-Robey	Formation and Structure-Property Relationships of Molecular Surface Architectures
Daniel Forciniti	Molecular Visualization and Modeling of Proteins at Interfaces
G. Charles Dismukes	Instrumentation for Ultra-Sensitive Detection of Oxygen and Fluorescence in Photosynthetic Bio-Materials
Masoud Ghandehari	Optical Chemo-Sensing for Civil and Mechanical Systems, An Interdisciplinary Exploratory Research Project
Challa V. Kumar	US-India Cooperative Research: Enzyme-Inorganic Materials: Peroxidase Behavior at Selected Interfaces
Gary J. Kirkpatrick	Development of Nested, Autonomous Phytoplankton Monitoring Technology
Peter Saggau	Adaptive Resolution Microscope for Fast Structural and Functional Optical Imaging
Rahmatallah A. Shoureshi Fu-Kuo Chang (CPIC) Darryll J. Pines (CPIC)	Creation of National Forum for Synergistic Program Development in Smart Structures and Sensor Technologies
Valerie J. Leppert	ADVANCE Fellow: Microscopy of Nanomaterials
Luke J. Mawst	Two-Dimensional Leaky-Mode VCSEL Arrays: Active Photonic Lattices

Investigator CPIC=Co-Principal Investigator Current	**Project Title**
Cheng S. Lee	Plastic Microfluidics-Based 2-D PAGE
Charles A. Schmuttenmaer	Terahertz Studies of Transient Photoconductivity in Quantum Dots and Electron Transfer in Bacterial Reaction Centers
C. Michael Elliott	Electrochemically Active Polymers - Designing Compositional Structures and Electronic Properties
Tianquan Lian	Femtosecond IR Probe of Ultrafast Dynamics of Molecular Adsorbates on Nanoparticles: Solvation and Electron Transfer
Paul A. Garris George Vincent Rebec (CPIC)	Real-Time Animal Telemetry
Xi-Cheng Zhang	Development of T-Ray Microscope
Sudipta Seal Lucille A. Giannuzzi (CPIC)	NSF REU Site in Nanomaterials Processing and Characterization (NANOPAC-SITE)
Deron A. Walters	CAREER: Enabling Nanoscale Science and Devices Using Selective Biomolecular-inorganic Interactions
Marshall I. Nathan P. P. Ruden (CPIC)	Uniaxial and Hydrostatic Stress on Group III-nitride Heterojunctions and Schottky Barriers
Zoe G. Cardon Francis Moussy (CPIC)	SGER: Developing a New Miniaturized Sensor for Detecting Glucose in Soil
Karen J. Burg Martine LaBerge (CPIC)	Innovations in Biomaterials
David A. Spivak	CAREER: Development of Polymerizable Diacetylene Surfactant Monomers for Two-Dimensional Imprinting and Sensors
Scott T. Sanders Xiaochun Li (CPIC)	Acquisition of Fiber-Optic Instrumentation for Innovative Spectroscopic Light Source to Advance Sensing Capabilities in Research and Education
Kevin J. Webb Andrew M. Weiner (CPIC)	Characterization of Scattering Media Using Vector Speckle and High-Order Speckle Correlations
Sankar Das Sarma Igor Zutic (CPIC)	Spin Electronics
Rebecca Richards-Kortum Carlyle B. Storm (CPIC)	Lasers in Medicine and Biology Gordon Conference - July 14–19, 2002 at Kimball Union Academy in Meriden, NH
Kirk V. Cammarata Joanna Mott (CPIC) Gregory W. Buck (CPIC) Patrick D. Larkin (CPIC) Lillian S. Waldbese	MRI/RUI: Acquisition of a Digital Imaging System to Support Research and Research Training in Applications of Molecular Biology

Investigator CPIC=Co-Principal Investigator Current	**Project Title**
Susan R. Stapleton	Oxidative Stress and G6PDH Expression
Louis A. Lyon	Stimuli-Sensitive Core/Shell Microgels
Seth D. Silverstein Yibin Zheng (CPIC)	Adaptive Digital Signal Processing for Spatial and Temporal Sampled Coherent Imaging Systems
Andrew M. Weiner Peter J. Miller (CPIC)	GOALI: Polarization Mode Dispersion Compensation in the Spectral Domain Using Liquid Crystal Modulator Arrays
Wolfgang Porod Gary H. Bernstein (CPIC)	NER: Computing Architectures for Coupled Nanomagnets
Asit K. Ray	Development of Optical Sensors Based on Spherical Microparticles
Yue Wu	U.S.-Korea Cooperative Research: Inorganic Nanostructure/Dye Molecules Hybrid Systems Studied by Nuclear Magnetic Resonance
Jeanne L. McHale	Spectroscopic Investigations of Interfacial Electron Transfer and Chromophore Aggregation
Navin Khaneja	CAREER: Optimal Control of Quantum Systems
John A. Marohn	CAREER: Variable Temperature Electric Force and Magnetic Resonance Force Microscopy Studies of Organic Electronic Materials
Richard L. Smith	CAREER: Chiral Ceramic Sensors
David D. Nolte	High-Speed Multi-Analyte Biosensor Using Adaptive Laser Interferometry
Jeffrey R. Mackey	SBIR Phase I: Force Transducer Based on Phase-Modulated Optical Polarimetry
Andrei M. Shkel	Feasibility Study of Polymer-based MEMS Low-Frequency Sensing Technology for Health Monitoring of Civil Structures
Eliza Hutter	International Research Fellowship Program: Electric Field Effects on the Self-Assembly and Hybridization of Functionalized Oligonucleotides
Alissa Fitzgerald	SBIR Phase I: Investigation of Charge Trapping in Plasma Enhanced Chemical Vapor Deposition (PECVD) Dielectrics Using Electrostatically Actuated Mechanical Resonators
Steven Cordero	SBIR Phase I: Optical Based Chemical Sensing Using Luminescent Nanomaterials
David D. Nelson	SBIR Phase I: Development of a High Precision, Autonomous Quantum Cascade Laser-Based Detector for Methane and Nitrous Oxide

Investigator CPIC=Co-Principal Investigator Current	**Project Title**
Wayne H. Richardson	SBIR Phase I: Software Tools for the Design of Nanoscale Electronic Devices and Circuits
Benaiah D. Schrag	SBIR Phase I: Scanning Magnetic Microscropy for Real-time Electromigration Imaging
Brian N. Strecker	SBIR Phase I: Microsphere-coupled Surface-enhanced Raman Spectroscopy Probe
Seong-Gi Baek	SBIR Phase I: An Innovative Normal Stress Sensor System for Complete Characterization of Polymer Shear Flow Properties
Albrecht Jander	SBIR Phase I: Delta-Sigma All-Digital Magnetometer
Peter-Patrick U. deGuzman	SBIR Phase I: Electrowetting Micro Optical Switch Array
Anuncia Gonzalez-Martin	SBIR Phase I: An Electrochemical Array-Based Nondestructive Evaluation System
Paul Shnitser	SBIR Phase I: Lobster-Eye X-Ray Imaging Sensor
Colleen M. Fitzpatrick	SBIR Phase I: Innovative Integrated Optical Circuit Fabrication and Processing Techniques
Jim Hang	SBIR Phase I: An Optical Sensor for Semiconductor Back-End Processes
Peggy Thompson	SBIR Phase I: Label-Free Biochip for Ultra-High Throughput Screening
Markus Erbeldinger	SBIR Phase I: Urea Sensing Biocatalytic Polymers
Jeff Lindemuth	SBIR Phase I: Improved Magneto-Optical Imaging Films Employing Surface Plasmon Resonance
Mourad Manoukian	SBIR Phase I: Solid State Electrochemical Carbon Dioxide Sensor
Wei Shi	SBIR Phase I: A Novel Coherent and Tunable Terahertz (THz) Module for Chemical Identification
Ewa Heyduk	SBIR Phase I: DNA Binding Proteins as Biosensors
Amy J. Hunter	SBIR Phase I: Fast Response Sensor for Airborne Biological Particles
S Sriram	SBIR Phase I: Photonic Band Gap Optical Waveguide Structures in Electro-optic Substrates
Ting Chen	SBIR Phase I: Optical Switch Manufactured Using Direct Write Method
Xingtao Wu	SBIR Phase I: Development of a Hybrid Microelectromechanical (MEMS) Driven Tunable Optical Filter Technology

Investigator CPIC=Co-Principal Investigator Current	**Project Title**
Rafael Perez-Reisler	SBIR Phase I: Development of a Novel Droplet Multi-Sensor
Phillip B. Danielson Robert M. Dores (CPIC) Thomas W. Quinn (CPIC) James C. Fogleman (CPIC) Egbert Schwartz (CPIC)	A WAVE Nucleic Acid Fragment Analysis System for Research and Education
Jack M. McCarthy Jody L. House (CPIC) John L. Freeouf (CPIC) C. Neil Berglund (CPIC)	NER: Massively Parallel Electron Photoemitter Micro-arrays for Nanoscale Lithography, Imaging, and Inspection Applications
Jun Hu Stephanie T. Lopina (CPIC)	NER: Bioengineering of Implantable Nanosize Optical Sensing Elements by Controlled Radical Polymerization
Robert Kurt Elaine R. Reynolds (CPIC) Shyamal K. Majumdar (CPIC)	MRI-RUI: Acquisition of the C1 Confocal Microscopy System for Cellular Analysis in the Biological Sciences
Salvador M. Fernandez	SBIR Phase I: Biosensor for Label-Free, Real-Time Monitoring of Environmental Pathogens
Vladimir M. Shalaev Alexander Wei (CPIC) Andrew M. Weiner (CPIC) Michael R. Melloch (CPIC)	NIRT: Plasmonic Nanophotonics and Optoelectronics
Sindee L. Simon Shubhra Gangopadhyay (CPIC)	NER: Supercritical Carbon Dioxide Extraction Process for Forming Nanoporous Materials for Low-k and Biosensor Applications
Vincent J. Fratello	SBIR Phase I: Liquid Phase Epitaxy of Potassium Tantalum Niobate on Low Dielectric Constant Substrates
Stephen R. Leone	Infrared Band-Specific Near Field Optical Microscopy Probing of Chemically Amplified Polymer Photoresists
Manish Gupta	SBIR Phase I: Cavity-Enhanced Capillary Electrophoresis
Doug Schulz Bruce Bishop (Princ. Invest. former)	SBIR Phase I: High-Temperature Gas Sensors with Enhanced Stability
Manuel Gamero	SBIR Phase I: Optical Detection and Sizing of Aerosol Nanoparticles (diameter detection limit below 2 nanometers)

Investigator CPIC=Co-Principal Investigator Current	**Project Title**
Margaret E. Kosal	SBIR Phase I: Colorimetric Sensor for Real-Time Detection of Nitroaromatic Explosives
Charles D. Pennington Shufang Luo (Princ. Invest. former)	STTR Phase I: Novel Lipid Deposition for Biosensor Surfaces
Nasser Peyghambarian Bruce S. Dunn (CPIC) Jeffrey I. Zink (CPIC) Ghassan E. Jabbour (CPIC) Michael R. Descour (CPIC)	NSF-EC Activity: Integration of New Hybrid Materials Containing Biomolecules for the Fabrication of Optical Sensor Systems
Thomas R. Kurfess Levent F. Degertekin (CPIC)	In-Line Optical Measurement of MicroElectroMechanical Systems (MEMS) Devices During Production
Pelagia Gouma pgouma@notes.cc.sunysb.edu	SGER: Bio-doped Electronic Ceramics for Use in Microsensors
Gregory Timp Klaus J. Schulten (CPIC) Alexey Bezryadin (CPIC) Jean-Pierre Leburton (CPIC)	NIRT: A Nanometer-Scale Gene Chip
Ann M. Anderson Mary K. Carroll (CPIC) Richard D. Wilk (CPIC) Michael E. Hagerman (CPIC)	RUI/MRI: Acquisition of Equipment to Establish an Aerogel Fabrication, Characterization and Applications Laboratory
Vladimir M. Shalaev	SGER: Fractal Surface Enhanced Chemical & Biological Sensors
Cyrus R. Safinya Philip A. Pincus (CPIC)	Biomolecular Materials: Structure, Phase Behavior, and Interactions
Peter S. Ungar Alan C. Walker (CPIC) Christopher A. Brown (CPIC)	Acquisition of a white light confocal microscope for quantitative characterization of dental microwear surfaces.
Ananth Dodabalapur John M. White (CPIC) Dim-Lee Kwong (CPIC) Micheal J. Krische (CPIC)	NIRT: Nanoscale Organic Circuits and Sensors
Jing Shi Zeev Valy Vardeny (CPIC)	NER: Nanoscale Molecular Spintronic Materials and Devices
Martin E. Huber	NER: High-Bandwidth Scanning DC SQUID Susceptometer for Characterization of Nanomagnetic Structures and Phenomena

Investigator CPIC=Co-Principal Investigator Current	**Project Title**
Levent F. Degertekin	NER: Acoustic Radiation Pressure Driven Atomic Force Microscope for Fast Imaging and Parallel Sensing of Biological and Chemical Processes at the Nanoscale
Stephen M. Wright Andrienne C. Friedli (CPIC) William M. Robertson (CPIC)	C-RUI: Development and Applications of a Novel Biosensor
Katharine Dovidenko	NER: Focused Ion Beam (FIB) Micromachining and Advanced Characterization of Carbon Nanotube-Metal Junctions
Hans D. Hallen	NER: Deposition of Molecular Nanostructures with Controlled in-plane Orientation
Bhubaneswar Mishra	Designer Molecules for Biosensor Applications
Keith E. Gubbins	NER: Molecular Modeling of Self-Assembled Nanostructures on Surfaces and in Narrow Pores
James W. Schneider	NER: Chemical Probing of Biosensor Nano-environments using Dynamic AFM
R. Fabian W. Pease	NIRT: Properties and Applications of Deformed Nanotubes
Seunghun Hong Peng Xiong (CPIC) Prescott B. Chase (CPIC) Stephan von Molnar (CPIC)	NIRT: Development, Functionalization, and Assembly of Nanoscale Biological Sensors
William D. Hunt	NER: Electron Beam Emitter SPR for Biosensor Applications Nanoscale Exploratory Research
Marc S. Levoy	ITR: High Performance Imaging Using an Array of Low-Cost Cameras
Lynn S. Penn Roderic P. Quirk (CPIC) Arthur W. Cammers-Goodwin (CPIC)	Design and Construction of Responsive Surfaces by Means of Tethered Chain Nanolayers
Dennis L. Matthews	Center for Biophotonics Science and Technology
Joseph Wang	Characterization of DNA-Linked Nanoparticle Networks for Advanced Genetic Testing
Erno Lindner	Current Polarized Ion-Selective Membranes for Enhanced Analytical Performances
Richard M. Crooks	Dendrimer-Encapsulated Metal Nanoparticles
Robert T. Kennedy	Affinity Interactions in Capillary Separations

Investigator CPIC=Co-Principal Investigator Current	**Project Title**
John F. Devlin	CAREER: Heterogeneous Reactions and Groundwater Flow in Reactive Porous Media
Amy J. Moll Harold D. Ackler (CPIC) William B. Knowlton (CPIC)	MRI: Acquisition of Materials Characterization Instrumentation
Robert W. Cohn Bruce W. Alphenaar (CPIC) Mahendra K. Sunkara (CPIC) Francis P. Zamborini (CPIC)	Acquisition of a Virtual Presence Surface Profiling Microscope for Nanomanipulation and Nanoassembly
Robert T. Kennedy	Affinity Interactions in Capillary Separations
John F. Rabolt D. Bruce Chase (CPIC)	Ultra-Fast Infrared Spectroscopy Using a Focal Plane Array for the Real Time Detection of Chemical and Biological Agents
Robert D. Grober James F. Cameron (CPIC)	Single Molecule Spectroscopic Imaging as an Optical Nanoprobe for Chemically Amplified Photoresists
Kevin K. Lehmann	An Optical Fiber Resonator for Cavity Ring-down Spectroscopic Detection and Measurement of Trace Species
Bassam A. Bamieh	SGER: Distributed Control of Capacitive Micro Cantilever Arrays
David C. Johnson	Acquisition of a Time-of-Flight SIMS System
Raghupathy Sivakumar Ian F. Akyildiz (CPIC)	Integrated Sensing: Communication Protocols and Testbed Development for Ad-Hoc Sensor Networks
Nathan S. Lewis	Achieving Molecular Level Control over the Chemical, Electrochemical, and Electrical Properties of Crystalline Si Surfaces
I. Charles Ume	Laser Ultrasound-Interferometric System for Packaged Electronic Devices Quality Evaluations
Yu Ding Feng Niu (CPIC)	Collaborative Research/GOALI: Analysis and Optimization Method for Distributed Sensor Systems in Electronics Assembly Processes Systems
Karsten Pohl	CAREER: Dynamics of Self-Assembly at Strained Metal Interfaces
Jia G. Lu	CAREER: Single Spin Transistors - Science, Application and Education
Rahul Simha	MRI: Acquisition of Research Infrastructure for Distributed Sensor Applications in the Home of the Future

Investigator CPIC=Co-Principal Investigator Current	**Project Title**
Ebtisam S. Wilkins Terry L. Yates (CPIC)	Miniaturized Portable Flow-Through Amperometric Immunosensor Device for Fast Field Analysis of Rodent Viruses
Ronald R. Hoy Elke Buschbeck (CPIC)	The Functional Organization and Evolution of a Novel Insect Visual System.
Gary C. Tepper	SGER: Biosensing in the Gas Phase: A New Approach Based on Imprinted Nanoparticles of a Linear Polymer
Mauricio Pereira da Cunha Paul Millard (CPIC)	SGER: Detection of Bioterrorism-Linked Microbial Pathogens Using Surface Acoustic Wave Liquid Sensors
Robin Shandas	Integrated Sensing: Non-Invasive Ultrasound-Based Micro-Flow Imaging System for Biomedical Applications
Franco Maloberti Jin Liu (CPIC) Murat Torlak (CPIC) Andrea Fumagalli (CPIC)	Integrated Sensing: Generic Autonomous Platform for Sensor Systems
Orlin D. Velev Peter K. Kilpatrick (CPIC)	NER: Bioelectronic Interfacing of Living Cells via Self-Assembled Microwires
Yoram Bresler	Fast Algorithms for 3D Cone-Beam Tomography
Sergey B. Mirov	International Cooperative Study of Multiphonon Relaxation of Mid IR Transitions in Laser Crystals with Short Phonon Spectra
Peter J. Hesketh James L. Gole (CPIC) Zhiping Zhou (CPIC)	Integrated Sensing Porous Silicon Integrated Sensor Arrays
Gunter Luepke	SPIN ELECTRONICS: Band-Offset and Time-Resolved Nonlinear-Optical Studies of Magnetic Heterostructure Interfaces
N R. Aluru	ITR: Computational Prototyping of Micro-Electro-Fluidic-Mechanical Systems
Gabriel P. Lopez Steven R. Brueck (CPIC)	Fluorescence Lifetime-Based Measurements of Biosensor Arrays Using Closed Loop Auto-Oscillating Systems
Sara M. Lindsay Paul D. Rawson (CPIC)	Linking Bioturbation and Sensory Biology: Chemoreception Mechanisms in Deposit-Feeding Polychaetes
Thomas L. Martin Mark T. Jones (CPIC)	ITR: Tailor-Made: Design of e-Textile Architectures for Wearable Computing
Wendell Lim	Engineering Protein-Based Logic Gates

Investigator CPIC=Co-Principal Investigator Current	**Project Title**
Milan N. Stojanovic Darko Stefanovic (CPIC)	Decision-Making Deoxyribozyme Networks
Jagannathan Sankar	Center for Advanced Materials and Smart Structures
Michael F. Rubner	MIT Materials Research Science and Engineering Center
Robert Reich	Collaborative Project: High-speed, Low-noise CCD Imaging Technology (Lincoln Project #10032).
Robert B. Barat Dale E. Gary (CPIC) John F. Federici (CPIC)	Terahertz Imaging System for Sensing of Chemical and Biological Agents
S. Michael Kilbey Scott M. Husson (CPIC) Richard V. Gregory (CPIC) Stephen E. Creager (CPIC)	Acquisition of a High-Speed, High-Sensitivity Ellipsometer for Materials Research and Education
Jaime F. Cardenas-Garcia	Development of a Bi-Axial Micro-Tensile Tester of MEMS Materials for Research and Student Training
Zeynep Celik-Butler	Micromachined Infrared Sensors on Flexible Substrates
Rebekka M. Wachter	Mechanism of Chromophore Formation in Green Fluorescent Protein
Peter N. Pintauro	Multicomponent Space-Charge Ion Uptake and Ion/Solvent Transport Models for Ion-Exchange Membranes
John L. Freeouf	Far UV Spectroscopic Ellipsometry of Electronic Materials
Martin E. Huber Kathryn A. Moler (CPIC)	Development of Wideband Scanning Superconducting Quantum Interference Device Susceptometers for Nanomagnetic Materials Research and Education
Mark S. Humayun James Weiland (CPIC)	Biocompatible Technology for a Light Sensitive Retinal Prosthesis
Elisabeth Smela Pamela A. Abshire (CPIC) Andreas G. Andreou (CPIC)	Integrated Sensing: Cell Clinics on a Chip
Sitharama S. Iyengar	Real Time Distributed Data Mining for Sensor Networks
Marvin H. White	Integrated Sensing: An Integrated Biosensor System for Cellular Studies
Cynthia G. Zoski Peixin He (CPIC)	GOALI: Addressable Multielectrode Arrays Based on Membrane Templates: Fabrication, Characterization, and Instrumentation

Investigator CPIC=Co-Principal Investigator Current	**Project Title**
Peter Rogan	HPNC: High Speed Networking for Automated Fluorescence Microscopy
Larry V. McIntire John W. Clark (CPIC)	Mini-symposia on New Technologies in Biomedical Optics and Recent Advances in Medical Imaging, Houston, Texas, October 23–26, 2002
Elizabeth J. Podlaha Julia Y. Chan (CPIC) David Young (CPIC) Wanjun Wang (CPIC) Michael C. Murphy (CPIC)	NIRT: Electrodeposition of Nanostructured Multilayers
Chang Liu Chryssostomos Chryssostomidis (CPIC)	Integrated Sensing: Biomimetic Sensors for Autonomous Underwater Vehicles
Ravindra B. Lal B. R. Reddy (CPIC) Anup Sharma (CPIC) Matthew E. Edwards (CPIC) Manmohan D. Aggarwal (CPIC)	Doctoral Research Capacity Building for Sensor Science Technology
Erik Rosenthal	SGER: Path Dissolution in Propositional Logic
Joda C. Wormhoudt	SBIR/STTR Phase II: Microchip-Laser-Based Optical Alloy Analysis Instrument
Frank L. Lewis David B. Wallace (CPIC) Khosrow Behbehani (CPIC)	GOALI: MEMS Based Sensors and Actuators for Medical and Biological Applications
William J. Kaiser Michael Fitz (CPIC) Gregory J. Pottie (CPIC)	Integrated Sensing: Energy-Aware Articulation in Sensor Networks
Farhad Ansari	Miniaturized MEMS Based Fiber Optic Distributed Health monitoring System for Civil Structures
Bruce W. Alphenaar Shi-Yu Wu (CPIC) Chakram S. Jayanthi (CPIC)	SPIN ELECTRONICS: Carbon Nanotube Based Spin Electronic Devices
Wijesuriya P. Dayawansa	Control of Patterns in Systems with Large Numbers of Actuators and Sensors
John Hetling Yang Dai (CPIC) Thomas C. Baker (CPIC)	Sensory Coding and Pattern Recognition with Hybrid Olfactory Biosensor
Ellen M. Arruda Karl Grosh (CPIC)	Biomicromechanics of Heart Muscle Tissue Function

Investigator CPIC=Co-Principal Investigator Current	**Project Title**
David J. Carlson Alan Fried (CPIC) James W. White (CPIC) Dirk Richter (CPIC) Frank K. Tittel (CPIC)	BIOCOMPLEXITY: High-Precision 13CO2/12CO2 Ratio Measurements Using an Optical Fiber Based Difference Frequency Generation Laser Source
Robert D. Throne	Data Fusion for Inverse Electrocardiography: Synthesis of Signals from Multiple Sensor Types and Locations
Mark R. De Guire Paul M. Kayima (CPIC)	Engineered Ceramic-Organic Interfaces: Properties and Applications
Richard L. McCreery	Raman Spectroscopy of Carbon-based Molecular Electronic Junctions
Peter T. Cummings	NIRT: Multiscale Simulation of the Synthesis, Assembly and Properties of Nanostructured Organic/Inorganic Hybrid Material
Shivshankar Sundaram	SBIR Phase II: Development of Integrated Fluid/Solid/Bio-Kinetic Simulation Software for the Characterization of Microsphere-based Bio-analytic Systems
Thomas A. DeFanti Oliver Yu (CPIC) Jason Leigh (CPIC) Peter C. Nelson (CPIC) Robert L. Grossman (CPIC)	CISE Research Resources: Matching Advanced Visualization and Intelligent Data Mining to High-Performance Experimental Networks
Nikolaos Papanikolopoulos Maria L. Gini (CPIC) Daniel L. Boley (CPIC) Bradley J. Nelson (CPIC) William K. Durfee (Co-	CISE Research Resources: Teams of Miniature Mobile Robots
John T. McDevitt	Conductive Polymer / Superconductor Nanocomposite Assemblies

APPENDIX F. DOD/DARPA PROGRAMS RELATED TO BIOSENSING

Excerpt from the Department of Defense's Fiscal Year (FY) 2004/FY 2005 Biennial Budget Estimates—*RESEARCH, DEVELOPMENT, TEST AND EVALUATION, DEFENSE-WIDE:* Volume 1 - Defense Advanced Research Projects Agency, page 1. Retrieved on 10/02/2003 from www.defenselink.mil/comptroller/defbudget/fy2004/budget_justification/pdfs/rdtande/DARPA_RDTE.pdf.

UNCLASSIFIED

RDT&E BUDGET ITEM JUSTIFICATION SHEET (R-2 Exhibit)								**DATE** February 2003
APPROPRIATION/BUDGET ACTIVITY RDT&E, Defense-wide BA1 Basic Research				**R-1 ITEM NOMENCLATURE** Defense Research Sciences PE 0601101E, R-1 #2				
COST (In Millions)	FY2002	FY2003	FY2004	FY2005	FY2006	FY2007	FY2008	FY2009
Total Program Element (PE) Cost	141.900	199.030	151.029	143.522	146.283	148.519	151.303	154.081
Bio/Info/MicroSciences BLS-01	72.657	85.631	87.861	82.099	82.679	84.029	83.948	84.843
Information Sciences CCS-02	8.318	24.094	16.325	15.791	18.592	18.565	18.547	18.528
Electronic Sciences ES-01	23.149	21.924	18.677	20.596	21.527	22.474	25.380	27.306
Materials Sciences MS-01	37.776	67.381	28.166	25.036	23.485	23.451	23.428	23.404

(U) Mission Description:

(U) The Defense Research Sciences Program Element is budgeted in the Basic Research Budget Activity because it provides the technical foundation for long-term National Security enhancement through the discovery of new phenomena and the exploration of the potential of such phenomena for Defense applications. It supports the scientific study and experimentation that is the basis for more advanced knowledge and understanding in information, electronic, biological and materials sciences.

(U) The Bio/Info/Micro Sciences project will explore and develop potential technological breakthroughs that exist at the intersection of biology, information technology and micro/physical systems to exploit advances and leverage fundamental discoveries for the development of new technologies, techniques and systems of interest to the DoD. The project will apply information and physical sciences to discover properties of biological systems that cross multiple length scales of biological architecture and function, from the molecular and genetic level through cellular, tissue, organ, and whole organisms' levels. Key focus areas include multidisciplinary programs in BioComputational Systems; Simulation of Bio-Molecular Microsystems; Bio Futures; Biological Adaptation, Assembly, and Manufacturing; Nanostructure in Biology; and Brain Machine Interface.

(U) The Information Sciences project supports basic scientific study and experimentation for national security requirements such as computational models, new mechanisms for performing computation and communication, innovative approaches to the composition of software, novel human computer interfaces, novel computing architectures, and automatic speech recognition research.

Excerpt from the Department of Defense's Fiscal Year (FY) 2004/FY 2005 Biennial Budget Estimates—*RESEARCH, DEVELOPMENT, TEST AND EVALUATION, DEFENSE-WIDE:* Volume 1 - Defense Advanced Research Projects Agency, page 129. Retrieved on 10/02/2003 from www.defenselink.mil/comptroller/defbudget/fy2004/budget_justification/pdfs/rdtande/DARPA_RDTE.pdf.

UNCLASSIFIED

RDT&E BUDGET ITEM JUSTIFICATION SHEET (R-2 Exhibit)								**DATE** February 2003
APPROPRIATION/BUDGET ACTIVITY RDT&E, Defense-wide BA2 Applied Research				**R-1 ITEM NOMENCLATURE** Biological Warfare Defense PE 0602383E, R-1 #14				
COST (In Millions)	FY2002	FY2003	FY2004	FY2005	FY2006	FY2007	FY2008	FY2009
Total Program Element (PE) Cost	171.878	161.956	137.254	138.533	139.975	147.104	145.888	145.745
Biological Warfare Defense Program BW-01	171.878	161.956	137.254	138.533	139.975	147.104	145.888	145.745

(U) Mission Description:

(U) DARPA's Biological Warfare Defense project is budgeted in the Applied Research Budget Activity because its focus is on the underlying technologies associated with pathogen detection and remediation. This project funds programs supporting revolutionary new approaches to biological warfare (BW) defense and does not duplicate efforts of other government organizations.

(U) Efforts to counter the BW threat include developing barriers to block entry of pathogens into the human body (including unique methods for rapid air and water purification), countermeasures to stop pathogen and chemical consequence and to modulate host immune response, medical diagnostics for the most virulent pathogens and their molecular mechanisms, biological and chemically-specific sensors, advanced decontamination and neutralization techniques and integrated defensive systems. Program development strategies include collaborations with pharmaceutical, biotechnology, government, and academic centers of excellence.

APPENDIX G. U.S. ARMY RESEARCH OFFICE-FUNDED PROJECTS RELATED TO BIOSENSING, ACTIVE AS OF MARCH 2004

Investigator	Project Title
O. Velev	Chemical and biological microassays in freely suspended droplets on novel fluidic chips
P. Treado	Development of novel spectroscopic techniques for the detection and identification of biological warfare agents
K. Spencer	Ultra-sensitive Raman detector
P. Barthelemy	Amphiphiles for DNA supramolecular assemblies
J. Yates, Jr.	Enzyme, antibody, and photocatalytically active nanoscale scavengers and sensors for CW and biological agents
M. Lean	High throughput sample preparation for detection of bioagents in water
K. Kishore	Compact submillimeter wave sources and detectors for biological and chemical spectroscopy
D. Porterfield	Compact submillimeter wave spectrometers for biological and chemical sensing
J. Seminario	Integrated molecular and nanoscale semiconductor devices: Applications to computing and biosensing
P. Burke	Active GHz nanobiosensor devices with chemical specificity
E. Brown	THz differential absorption radar for bioparticulate detection
E. Brown	Remote detection of bioparticles by Raman lidar
A. Marakelz	Terahertz time domain spectroscopy of conformational dynamics of sensor proteins: Basic research and pathogen sensor development
D. Van der Weide	Biomolecular interaction sensing with sub-terahertz fields
T. Crowe	Science and technology of chemicals and biological sensing at terahertz frequencies
M. Norton	DNA nanostructures for surface pattering
G. Hitchens	New DNA/RNA sequencer for rapid assessment of exposure to infectious agents
V. Fischetti	Using bacteriophage lytic enzymes to specifically destroy BW bacteria
E. Wang	Generation of advanced diagnostics and counter measures for individuals most vulnerable to biothreats
D. Gorenstein	A thioaptamer chip for diagnostics and therapeutic targeting of pathogenic and human proteomes

Investigator	Project Title
S. Summers	Amplification of molecular signal using highly stabilized acoustic wave devices
S. Paik	Molecular signatures of biological pathogens
S. Iadonato	Molecular signatures of biological pathogens
S. Weaver	Automated sequencing for biological defense research
T. Scofield	Epidemic outbreak surveillance/Lackland test bed project
A. Lapidus	Bioengineered proteins for chemical/biological defense, protection, and decontamination
J. Tabb	Phage array biosensor for detection of biowarfare agents
K. DeBoer	Bioengineered proteins for chemical/biological defense, protection and decontamination
R. Rohwer	The BugID system for discovering optimal nucleotide probes
A. Ferrante	Cellular persistence and stability (CEPAS)
D. Morse	Institute for Collaborative Biotechnologies
E. Kool	Use of multiple fluorescent labels in biological sensing
K. Clinkenbeard	Development of aptamer beacons to lipopolysaccharide for the real-time sensing of biological warfare agents
S. Stupp	Infrastructure of Institute for Bioengineering and nanoscience in advanced medicine
V. Petrenko	Phage landscape libraries as a source of substitute antibodies for detector platforms
P. Cremer	Designing lithographically patterned phospholipid bilayer arrays for next-generation biosensors and immunoassays
J. Currie	Bio-fluidic chip technology for chemical/biological microsystems
P. Gascoyne	A general-purpose analysis system based on a programmable fluid processor
A. Scherer	Monolithic integration of microfluidics and optoelectronics for biological analysis
C. Meinhart	An integrated tunable laser cavity sensor for immunoassay analysis and molecular diagnostics
C. Turnbough	Peptide ligands for the detection of spore-forming bio-agents
A. Ellington	Texas consortium for the development of biological sensors
T. Haddock	MEMS water safety monitor
J. Wormhoudt	Portable laser induced breakdown spectroscopy sensor for detection of biological agents
S. Palamakumbura	Development of a detector using fluorescent coated filters

Investigator	Project Title
R. VanTassell	Fluorescent, polymerized affinity liposomes for the detection of bacterial toxins
R. Deans	Detection of infectious bacteria in water
R. VanTassell	Viability assay for monitoring decontamination of pathogenic bacteria
E. Thomas	Proposal to host the Institute for Soldier Nanotechnologies at MIT
H. Rabitz	Optimum quantal discrimination of chemical and biological agents

APPENDIX H. U.S. DEPARTMENT OF ENERGY RESEARCH RELATED TO BIOSENSING (1999)

Excerpted March 2003 from DOE (1999) *Biomedical Engineering Research at DOE National Labs*, available online, www.osti.gov/sc73/doe-sc-1999-1.pdf.

Laboratory	Project Title	Name
Ames Laboratory	Technique Measures DNA Damage from Carcinogen	Gerald J. Small
	Analytical Techniques Measure Trace Components in Cells	Edward S. Yeung
	Genetic Reader	Edward S. Yeung
Argonne National Laboratory	Biochips for Gene Research	Andrei Mirzabekov
	Immunoassays	Fred Stevens
	Gene Expression and Protein Function	Gayle Woloschak
	Biophysics of Myeloma Pathology	Fred Stevens
	Motor Neuron Diseases	Gayle Woloschak
Brookhaven National Laboratory	Sensitive Detection and Rapid Identification of Biological Agents by Single Molecule Detection	M. Wu
	Development of Diagnostics for Lyme Borreliosis	J.J. Dunn
	Methyl Histidine Kinetics as an Indicator of Muscle Mass and Metabolism	P. Molina
Idaho National Engineering and Environmental Laboratory	Analysis of Biological Fluids by Ion Mobility Spectrometry	Dave Atkinson
	SIMS technology for the Study of Microsurface Chemistry	Jim Delmore
	Microbiological Identification from Cell Membrane	Jani Ingram
Lawrence Livermore National Laboratory	Development of a Hand held MiniPCR Instrument	Fred Milanovich
	Functional Gene Expression Microarrays	Andrew J. Wyrobek
	DNA Chip Analysis	Gary Andersen
Los Alamos National Laboratory	Rapid Identification of Microbial Species	Paul Jackson
	Noninvasive Intercranial Pressure Measurement System	William O. Wray

Laboratory	Project Title	Name
	Noninvasive Measurement of Drug Concentrations in Tissues	Irving J. Bigio
	Low Frequency Impedance Spectroscopy for Biomolecular Characterizations	Benno P. Schoenborn
	Beryllium Health Effects	Babs Marrone
National Renewable Energy Laboratory	Regenerable Enzyme Electrodes	Paul Weaver
Oak Ridge National Laboratory	Monitoring Inflammatory Cytokines Using Maldi Mass Spectrometry	S.J. Kennel
	Electrospray Ionization and Ion/Ion Chemistry for Rapid ID of Pathogens	Scott A. McLuckey
	New Approaches for Monitoring of Trace Compounds in Physiological Media	G.J. Van Berkel
	The Molecular Analysis of Genomes by AFM	D.P. Allison
	Flowthrough Genosensor Chips	K.L. Beattie
	Rapid Screening of DNA Using Maldi Mass Spectrometry	M.V. Buchanan
	Micromachined Biosensor Arrays	Mitchel J. Doktycz
Oak Ridge National Laboratory (con't.)	Medical Telesensor Application-Specific Integrated Circuits (ASICs)	T.L. Ferrell
	Lab-on-a-Chip Technologies for Medical Diagnostics and Drug Discovery	J. Michael Ramsey
	Integrated Biochip for Medical Diagnosis	Tuan Vo-Dinh
	Electronic Nose on a Chip	R.J. Warmack
Pacific Northwest National Laboratory	Boron Neutron Capture Therapy (BNCT) Real Time Dosimetry	Mary Bliss

Laboratory	Project Title	Name
Sandia National Laboratories	Accelerated Molecular Discovery Arrays	Deon Anex
	Measuring Blood Rheology Using Thickness-Shear-Mode Resonators	Richard W. Cernosek
	Biological Weapon Detector Using Bioaffinity Array Impedance Analysis with Chemical Amplification Through Redox Recycling – BioCCD	Albert William Flounders
	Combinatorial BioFET Microsensor Arrays	Albert William Flounders
	Non-Invasive Biomedical Monitoring	David Haaland
	Investigation of Technologies to Improve Fiber Optic Biosensor for Counter-Proliferation Purposes	Colin E. Hackett
	Parallel Microseparations-Based Detection of Biological Toxins	Joe Schoeniger
	Miniature UV Fluorescence Based Biological Agent Sensors	Kevin Schroder
	Optical Detection of Biologicals	John S. Wagner
	Optical Detection of Pharmaceuticals in Optically Dense Media	John S. Wagner
	Optical Detection of PrpSc	John S. Wagner

APPENDIX I. EUROPEAN UNION 6TH FRAMEWORK PROGRAMME (2002–2006)

(Excerpts from www.cordis.lu/fp6/)

Genomics and Biotechnology for Health

Advanced genomics and its application for health

Fundamental knowledge and basic tools for Functional Genomics in all organisms

Programme objectives:

To foster the basic understanding of genomic information, by developing the knowledge base, tools and resources needed to decipher the function of genes and gene products relevant to human health and to explore their interactions with each other and with their environment.

Research actions:

- Gene expression and proteomics to enable researchers to better decipher the functions of genes and gene products as well as to define the complex regulatory networks (biocomplexity) that control fundamental biological processes. Research will focus on: developing high throughput tools and approaches for monitoring gene expression and protein profiles and for determining protein function and protein interactions.

* Structural genomics to enable researchers to determine, more effectively and at a higher rate than is currently feasible, the 3-D structure of proteins and other macromolecules, which is important for elucidating protein function and essential for drug design. Research will focus on: developing high throughput approaches for determining high-resolution 3-D structures of macromolecules.

* Comparative genomics and population genetics to enable researchers to use well characterised model organisms for predicting and testing gene function and to take full advantage of specific population cohorts available in Europe to determine the relationship between gene function and health or disease. Research will focus on: developing model organisms and transgenic tools; developing genetic epidemiology tools and standardised genotyping protocol.

* Bioinformatics to enable researchers to access efficient tools for managing and interpreting the ever increasing quantities of genome data and for making it available to the research community in an accessible and usable form. Research will focus on developing bioinformatic tools and resources for data storage, mining and processing; developing compu-

tational biology approaches for in silico prediction of gene function and for the simulation of complex regulatory networks.

Information Society Technologies

Intelligent systems for the monitoring of health status

Objectives: To improve early illness detection and medical intervention by carrying out medium to long term multidisciplinary research on IST health application systems. The aim is to foster closer collaboration between research activities in areas such as health telemetric, biomedical engineering and advanced communication technologies. Longer term work is also expected on new systems that take into account the results of functional genomics research. Activities will complement the existing clusters on "Ambient intelligence-based systems for health promotion, illness prevention and patient treatment."

Focus: Minimally invasive personal health systems for illness prevention and/or for health status monitoring of patients including systems based on flexible and smart technologies adaptable to the human body and integrating the possibilities of electrical, optical, chemical, & mechanical sensors. These systems monitor various parameters (bio-signals, location, etc), and when needed, communicate securely with health professionals as well as with intelligent support systems. The focus is on development of new sensor technologies as well as intelligent decision support systems. Research on knowledge technologies for access and delivery of *personalized* health promotion material based on the current health status and including, where appropriate, health and genetic profile. The problems to address include interoperability of databases containing individual's health information, semantic based knowledge representation, knowledge capturing and retrieval which facilitate compliance with data protection, electronic signature and other information

Society related legislation

<u>Systems for health professionals: creating a "Health knowledge info-structure</u>"

Objective: To allow health professionals timely interaction with heterogeneous, redistributed, medical and other health related databases. Work will consist of medium to long term research on the development of more efficient and secure "Health Knowledge info-structure", (i.e. a network of interactive and secure medical and health systems). This will complement the existing cluster on "Ambient intelligence-based system for health professionals"

Focus:

- Advanced navigation tools for health professionals for timely retrieval of vital information including health info-structure tools such as user friendly systems and interfaces as well as mobile systems for ubiquitous, timely and secure access to medical data at the point of care. A midterm strategy is the fostering of closer collaboration between the bio-informatics community and medical informatics researchers in order to accelerate and validate the results of functional genomics and develop the future forms of clinical systems that will incorporate genetic information.
- Medical knowledge and evidence management, data mining, capturing and retrieval, intelligent interactive environments and interoperability of large health databases, using open source where appropriate. All systems handling person identifiable data must comply with the requirements of the information society related legislation.

Systems for independent living

Objective: To provide people with disabilities and their careers, and the elderly, with systems based on the ambient intelligence concept in order to facilitate employment, education, and full participation in society. Work will include innovative applications and services to facilitate citizens' civic involvement through enhanced remote access to general interest services. This will require the availability of new interoperable tools influencing the design, and content authoring of publicly accessible web sites. Medium and long term research and technological development will concentrate on the rehabilitative and interface aspects of advanced, systems that could be embedded in our surroundings or in the case of medical implants inside our bodies. Using a design-for-all approach, work will be undertaken principally with mainstream industries to find solutions to the needs of people with disabilities that can provide links and interfaces with assistive technologies.

Focus:

- The work will cover innovative IST-based assistive systems for supporting cognition, mobility, orientation, and sensory abilities, and secure living conditions in the home. r- Longer term research will address advanced interfaces for compensating the effects of impaired functionalities and individual performance using an up-to-date understanding of cognitive, behavioural, and sensory processes to meet the requirements of the target groups, within the research and development work, particular attention should be paid to covering the legal, regulatory, financial, ethical and societal aspects to understand better the pre-requisites for successful dissemination of results.

Intelligent systems and services for civilian and environmental crises management

Objectives: The objective is to assist public administrations and emergency services in the management of specific emergency scenarios by funding research into intelligent decision support systems for the risk assessment and post crisis management of natural or man made risks, including their impact on the structural integrity of large infrastructures. .

A secondary objective is the improvement of the associated risk management systems dependent on the integration and management of multiple data sources (including the decision support systems cited above). For many such applications, a combined access to, and a more intelligent use of heterogeneous, multi-source data, is required to reduce the risk and adverse economic or societal impacts related to environmental emergencies, and to deliver effective, sustainable, high quality information services. The more efficient provision of harmonised, accurate information combined with easy access across borders and nations, in line with the GMES objectives, is one of the main challenges for the services to be developed.

Focus:

- Intelligent systems combining data from earth observation, satellite positioning systems and in-situ sensors with geo-referenced information, and advanced methods and technologies for extracting knowledge from environmental data, that contribute to effective decision support in the form of improved risk and damage assessment, prevention and response actions including emergency telemedicine.
- In-depth analysis and comparative assessment of the performance, scalability and effectiveness of existing risk assessment tools, methods and systems to meet with the challenges foreseen over the next ten years.
- Pre-standardisation activities leading to harmonised data models, metadata, functional architecture and harmonised approach to services relevant to risk management.

Nanotechnologies and nanosciences, knowledge-based multifunctional materials, and new production processes and devices

Molecular and bio-molecular mechanisms and engines

The objective will be to develop new concepts and technologies for further developments with substantial breakthrough potential applications. Research may address a vast variety of areas, such as molecular electronics, artificial photosynthesis and molecular motors.

Interfaces between biological and non biological systems -

The objective is to realise novel forms of integration of biological and non-biological systems at the nano-level. Research may include bio-molecular, chemical and physical

modifications at the substrate surface, including patterning or growth of cells, enabling specific bioactivity/biomimetic performance and integration in devices with new potential applications. Health and environmental risks should be addressed.

Handling and control instrumentation at the level of single atoms or molecules and/or < 10 nm

The objective is to develop instrumentation and methods, for manipulation and manufacture at the nano-scale, supported by appropriate analysis and control, including benchmarking of efficient and cost effective instrumentation, and nano-metrology. Research at frontiers of knowledge may include the study of a variety of advanced techniques for nano-scale manufacture; the development of breakthrough technologies and methodologies exploiting the self-assembling properties of matter.

Applications in areas such as health and medical systems, chemistry, energy, optics, food and the environment

Nanosciences and nanotechnologies are fast developing domains with great potential, both in terms of improving the quality of life of all people and of creating wealth though novel knowledge-based and sustainable processes. The goal is to foster the potential nano-technologies in breakthrough applications through the integration of research developments in materials and technological devices in an industrial context. The development of new, higher performance "nano-enabled" services, products, components, devices, systems and processes still requires long term research efforts. The availability of up-to-date information and the development of realistic scenarios are key elements for elaborating possible forms and scope for the intervention of public funds.

Development of fundamental knowledge

There is a pressing industrial need to better understand complex physico-chemical and biological phenomena relevant to the mastering and processing of multifunctional and eco-efficient materials providing the basis for developing novel materials with predefined physical, chemical or biological characteristics.

Understanding materials phenomena

Research will focus on materials phenomena offering new options for the long-term. Research projects should support high-risk activities to design and develop new structures with defined characteristics, which can lead to new industrial applications. The activities should address the understanding of properties, behaviour and synthesis of materials in order to exploit the potential use of novel highly complex composite systems, molecules and new multi-functional materials derived from them.

Computational strategies, experimental, theoretical, simulation and modelling are key tools to be considered.

New materials by design

The main objective is to develop novel multi-functional materials for multisectoral applications by providing new materials processing solutions and encouraging new approaches, such as "learning from nature" or materials "made to measure", using whenever appropriate the potential of nanotechnology. Emphasis should be put on developing novel materials by means of "design approaches", including prediction and modelling, on exploring new complex multi-functional properties of materials and on tailoring the materials in order to obtain a desired set of properties suitable for given applications and respecting consumer needs and perceptions. In using the potential of nanotechnology, a particular attention should be given to self-repairing materials.

Integration of nanotechnologies, new materials, and new production technologies for improved security and quality of life

This area has been added to the three first areas, as defined in the specific programme, due to the "integrating" challenge of the expected output and due also to the number of EoIs received on the subject. A specific target should indeed be to put materials science and advanced industrial technologies at the service of health. In this context, integration of technological developments, and in particular of the new generation of smart and hybrid materials interacting with their surrounding and related manufacturing equipment, is bringing huge potential for the development of sensors, actuators and devices, leading to a greater security and safety of people and the environment.

Systems, instruments and equipment for better diagnosis and/or surgery, including for remote operations

The long-term objective is the development of remote surgically precise systems, new medical instruments and/or intelligent diagnosis equipment and systems, supporting challenges such as the development of health care for the future. A specific technical goal should be the miniaturisation of systems and instruments, including sterilisation aspects. The advances in biosensors should also be considered here.

Tissue engineering, new biomimetic and bio-hybrid systems

The new developments in new materials and industrial processes for health will strongly boost treatment and healing, in fields such as artificial organs. Research should encompass the understanding, modelling and development of biomaterials through new bioreactor developments including adult stem cell research. The final goal should be the development of advanced intelligent bio-hybrid systems and their production lines.

New generation of sensors, actuators and systems for health, safety and security of people and environment

The target is to support technological platforms for the development of novel, low cost and highly reliable sensors and actuators, in particular those based on nano or microtechnologies, in combination with signal treatment. The resulting systems will enable the real-time detection of hazards and species from various origins, to monitor quality, reliability and safety of products and systems and to provide early feedback to protect people and the environment. The long-term objective is the development of stable, multifunctional, precise, small and low-cost systems for optimised use, as well as of an efficient related metrology infrastructure.

Food

Traceability processes all along the production chain

The objective is to increase consumer confidence in the food supply by strengthening the scientific and technological basis for ensuring complete traceability along the entire food chain including animal feed. It will ensure that products can be linked to their source while also protecting products of declared origin (both geographical and production system). It will also assure traceability of genetically modified organisms, and other products based on recent biotechnology developments, from raw material origin to purchased food products.

Methods of analysis, detection and control

The objective is to contribute to the development, improvement, validation and harmonisation of reliable and cost-effective sampling and measurement strategies for chemical contaminants and existing or emerging pathogenic micro-organisms (such as viruses, bacteria, yeasts, fungi, parasites, and new agents of the prion type including development of ante mortem diagnostic tests for BSE and scrapie) so as to control the safety of the food and feed supply and ensure accurate data for risk analysis.

> With changes in production methods, processing technologies and distribution systems, many pathogens and contaminants are controlled ever more rigorously today. However, new pathogens or food safety issues may arise as a consequence of factors outside the control of the food producer. Increasingly, foods do not come from one source or one country, but are a combination of raw materials coming from many diverse countries and very different production systems. The aim will be to improve detection and control techniques along the food production chain, using powerful new and more sophisticated technologies linked to primary production, ensuring that the original

contamination does not enter the chain at critical points. Particular attention will be given to possible anticipation and control of emerging risks in food and feed including new contaminants and pathogens, non-conventional agents and stress adaptation of pathogens. Projects should take account of aspects of communication with stakeholders, especially consumers.

APPENDIX J. EUROPE AND JAPAN PATENTS RELATED TO BIOSENSING, 1999–2003

In scientometrics, the quantity of patents and the rate of appearance of new patents have been established as indicators of the progress of science toward technological application. Hence, the WTEC Biosensing Panel gathered patent information for the sites it visited in Europe and Japan as an indicator of progress in development of biosensing applications. This appendix contains two tables, I.1 and I.2, showing patents that have been registered for the sites the panel visited in Europe and in Japan, respectively. The information was collected for the complete years 1999 through 2002, and includes patents registered in the first two months of 2003.

Figure I.1 shows the total number of patents in biosensing combining those in both Europe and Japan. Europe has 45 percent of the patents, while Japan has 55 percent; Matsushita alone accounts for 21 percent of the recorded patents. The patents in biosensing tend to corroborate the perception that Japan aggressively pursues the commercialization of new technologies. The patents from all sources show the greatest emphasis for biosensing is in new devices and new chemical analysis methods.

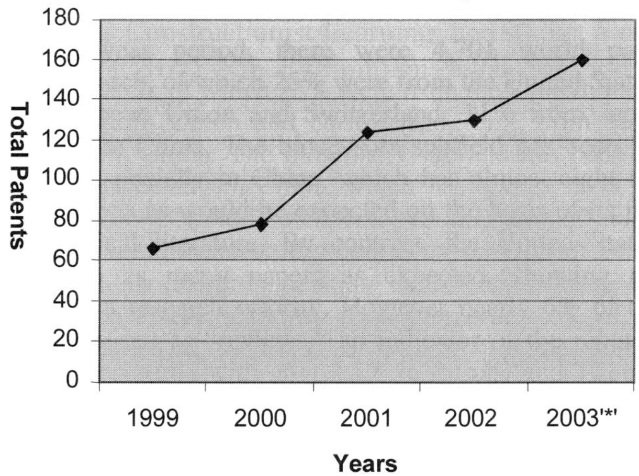

Fig. I.1. Number of overall biosensing-related patents recorded in the sites the WTEC panel visited in both Europe and Japan for the period 1999–2003.

Although the events of 9/11/2001 have accelerated interest in the detection of toxic chemical and biological agents, the increase in patents in 2001 was spread relatively uniformly throughout the year, so that the patent data do not necessarily demonstrate a response to 9/11. The data for 2003 in Figure I.1 is an estimate by projecting the first two months of the year at a constant rate for the full year.

Under a program different from the Biosensing Study, WTEC is analyzing U.S. patents. It will be interesting to compare the findings about the U.S. patents with the European and Japanese patents listed in this appendix.

REFERENCES

Verbeek, Debachere, Luwel. 2003. Science cited in patents: A geographic "flow" analysis of bibliographic patterns in patents. *Scientometrics* 58(2):241–263.

Table J.1.
European Patents Related to Biosensing in Sites Visited by WTEC Panel (1999–2003).

Researcher	Organization	Patent number	Patent Title	Year issued
Anthony Turner	Cranfield University	GB2337332	Affinity electrode for electrochemical analysis	1999-11-17
Anthony Turner	Cranfield University	WO9910736	Protein sensor	1999-03-04
Anthony Turner	Cranfield University	US5922616	Biochemical sensor and novel media for bioelectrochemical reactions	1999-07-13
Anthony Turner	Cranfield University	WO0106256	Polymer for binding amine containing ligands and uses thereof	2001-01-25
Anthony Turner	Cranfield University	WO0130856	Molecularly imprinted polymers produced by template polymerisation	2001-05-03
Anthony Turner	Cranfield University	WO0155235	Molecularly imprinted polymer	2001-08-02
Anthony Turner	Cranfield University	WO0166567	Design, synthesis and use of affinity ligands	2001-09-13
Anthony Turner	Cranfield University	WO0184363	Remote selection of data resource	2001-11-08
Anthony Turner	Cranfield University	GB2364571	Diagnosing and/or monitoring urinary tract infection	2002-01-30
Anthony Turner	Cranfield University	WO0229412	Selective binding materials	2002-04-11
David Reinhoudt	University of Twente	WO0029337	A system and a method for removing ions from aqueous liquid streams	2000-05-25

Table J.1.
European Patents Related to Biosensing in Sites Visited by WTEC Panel (1999–2003).

Researcher	Organization	Patent number	Patent Title	Year issued
David Reinhoudt	University of Twente	EP0907641	Fluoroionophores and their use in optical ion sensors	1999-04-14
David Reinhoudt	University of Twente	EP1019401	Complex comprising a rare earth metal ion and a complexing moiety	2000-07-19
David Reinhoudt	University of Twente	US6294390	Covalently immobilized fluoroionophores for optical ion sensors	2001-09-25
David Reinhoudt	University of Twente	US6417005	Covalently immobilized fluoroionophores as optical ion sensors	2002-07-09
David Reinhoudt	University of Twente	US6468406	Anion-complexing compound, method of preparing the same, an ion-selective membrane and a sensor provided with such a compound or membrane	2002-10-22
Dermot Diamond	Dublin City University	IE980221	Separation of enantiomers	1999-10-20
Brian MacCraith	Dublin City University	US6137117	Integrating multi-waveguide sensor	2000-10-24
Brian MacCraith	Dublin City University	GB2355524	Device for measuring colour and turbidity in a liquid sample	2001-04-25
Brian MacCraith	Dublin City University	WO0129541	Device for measuring water quality	2001-04-26
Brian MacCraith	Dublin City University	WO02059583	A luminescence based sensor	2002-08-01
Brian MacCraith	Dublin City University	EP1241464	Non-contact optical monitor	2002-09-18
Horst Vogel	Ecole Polytechnique Fédérale de Lausanne	WO9905509	Detection and investigation of biological molecules by fourier transform infra-red spectroscopy	1999-02-04

Table J.1.
European Patents Related to Biosensing in Sites Visited by WTEC Panel (1999–2003).

Researcher	Organization	Patent number	Patent Title	Year issued
Horst Vogel	Ecole Polytechnique Fédérale de Lausanne	CA2316966	Positioning and electrophysiological characterization of individual cells and reconstituted membrane systems on microstructured carriers	1999-06-24
Horst Vogel	Ecole Polytechnique Fédérale de Lausanne	WO0073798	Vesicle containing polymers and sensor detection methods based thereon	2000-12-07
Horst Vogel	Ecole Polytechnique Fédérale de Lausanne	WO0179820	Assembly and method for a correlator structure	2001-10-25
Horst Vogel	Ecole Polytechnique Fédérale de Lausanne	WO0246766	Bioanalytical reagent, method for production thereof, sensor plat-forms and detection methods based on use of said bioanalytical reagent	2002-06-13
Andreas Hierlemann	ETH Zurich	WO0250528	Microsensor and single chip integrated microsensor system	2002-06-27
Dr. Zenobi	Eidgenössische Technische Hochschule	AU7658401	Holder device for smoke products, particularly cigarettes	2002-02-13
Dr. Zenobi	Eidgenössische Technische Hochschule	EP1193730	Atmospheric-pressure ionization device and method for analysis of a sample	2002-04-03
Dr. Zenobi	Eidgenössische Technische Hochschule	WO0209540	Holder device for smoke products, particularly cigarettes	2002-02-07
Henry Baltes	Eidgenössische Technische Hochschule	WO0250528	Microsensor and single chip integrated microsensor system	2002-06-27
Gordon Holt		US2002115667	Methods for therapeutic use of glucosylceramide synthesis inhibitors and composition thereof	2002-08-22

Table J.1.
European Patents Related to Biosensing in Sites Visited by WTEC Panel (1999–2003).

Researcher	Organization	Patent number	Patent Title	Year issued
Gordon Holt		WO02054081	Proteins, genes and their use for diagnosis and treatment of kidney response	2002-07-11
Gordon Holt		WO02054079	Proteins, genes and their use for diagnosis and treatment of cardiac response	2002-07-11
Gordon Holt		WO0062780	Use of glucosylceramide synthesis inhibitors in therapy	2000-10-26
Gunter Gauglitz	Eberhard Karls University	AU1364300	Quantitative determination of analytes in a heterogeneous system	2000-08-24
Gunter Gauglitz	Eberhard Karls University	GB2340231	Optical transducers based on liquid crystalline phases	2000-02-16
Gunter Gauglitz	Eberhard Karls University	GB2334581	Microtitre plate	1999-08-25
Hubert Girault	EPFL	WO02090112	Polymer bonding by means of plasma activation	2002-11-14
Hubert Girault	EPFL	WO02080222	Apparatus and method for dispensing a sample	2002-10-10
Hubert Girault	EPFL	FR2775007	Electrode array useful for waste water treatment, electrolysis of sea water, and electrosynthetic organic reactions	1999-08-20
Hubert Girault	EPFL	EP0944834	Surface patterning of affinity reagents using photoablation	1999-09-29
Hubert Girault	EPFL	EP0968416	Methods of fabricating chemical sensors	2000-01-05
Hubert Girault	EPFL	WO0058724	Microscale total analysis system	2000-10-05
Hubert Girault	EPFL	WO0186279	Electrophoretic separation of compounds	2001-11-15
Hubert Girault	EPFL	US2002130044	Mechanical control of fluids in micro-analytical devices	2002-09-19

Table J.1.
European Patents Related to Biosensing in Sites Visited by WTEC Panel (1999–2003).

Researcher	Organization	Patent number	Patent Title	Year issued
Andreas Manz	Imperial College	EP1287365	Sampling system	2003-03-05
Andreas Manz	Imperial College	EP1287347	Sampling system for a separation channel	2003-03-05
Andreas Manz	Imperial College	WO02059013	An analysable pacage, an analysis system and a method for analysing a packed product	2002-08-01
Andreas Manz	Imperial College	AU7421901	Chemical sensor for wellbore applications	2002-01-02
Andreas Manz	Imperial College	WO0198630	Chemical sensor for wellbore applications	2001-12-27
Andreas Manz	Imperial College	GB2363809	Chemical sensor for wellbore applications	2002-01-09
Andreas Manz	Imperial College	CA2336896	Electrochemiluminescence cell with floating reaction electrodes	2000-01-20
Andreas Manz	Imperial College	AU4919199	Electrochemiluminescence cell with floating reaction electrodes	2000-02-01
Andreas Manz	Imperial College	EP1223426	Fluid transport apparatus and method	2002-07-17
Andreas Manz	Imperial College	US2002027075	Method for controlling sample introduction in microcolumn separation techiques and samplinh device	2002-03-07
Andreas Manz	Imperial College	US6423198	Method for controlling sample introduction in microcolumn separation techniques and sampling device	2002-07-23
Andreas Manz	Imperial College	US2002036140	Method for controlling sample introduction in microcolumn separation techniques and sampling device	2002-03-28

Table J.1.
European Patents Related to Biosensing in Sites Visited by WTEC Panel (1999–2003).

Researcher	Organization	Patent number	Patent Title	Year issued
Andreas Manz	Imperial College	US2001025793	Method for controlling sample introduction in microcolumn separation techniques and sampling device	2001-10-04
Andreas Manz	Imperial College	US2001023824	Method for controlling sample introduction in microcolumn separation techniques and sampling device	2001-09-27
Andreas Manz	Imperial College	US2001008213	Method for controlling sample introduction in microcolumn separation techniques and sampling device	2001-07-19
Andreas Manz	Imperial College	US2001004964	Method for controlling sample introduction in microcolumn separation techniques and sampling device	2001-06-28
Andreas Manz	Imperial College	US2001004963	Method for controlling sample introduction in microcolumn separation techniques and sampling device	2001-06-28
Andreas Manz	Imperial College	WO9908791	Method for producing catalysts containing metal nanoparticles on a porous support, especially for gas phase oxidation of ethylene and acetic acid to form vinyl acetate	1999-02-25
Andreas Manz	Imperial College	WO0032017	Plasma generator	2000-06-02
Andreas Manz	Imperial College	GB2344212	Plasma generator	2000-05-31

Table J.1.
European Patents Related to Biosensing in Sites Visited by WTEC Panel (1999–2003).

Researcher	Organization	Patent number	Patent Title	Year issued
Andreas Manz	Imperial College	US6074979	Polybetaine-stabilized, palladium-containing nanoparticles, a process for preparing them and also catalysts prepared from them for producing vinyl acetate	2000-06-13
Andreas Manz	Imperial College	WO0142774	Potentiometric sensor	2001-06-14
Andreas Manz	Imperial College	DE19734974	Production of supported catalyst for vinyl acetate production	1999-02-25
Andreas Manz	Imperial College	GB2362712	Sampling fluids	2001-11-28
Andreas Manz	Imperial College	AU5864201	Sampling system	2001-12-03
Andreas Manz	Imperial College	WO0190740	Sampling system for a separation channel	2001-11-29
Andreas Manz	Imperial College	GB2362713	Sampling system for gas	2001-11-28
Andreas Manz	Imperial College	WO9908790	Shell catalyst, method for its production and use, in particular for gaseous phase oxidation of ethylene and acetic acid into vinyl acetate	1999-02-25
Andreas Manz	Imperial College	DE19734975		1999-03-11
Loic Blum		US6124109	System for qualitatively and/or quantitatively analyzing preferably biological substances using enhanced chemiluminescence, and method and analysis kit using same	2000-09-26
Klaus Mosbach	Lund University	GB2337332	Affinity electrode for electrochemical analysis	1999-11-17

Table J.1.
European Patents Related to Biosensing in Sites Visited by WTEC Panel (1999–2003).

Researcher	Organization	Patent number	Patent Title	Year issued
Klaus Mosbach	Lund University	US5872198	Molecularly imprinted beaded polymers and stabilized suspension polymerization of the same in perfluorocarbon liquids	1999-02-16
Klaus Mosbach	Lund University	WO9933768	Materials for screening of combinatorial libraries	1999-07-08
Klaus Mosbach	Lund University	US5959050	Supports useful for molecular imprinting technology	1999-09-28
Klaus Mosbach	Lund University	US5994110	Methods for direct synthesis of compounds having complementary structure to a desired molecular entity and use thereof	1999-11-30
Klaus Mosbach	Lund University	EP0982591	Imprints formed using functionally complementary monomers	2000-03-01
Klaus Mosbach	Lund University	WO0041723	Molecularly imprinted microspheres prepared using precipitation polymerisation	2000-07-20
Klaus Mosbach	Lund University	US6127154	Methods for direct synthesis of compounds having complementary structure to a desired molecular entity and use thereof	2000-10-03
Klaus Mosbach	Lund University	US6255461	Artificial antibodies to corticosteroids prepared by molecular imprinting	2001-07-03
Klaus Mosbach	Lund University	US6274686	Amide containing molecular imprinted polymers	2001-08-14
Klaus Mosbach	Lund University	US6316235	Preparation and use of magnetically susceptible polymer particles	2001-11-13
Klaus Mosbach	Lund University	AU6092101	Molecular imprinting	2001-12-03

Table J.1.
European Patents Related to Biosensing in Sites Visited by WTEC Panel (1999–2003).

Researcher	Organization	Patent number	Patent Title	Year issued
Klaus Mosbach	Lund University	US2002001821	Methods for direct synthesis of compounds having complementary structure to a desired molecular entity and use thereof	2002-01-03
Klaus Mosbach	Lund University	WO0222846	Process	2002-03-21
Klaus Mosbach	Lund University	WO0237100	Novel applications of nickel nitrilotriacetic acid (ni-nta) resin: hemeprotein removal, recovery, and purification from biological samples	2002-05-10
Klaus Mosbach	Lund University	AU1826802	Novel applications of nickel nitrilotriacetic acid (ni-nta) resin: hemeprotein removal, recovery, and purification from biological samples	2002-05-15
Klaus Mosbach	Lund University	WO02068958	Molecularly imprinted scintillation polymers	2002-09-06
Klaus Mosbach	Lund University	US6489418	Preparation and application of artificial anti-idiotypic imprints	2002-12-03
Albert van den Berg	University of Twente	WO0144575	Sizing composition	2001-06-21
Albert van den Berg	University of Twente	US6040385	Adhesion promoters for plastisols	2000-03-21
Albert van den Berg	University of Twente	US6184399	Process for preparing a fatty acyl isethionate salt	2001-02-06
Michael Grunze	Ruprecht-Karls-University	US5922550	Biosensing devices which produce diffraction images	1999-07-13
Michael Grunze	Ruprecht-Karls-University	WO0123962	Surface-modified layer system	2001-04-05
Michael Grunze	Ruprecht-Karls-University	WO0144813	Use of wicking agent to eliminate wash steps for optical diffraction-based biosensors	2001-06-21

Table J.1.
European Patents Related to Biosensing in Sites Visited by WTEC Panel (1999–2003).

Researcher	Organization	Patent number	Patent Title	Year issued
Michael Grunze	Ruprecht-Karls-University	WO0170296	Polyphosphazene derivatives	2001-09-27
Michael Grunze	Ruprecht-Karls-University	DE10019982	Use of antithrombogenic phosphazene polymer films or coverings on stents, catheters or other implants to reduce cell proliferation and hence to limit restenosis	2001-10-25
Michael Grunze	Ruprecht-Karls-University	WO0180919	Poly-tri-fluoro-ethoxypolyphosphazene coverings and films	2001-11-01
Michael Grunze	Ruprecht-Karls-University	DE10052823	Method for topographical analysis of molecular structures, e.g. Multi-enzyme complexes using optical microscope, comprises arranging molecular structures in specific pattern with calibrating materials on reactive carrier	2002-05-02
Michael Grunze	Ruprecht-Karls-University	US2002054851	32p-polyphosphazenes	2002-05-09
Michael Grunze	Ruprecht-Karls-University	US6399295	Use of wicking agent to eliminate wash steps for optical diffraction-based biosensors	2002-06-04
Michael Grunze	Ruprecht-Karls-University	DE10059349	Ultrastructure analysis of nano-structures, comprises forming grid on carrier with reactive features formed by nano-lithography to give set binding sites and specific orientations for analysis by photon or particle beams	2002-06-06
Michael Grunze	Ruprecht-Karls-University	WO02064666	Substrates containing polypho-sphazene as matrixes and substrates containing polyphosphazene with a microstructured surface	2002-08-22

Table J.1.
European Patents Related to Biosensing in Sites Visited by WTEC Panel (1999–2003).

Researcher	Organization	Patent number	Patent Title	Year issued
Michael Grunze	Ruprecht-Karls-University	WO02077073	Plastic articles having a polyphosphazene coating	2002-10-03
Michael Grunze	Ruprecht-Karls-University	WO03015719	Device based on nitinol with a polyphosphazene coating	2003-02-27
Nico de Rooij	University of Neuchâtel	EP0923957	Liquid droplet spray device for an inhaler suitable for respiratory therapies	1999-06-23
Nico de Rooij	University of Neuchâtel	EP0988527	Electrochemoluminescent detector	2000-03-29
Nico de Rooij	University of Neuchâtel	EP1129741	Spray device for an inhaler	2001-09-05
Nico de Rooij	University of Neuchâtel	EP1149602	Spray device for an inhaler suitable for respiratory therapies	2001-10-31
Nico de Rooij	University of Neuchâtel	EP1236974	Apparatus for measuring a pressure at two points of a flowing fluid	2002-09-04
Nico de Rooij	University of Neuchâtel	WO02071001	Device for measuring pressure in two points of a fluid flow	2002-09-12
Nico de Rooij	University of Neuchâtel	EP1273346	Multi-channel fluid dispensing apparatus	2003-01-08
Nico de Rooij	University of Neuchâtel	WO03004163	Multi-channel fluid dispenser	2003-01-16
Nico de Rooij	University of Neuchâtel	US6509195	Electrochemoluminescent detector	2003-01-21
Dr. Livache	University of Grenobel	EP0890651	Analysing chip with local heating electrodes	1999-01-13
Dr. Livache	University of Grenobel	JP11127900	Analyzer of chip base comprising electrode equipped with partial heating means	1999-05-18
Dr. Livache	University of Grenobel	JP11148911	Supporting body or structural body for electrode, etc.	1999-06-02
Dr. Livache	University of Grenobel	FR2772926	Testing method for electronic integrated circuit	1999-06-25

Table J.1.
European Patents Related to Biosensing in Sites Visited by WTEC Panel (1999–2003).

Researcher	Organization	Patent number	Patent Title	Year issued
Dr. Livache	University of Grenobel	WO9934227	Device and method for testing an electronic chip sensitive element	1999-07-08
Dr. Livache	University of Grenobel	WO0047317	Method for producing addressed ligand matrixes on a support	2000-08-17
Dr. Livache	University of Grenobel	US6187914	Nucleoside derivatives, and their use in oligonucleotide synthesis	2001-02-13
Dr. Livache	University of Grenobel	US6197949	Electronically conductive polymer/nucleotide copolymer. Preparation method therefore and use thereof	2001-03-06
Dr. Livache	University of Grenobel	US6207797	Method for reducing the surface reactivity of copolymers produced by electrochemical polymerization	2001-03-27
Dr. Livache	University of Grenobel	US6255677	Chip-based analysis device compri-sing electrodes with localized heating	2001-07-03
Serge Cosnier	University of Grenobel	US6197881	Electrically conductive copolymers and their preparation	2001-03-06
Serge Cosnier	University of Grenobel	FR2798145	Electrically conductive polymers with light-activatable groups which can be grafted on with biomolecules, e.g. Proteins or enzymes, used for the production of electronic biosensors	2001-03-09
Serge Cosnier	University of Grenobel	WO0112699	Electrically conductive polymers capable of being covalently grafted on by light, method for obtaining same and uses as supports in probes for specific identification in electronic biosensors	2001-02-22

Table J.1.
European Patents Related to Biosensing in Sites Visited by WTEC Panel (1999–2003).

Researcher	Organization	Patent number	Patent Title	Year issued
Frieder Scheller	Potsdam University	US2003029231	Mobile hand-held unit comprising a reusable biosensor	2003-02-13
Frieder Scheller	Potsdam University	WO0150126	Mobile hand-held unit comprising a reusable biosensor	2001-07-12
Frieder Scheller	Potsdam University	US6171238	Portable hand-held device with a biosensor	2001-01-09
Frieder Scheller	Potsdam University	DE19938369	Detecting molecular interactions, useful e.g., for screening or diagnosis, based on variation in movement of particles loaded with motor proteins	2001-03-01
Otto Wolfbeis	University of Regensburg	JP11005796	Indacene derivative	1999-01-12
Otto Wolfbeis	University of Regensburg	US5942189	Luminescence-optical method and sensor layer for quantitative determination of at least one chemical component of a gaseous or liquid sample	1999-08-24
Otto Wolfbeis	University of Regensburg	US5981746	Luminescence indicator	1999-11-09
Otto Wolfbeis	University of Regensburg	EP0973033	Ion sensor	2000-01-19
Otto Wolfbeis	University of Regensburg	JP2000028532	Ion sensor	2000-01-28
Otto Wolfbeis	University of Regensburg	US6046055	Luminescence-optical method and sensor layer for quantitative determination of at least one chemical component of a gaseous or liquid sample	2000-04-04
Otto Wolfbeis	University of Regensburg	DE19856152	New heterocyclylalkene substituted quinolinium and pyridinium derivatives useful for labelling of biomolecules, particles and pharmaceuticals	2000-06-08

Table J.1.
European Patents Related to Biosensing in Sites Visited by WTEC Panel (1999–2003).

Researcher	Organization	Patent number	Patent Title	Year issued
Otto Wolfbeis	University of Regensburg	WO0034394	Pyridine dyes and quinoline dyes used as markers for biomolecules, polymers, medicaments, and particles	2000-06-15
Otto Wolfbeis	University of Regensburg	WO0042438	Optical-chemical sensor for detecting chloride	2000-07-20
Otto Wolfbeis	University of Regensburg	WO0136973	Method for solubilising optical markers	2001-05-25
Otto Wolfbeis	University of Regensburg	US2002034826	Optical-chemical Sensor	2002-03-21

Table J.2.
Japanese Patents Related to Biosensing in Sites Visited by WTEC Panel (1999–2003).

Research er	Organization	Patent Title	Year issued
Naoya Ogata		Ion conductor and manufacturing method of the same	2002-04-26
Naoya Ogata		Organic electroluminescent element material, organic electroluminescent element using the same and method for producing the organic electr…	2002-02-26
Naoya Ogata		Secondary battery using polymer electrolyte	2002-02-08
Naoya Ogata		Highly heat-resistant coating composition, organic- solvent-soluble polyimide, and highly heat-resistant film and its production	2001-07-03
Naoya Ogata		Lithium ion conductive polymer film using DNA	2001-04-13
Naoya Ogata		Chemical substance capturing agent, resin composition, resin molded body, and tableware for baby and infant	2001-04-10
Naoya Ogata		Functionalized nanotubes	2001-03-20
Naoya Ogata		Method of making functionalized nanotubes	2001-03-20
Naoya Ogata		Anisotropic film and its production	2001-02-20
Naoya Ogata		Proton conductive polymer film using DNA	2000-10-20
Naoya Ogata		Composite of polysilamine and strong acid	1999-11-03
Naoya Ogata		Organic electroconductive material	2000-06-13
Naoya Ogata		Culture, culture bed and coating agent enable formation of spheroid and culture for long time of primary hepatic cell by using culture be	2000-05-23
Naoya Ogata		High molecular solid electrolyte	2000-04-07
Naoya Ogata		Functionalized nanotubes	1999-01-19

Table J.2.
Japanese Patents Related to Biosensing in Sites Visited by WTEC Panel (1999–2003).

Researcher	Organization	Patent Title	Year issued
Naoya Ogata		Composite of polycyclamin and strong acid	1999-11-03
Naoya Ogata		Composite of polysilamine and strong acid	1999-11-03
Naoya Ogata		Phase transition type optically active polymer and its production	1999-05-11
Eiichi Tamiya	JATST	Method of manufacturing a microfluidic structure, in particular a biochip, and structure obtained by said method	2002-12-19
Eiichi Tamiya	JATST	Method and device for transferring chemical substance	2002-05-21
Eiichi Tamiya		Novel biochip and its making method	2002-05-09
Eiichi Tamiya	JATST	Eccentric rotary table device and treatment device using the same	2002-05-08
Eiichi Tamiya	JATST	New microorganism	2001-12-25
Eiichi Tamiya	JATST	Deodorization method	2002-01-08
Eiichi Tamiya	JATST	Dioxin measuring method and device	2001-08-17
Eiichi Tamiya	JATST	Method and apparatus for measuring concentration of organic substance	2001-03-23
Eiichi Tamiya	JATST	Biosensor for measuring influence on metabolic activity of cell	2001-03-27
Eiichi Tamiya	JATST	Cold-active protease cp70	2001-03-13
Eiichi Tamiya	JATST	Polymerase chain reaction device having integrated microwell	2000-09-05
Eiichi Tamiya	JATST	Biosensor and fixation method for biological substance	2000-09-29
Eiichi Tamiya	JATST	Partially acylated chitosan particle and its preparation	2000-09-19
Eiichi Tamiya	JATST	New biological chip and analytical method	2000-09-14
Eiichi Tamiya	JATST	Method and device for measuring antigen	2000-09-08

Table J.2.
Japanese Patents Related to Biosensing in Sites Visited by WTEC Panel (1999–2003).

Researcher	Organization	Patent Title	Year issued
Eiichi Tamiya	JATST	Formaldehyde dehydrogenase immobilized filter	2000-08-15
Eiichi Tamiya	JATST	Measurement of concentration of organic substance	2000-08-02
Eiichi Tamiya	JATST	Colored polymer particle and its production	2000-06-27
S. Ramachandra Rao	JATST	Production of zinc oxide from acid soluble ore using precipitation method	2003-02-13
S. Ramachandra Rao	JATST	Production of zinc oxide from acid soluble ore using precipitation method	2002-01-24
S. Ramachandra Rao	JATST	Production of zinc oxide from acid soluble ore using precipitation method	2002-01-24
S. Ramachandra Rao	JATST	Platenolide synthase gene	1999-08-31
Shigeori Takenaka	Kyushu University	Water-soluble flourescent intercalator compound	2001-08-16
Shigeori Takenaka	Kyushu University	Detecting reagent for double-stranded nucleic acid and double-stranded nucleic acid detecting method	2002-07-10
Shigeori Takenaka	Kyushu University	Gene detection method, detection device, and detection chip	2002-07-25
Shigeori Takenaka	Kyushu University	Novel ferrocene-type polycyclic hydrocarbon derivatives, novel ferrocene-type naphthalenediimide derivatives, process for producing the same	2002-07-11
Shigeori Takenaka	Kyushu University	Detecting reagent for double-stranded nucleic acid and double-stranded nucleic acid detecting method	2002-07-10
Shigeori Takenaka	Kyushu University	Biomolecule microarray support, biomolecule microarray using the support, and method of fabricating the support	2002-05-29
Shigeori Takenaka	Kyushu University	Method for detecting nucleic acids	2002-06-20

Table J.2.
Japanese Patents Related to Biosensing in Sites Visited by WTEC Panel (1999–2003).

Researcher	Organization	Patent Title	Year issued
Shigeori Takenaka	Kyushu University	Probe for detecting a highly ordered structural site of a single stranded nucleic acid of a gene, and a method and a device for detecting	2002-06-06
Shigeori Takenaka	Kyushu University	Biomolecule microarray	2002-05-29
Shigeori Takenaka	Kyushu University	Protection of partial complementary nucleic acid fragment using a electroconductive chip and intercalator	2001-01-03
Shigeori Takenaka	Kyushu University	Biomolecule microarray support, biomolecule microarray using the support, and method of fabricating the support	2002-05-29
Shigeori Takenaka	Kyushu University	Threading intercalator having oxidation-reduction activity	2001-06-13
Shigeori Takenaka	Kyushu University	Gene detecting chip, detector, and detecting method	2002-03-20
Shigeori Takenaka	Kyushu University	Analysis of expression of gene using plural potentials	2002-06-20
Shigeori Takenaka	Kyushu University	Threading intercalator having oxidation-reduction activity	2001-06-13
Shigeori Takenaka	Kyushu University	Method for assaying complementarity of sample nucleic acid fragment	2001-11-20
Shigeori Takenaka	Kyushu University	Fluorescent intercalator and detection method for complementary nucleic acid fragment	2001-10-19
Shigeori Takenaka	Kyushu University	Method for detecting single nucleotide polymorphism (SNP) and point mutation in gene, detection apparatus and detection chip	2001-10-31
Shigeori Takenaka	Kyushu University	Gene detecting chip, detector, and detecting method	2001-10-17
Shigeori Takenaka	Kyushu University	Stitch in-type intercalator having oxidation-reduction activity	2001-08-21
Shigeori Takenaka	Kyushu University	Chip for detection of gene, detector and detection method	2001-09-07
Shigeori Takenaka	Kyushu University	Protein chip and method for detecting protein	2001-09-07

Table J.2.
Japanese Patents Related to Biosensing in Sites Visited by WTEC Panel (1999–2003).

Researcher	Organization	Patent Title	Year issued
Shigeori Takenaka	Kyushu University	Probe for detecting a highly ordered structural site of a single stranded nucleic acid of a gene, and a method and a device for detecting…	2001-09-25
Shigeori Takenaka	Kyushu University	Method for testing complementation of nucleic acid	2001-09-12
Shigeori Takenaka	Kyushu University	Method for testing complementation of nucleic acid	2001-09-12
Shigeori Takenaka	Kyushu University	Chemically modified supports and process for producing the same	2001-05-17
Shigeori Takenaka	Kyushu University	Fluorescent intercalator compound	2001-08-16
Shigeori Takenaka	Kyushu University	Gene detecting chip, detector, and detecting method	2001-10-17
Shigeori Takenaka	Kyushu University	Method of manufacturing n,n-disubstituted- naphthalenediimide	2001-06-19
Shigeori Takenaka	Kyushu University	Sewing-in type intercalator, detecting method for nucleic acid fragment, and detecting kit therefore	2001-06-22
Shigeori Takenaka	Kyushu University	Chemically modified border brim and method of manufacturing for the same	2001-05-17
Shigeori Takenaka	Kyushu University	Method for detecting partial complementary nucleic acid fragment	2001-04-27
Shigeori Takenaka	Kyushu University	Method for detecting single nucleotide polymorphism (SNP) and point mutation in gene, detection apparatus and detection chip	2001-08-01
Shigeori Takenaka	Kyushu University	Modification of oligonucleotide	2001-03-13
Shigeori Takenaka	Kyushu University	Quantitative analysis of biochemical compound utilizing electrochemical reaction	2000-07-12
Shigeori Takenaka	Kyushu University	Threading intercalator having oxidation-reduction activity	2001-06-13
Shigeori Takenaka	Kyushu University	Chemically modified supports and process for producing the same	2001-05-17
Shigeori Takenaka	Kyushu University	Gene detecting chip, detector, and detecting method	2001-10-17

Table J.2.
Japanese Patents Related to Biosensing in Sites Visited by WTEC Panel (1999–2003).

Researcher	Organization	Patent Title	Year issued
Shigeori Takenaka	Kyushu University	DNA analyzing element, PNA analyzing element, highly sensitive method for determining sample nucleic acid piece with complementarity	2001-02-27
Shigeori Takenaka	Kyushu University	Method and device for detecting one base substitution SNP and point mutation of gene and detection chip	2001-08-01
Shigeori Takenaka	Kyushu University	Method for detecting and quantitatively determining sample nucleic acid fragment by scanning electrochemical microscope	2001-01-19
Shigeori Takenaka	Kyushu University	Method for detecting single nucleotide polymorphism (SNP) and point mutation in gene, detection apparatus and detection chip	2001-08-01
Shigeori Takenaka	Kyushu University	DNA chip, PNA chip, and their preparation methods	2001-02-21
Shigeori Takenaka	Kyushu University	Method for determining cholesterol using sensitization- type current-measuring tool	2000-11-30
Shigeori Takenaka	Kyushu University	Detection of partly complementary nucleic acid fragment	2001-01-03
Shigeori Takenaka	Kyushu University	Double-stranded DNA fragment having electroconductivity and water-soluble fullerene derivative	2000-10-17
Shigeori Takenaka	Kyushu University	Method for determining analyte using intensifying current-measuring tool	2000-07-18
Shigeori Takenaka	Kyushu University	Sensitized type detecting method for DNA	2000-05-26
Shigeori Takenaka	Kyushu University	Quantitative analysis of biochemical compound utilizing electrochemical reaction	2000-07-12
Shigeori Takenaka	Kyushu University	DNA sensor and detection of DNA	2000-05-09
Shigeori Takenaka	Kyushu University	Probe for detecting specific single-stranded nucleic acid site of gene, detection of specific single-stranded nucleic acid site of gene,	2001-09-25
Fumiaki Emoto	Matsushita	Fluorescence detecting device	2002-12-26

Table J.2.
Japanese Patents Related to Biosensing in Sites Visited by WTEC Panel (1999–2003).

Researcher	Organization	Patent Title	Year issued
Fumiaki Emoto	Matsushita	Fluorescence detecting device, method for producing the same, and fluorescence detecting method employing the same	2002-12-26
Fumiaki Emoto	Matsushita	Device for measuring extracellular potential, method of measuring extracellular potential by using the same and apparatus for quickly screen	2002-07-18
Fumiaki Emoto	Matsushita	Thin-film transistor array	2001-12-21
Fumiaki Emoto	Matsushita	Transmission type liquid crystal display device	2001-08-03
Fumiaki Emoto	Matsushita	Active matrix type thin film transistor board	2000-11-07
Fumiaki Emoto	Matsushita	Liquid crystal display device and manufacture thereof	2000-10-24
Fumiaki Emoto	Matsushita	Driving circuit of liquid crystal display device	2000-10-20
Fumiaki Emoto	Matsushita	Liquid crystal display element and liquid crystal display device	2000-09-08
Fumiaki Emoto	Matsushita	Manufacture of thin film transistor	2000-09-08
Fumiaki Emoto	Matsushita	Liquid crystal display device	1999-08-27
Fumiaki Emoto	Matsushita	Active matrix display device	1999-07-21
Fumiaki Emoto	Matsushita	Liquid crystal display device	1999-07-30
Fumiaki Emoto	Matsushita	Image display device	1999-03-05
Hiroaki Oka	Matsushita	Cell diagnosing method and device and apparatus used for it	2003-02-27
Hiroaki Oka	Matsushita	Cell potential measuring electrode and measuring apparatus using the same	2003-02-04
Hiroaki Oka	Matsushita	Method and apparatus for detecting physicochemical changes emitted by biological sample	2003-02-05

Table J.2.
Japanese Patents Related to Biosensing in Sites Visited by WTEC Panel (1999–2003).

Researcher	Organization	Patent Title	Year issued
Hiroaki Oka	Matsushita	Method and apparatus for detecting physicochemical changes in a biological sample	2003-02-05
Hiroaki Oka	Matsushita	Extracellular recording electrode	2003-01-03
Hiroaki Oka	Matsushita	Signal detecting sensor provided with multi "electrode…	2002-12-12
Hiroaki Oka	Matsushita	Significant signal extracting method, recording medium, and program	2002-11-07
Hiroaki Oka	Matsushita	Device for measuring extracellular potential, method of measuring extracellular potential by using the same and apparatus for quickly screen…	2002-08-28
Hiroaki Oka	Matsushita	Apparatus and method for screening, olfactory mucosa stimulating compound found by the screening method, and therapeutic apparatus and elect	2002-08-28
Hiroaki Oka	Matsushita	Cell potential measuring electrode and measuring apparatus using the same	1999-07-08
Hiroaki Oka	Matsushita	Process and device for producing silicon for solar cell	2002-01-29
Hiroaki Oka	Matsushita	Integral multiple electrode for extracellular recording	2002-08-22
Hiroaki Oka	Matsushita	Carbonizing furnace	2002-01-23
Hiroaki Oka	Matsushita	Extracellular recording integrated composite electrode	2002-08-22
Hiroaki Oka	Matsushita	Apparatus and method for screening, olfactory mucosa stimulating compound found by the screening method, and therapeutic apparatus and elect	2002-08-28
Hiroaki Oka	Matsushita	Induction heating furnace and heat processing apparatus	2001-11-22
Hiroaki Oka	Matsushita	Electrode for extracellular recording	2001-10-10
Hiroaki Oka	Matsushita	Cell potential measuring electrode and measuring apparatus using the same	1999-07-08

Table J.2.
Japanese Patents Related to Biosensing in Sites Visited by WTEC Panel (1999–2003).

Researcher	Organization	Patent Title	Year issued
Hiroaki Oka	Matsushita	Method and apparatus for screening laminated ceramic electronic component	2001-02-09
Hiroaki Oka	Matsushita	Monitor system	2000-09-29
Hiroaki Oka	Matsushita	Cell potential measuring electrode and measuring apparatus using the same	2000-10-17
Hiroaki Oka	Matsushita	Cell potential measuring electrode and measuring apparatus using the same	1999-07-08
Hiroaki Oka	Matsushita	Brake pedal actuating force detector	1999-09-21
Hiroaki Oka	Matsushita	Data gathering and recording device	1999-09-17
Hiroaki Oka	Matsushita	Electrode for measuring cell potential and measuring apparatus by using the same	1999-07-08
Hiroaki Oka	Matsushita	Cell potential measuring electrode and measuring apparatus using the same	1999-07-08
Hirokazu Sugihara	Matsushita	Cell potential measuring electrode and measuring apparatus using the same	2003-02-04
Hirokazu Sugihara	Matsushita	Method and apparatus for detecting physicochemical changes emitted by biological sample	2003-02-05
Hirokazu Sugihara	Matsushita	Methods and device for in vitro detection and characterization of psychoactives using analysis of repetitive electrical activity in a neurona	2003-01-28
Hirokazu Sugihara	Matsushita	Method and apparatus for detecting physicochemical changes in a biological sample	2003-02-05
Hirokazu Sugihara	Matsushita	Extracellular recording electrode	2003-01-03
Hirokazu Sugihara	Matsushita	Signal detecting sensor provided with multi "electrode…	2002-12-12
Hirokazu Sugihara	Matsushita	Significant signal extracting method, recording medium, and program	2002-11-07
Hirokazu Sugihara	Matsushita	Device for measuring extracellular potential, method of measuring extracellular potential by using the same and apparatus for quickly scree…	2002-07-18

Table J.2.
Japanese Patents Related to Biosensing in Sites Visited by WTEC Panel (1999–2003).

Researcher	Organization	Patent Title	Year issued
Hirokazu Sugihara	Matsushita	Methods and device for in vitro detection and characterization of psychoactives using analysis of repetitive electrical activity in a...	2000-12-28
Hirokazu Sugihara	Matsushita	Cell potential measuring electrode and measuring apparatus using the same...	1999-07-08
Hirokazu Sugihara	Matsushita	Integral multiple electrode for extracellular recording	2002-08-22
Hirokazu Sugihara	Matsushita	Extracellular recording integrated composite electrode	2002-08-22
Hirokazu Sugihara	Matsushita	Electrode for extracellular recording	2001-10-10
Hirokazu Sugihara	Matsushita	Measurement of complete electrical waveforms of tissue or cells	2001-10-02
Hirokazu Sugihara	Matsushita	Methods and device for in vitro detection and characterization of psychoactives using analysis of repetitive electrical activity in a neuron	2000-12-28
Hirokazu Sugihara	Matsushita	Delivery device for motor-bicycle	2000-11-28
Hirokazu Sugihara	Matsushita	Delivery device for motor-bicycle	2000-11-28
Hirokazu Sugihara	Matsushita	Methods and device for in vitro detection and characterization of psychoactives using analysis of repetitive electrical activity in a...	2000-12-28
Hirokazu Sugihara	Matsushita	Planar electrode	2000-11-21
Hirokazu Sugihara	Matsushita	Cell potential measuring electrode and measuring apparatus using the same...	2000-10-17
Hirokazu Sugihara	Matsushita	Cell potential measuring electrode and measuring apparatus using the same...	1999-07-08
Ichiro Yamashita	Matsushita	Solid electrolytic capacitor and production method thereof, and conductive polymer polymerizing oxidizing agent solution	2003-02-11
Ichiro Yamashita	Matsushita	Method for precisely machining microstructure	2002-12-19

Table J.2.
Japanese Patents Related to Biosensing in Sites Visited by WTEC Panel (1999–2003).

Researcher	Organization	Patent Title	Year issued
Ichiro Yamashita	Matsushita	Nucleotide detector, process for producing the same and process for forming fine particle membrane	2002-03-13
Ichiro Yamashita	Matsushita	Nucleotide detector, process for producing the same and process for forming fine particle membrane	2002-03-13
Ichiro Yamashita	Matsushita	Nucleotide detector, process for producing the same and process for forming fine particle membrane	2002-03-13
Ichiro Yamashita	Matsushita	Nucleotide detector, process for producing the same and process for forming fine particle membrane	2002-03-13
Ichiro Yamashita	Matsushita	Gas concentration measuring apparatus and combustion furnace	2001-03-23
Ichiro Yamashita	Matsushita	Method and apparatus for treating harmful substance	2001-03-06
Ichiro Yamashita	Matsushita	Transmitter, receiver and multi-rate transmission system using them	2000-02-18
Nobuhiko Ozaki	Matsushita	Method and apparatus for detecting physicochemical changes emitted by biological sample	2003-02-05
Nobuhiko Ozaki	Matsushita	Method and apparatus for detecting physicochemical changes in a biological sample	2003-02-05
Nobuhiko Ozaki	Matsushita	Device for measuring extracellular potential, method of measuring extracellular potential by using the same and apparatus for quickly screen...	2002-11-07
Nobuhiko Ozaki	Matsushita	Attachment for display rack for books	2001-02-13
Nobuhiko Ozaki	Matsushita	Capacitive force measuring apparatus	2001-02-21
Tetsuo Yukimasa	Matsushita	Extracellular recording electrode	2003-01-03

**Table J.2.
Japanese Patents Related to Biosensing in Sites Visited by WTEC Panel (1999–2003).**

Researcher	Organization	Patent Title	Year issued
Tetsuo Yukimasa	Matsushita	Apparatus and method for screening, olfactory mucosa stimulating compound found by the screening method, and therapeutic apparatus and ele...	2002-08-21
Tetsuo Yukimasa	Matsushita	Extracellular recording integrated composite electrode	2002-08-22
Tetsuo Yukimasa	Matsushita	Apparatus and method for screening, olfactory mucosa stimulating compound found by the screening method, and therapeutic apparatus and electr...	2002-08-28
Tetsuo Yukimasa	Matsushita	Scanning type probe microscope probe and method of producing the same, and a scanning type probe microscope having this probe and polymer pro...	2002-08-21
Tetsuo Yukimasa	Matsushita	Device for measuring extracellular potential, method of measuring extracellular potential by using the same and apparatus for quickly screen...	2002-07-18
Tetsuo Yukimasa	Matsushita	Apparatus and method for screening, olfactory mucosa stimulating compound found by the screening method, and therapeutic apparatus and electr...	2002-08-28
Tetsuo Yukimasa	Matsushita	Scanning type probe microscope probe and method of producing the same, and a scanning type probe microscope having this probe and polymer pro	2002-08-21
Tetsuo Yukimasa	Matsushita	Integral multiple electrode for extracellular recording	2002-08-22
Tetsuo Yukimasa	Matsushita	Electrode for extracellular recording	2002-08-22
Tetsuo Yukimasa	Matsushita	Manufacturing method of electric double-layer capacitor and manufacturing device thereof	2001-09-14
Tetsuo Yukimasa	Matsushita	Electric double-layered capacitor	1999-08-31
Kenji Yokoyama	NIAIST	5_"amidino_"n_"2_"aminophenethyl_"n_ " hydroxybenzenesulfonamide derivative, medicinal composi..	2003-02-27

Table J.2. Japanese Patents Related to Biosensing in Sites Visited by WTEC Panel (1999–2003).

Researcher	Organization	Patent Title	Year issued
Norihiko Minoura	NIAIST	Novel sulfated saccharide and process for producing the same	2002-12-27
Norihiko Minoura	NIAIST	Detection sensor of vero toxin produced by e-sherichia coli o-157 and detection method thereof	2002-01-23
Norihiko Minoura	NIAIST	Sugar chain derivative and method for producing the same	2001-12-11
Norihiko Minoura	NIAIST	Discrimination of protein	2001-02-27
Norihiko Minoura	NIAIST	Amino acid polymer having carboxybetaine type structure in side chain and its production	2000-03-28
Norihiko Minoura	NIAIST	Material capable of selectively adsorbing and desorbing protein and its production	2000-07-25
Norihiko Minoura	NIAIST	Sulfated galactose compound, its intermediate and production of sulfated galactose compound	1999-11-16
Norihiko Minoura	NIAIST	Sulfated galactose polymer	1999-11-16
Norihiko Minoura	NIAIST	Sulfated galactose compound, its intermediate and production of sulfated galactose compound	1999-11-16
Norihiko Minoura	NIAIST	Contact lens containing natural biopolymer and its production	1999-02-26
Norihiko Minoura	NIAIST	Substance having protein molecule-discriminating function and its production	1999-01-26
Soichi Yabuki	NIAIST	Enzyme electrode	2001-08-03
Soichi Yabuki	NIAIST	Oxygen electrode	2001-08-03
Yukari Sato	NIAIST	Designing method, cad device, computer program, and storage medium	2002-05-17
Yukari Sato	NIAIST	Designing method, cad apparatus and storage medium	2002-02-28
Yukari Sato	NIAIST	New adjuvant and vaccine using the same	2000-06-20

Table J.2.
Japanese Patents Related to Biosensing in Sites Visited by WTEC Panel (1999–2003).

Researcher	Organization	Patent Title	Year issued
Shigeru Toyama	National Rehabilitation Center for the Disabled (NRCD)	Solid state imaging device and its manufacturing method	2001-08-17
Shigeru Toyama	NRCD	Solid-state image pick-up device and its manufacturing method	2001-06-22
Shigeru Toyama	NRCD	Solid-state image pickup element and manufacture thereof	2000-08-29
Shigeru Toyama	NRCD	Solid-state image sensing element and manufacture thereof	2000-04-07
Shigeru Toyama	NRCD	Flue gas treatment equipment capable of treating dioxin	2000-01-11
Shigeru Toyama	NRCD	Schottky barrier type solid-state image pickup element and image pickup device using it	1999-10-19
Shigeru Toyama	NRCD	Surface plasmon resonance system immunoassay device	1999-10-05
Shigeru Toyama	NRCD	Measuring system of physiological phenomenon by sensor fusion	1999-10-05
Shigeru Toyama	NRCD	Rear surface irradiation solid state image sensor and its fabrication	1999-05-25
Masahiko Hara	RIKEN	Hybridization substrate, method of manufacturing same, and method of use for same	2003-01-29
Masahiko Hara	RIKEN	Hybridization substrate, method of manufacturing same, and method of use for same	2003-01-29
Masahiko Hara	RIKEN	Substrate for detecting base sequences, method of manufacturing the substrate, and method of detecting base sequences using the substrate	2001-12-27
Masahiko Hara	RIKEN	Antiviral mask	2002-05-14
Masahiko Hara	RIKEN	Laminated molding and its molding method	2002-05-08
Masahiko Hara	RIKEN	Liquid gun and liquid shot	2002-04-16

Table J.2.
Japanese Patents Related to Biosensing in Sites Visited by WTEC Panel (1999–2003).

Researcher	Organization	Patent Title	Year issued
Masahiko Hara	RIKEN	Substrate for detecting base sequence, method for producing substrate for detecting base sequence and method for detecting base sequence	2001-12-27
Masahiko Hara	RIKEN	Substrate for detecting base sequences, method of manufacturing the substrate, and method of detecting base sequences using the substrate	2001-12-27
Masahiko Hara	RIKEN	Method for molding resin molded object	2001-12-25
Masahiko Hara	RIKEN	Voice monitoring system using laser beam	2001-11-13
Masahiko Hara	RIKEN	Virus infection-preventive agent for domestic animal	2000-02-15
Masahiko Hara	RIKEN	Glucose absorption inhibitor	1999-11-02
Masahiko Hara	RIKEN	Identification of yeast-like fungus by biotechnological method	1999-06-29
Masahiko Hara	RIKEN	Oligonucleotide for identifying microbe and identification of microbe by using the same	1999-05-25
Masahiko Hara	RIKEN	Voice monitoring system using laser beam	2001-11-13
Eiry Kobatake	Tokyo Inst. Tech.	Light emitting method of acridinium derivative and method of detecting substance to be examined using same	1999-03-09
Masuo Aizawa	Tokyo Inst. Tech.	Deep water organism-carrying and raising vessel	2002-05-28
Masuo Aizawa	Tokyo Inst. Tech.	Neutral fat measuring sensor	2001-12-14
Masuo Aizawa	Tokyo Inst. Tech.	High-pressure culture apparatus	2001-09-25
Masuo Aizawa	Tokyo Inst. Tech.	Environmental equipment for biology experiment	2000-12-12
Masuo Aizawa	Tokyo Inst. Tech.	Biosensor using dehydrogenase and coenzyme	2000-02-02

Table J.2.
Japanese Patents Related to Biosensing in Sites Visited by WTEC Panel (1999–2003).

Researcher	Organization	Patent Title	Year issued
Masuo Aizawa	Tokyo Inst. Tech.	Light emitting method of acridinium derivative and method of detecting substance to be examined using same	1999-04-23
Koji Sode	TUAT	Oxygen electrode	2002-09-19
Koji Sode	TUAT	Glucose dehydrogenase	2002-09-19
Koji Sode	TUAT	Novel glucose dehydrogenase and process for producing the dehydrogenase	2002-05-10
Koji Sode	TUAT	Enzyme-mimicking polymers	2002-03-21
Koji Sode	TUAT	Glucose dehydrogenase	2002-06-12
Koji Sode	TUAT	Glucose dehydrogenase	2002-05-22
Koji Sode	TUAT	Novel glucose dehydrogenase and process for producing the dehydrogenase	2002-05-10
Koji Sode	TUAT	Enzyme-mimicking polymers	2002-03-21
Koji Sode	TUAT	Kit for assaying saccharified protein	2001-11-29
Koji Sode	TUAT	Glucose dehydrogenase	2002-01-30
Koji Sode	TUAT	Glucose dehydrogenase	2002-01-02
Koji Sode	TUAT	3,3'-diketotrehalose	2000-09-21
Koji Sode	TUAT	Glucose dehydrogenase	2002-01-30
Koji Sode	TUAT	Glucose dehydrogenase	2002-01-02
Koji Sode	TUAT	3,3'-diketotrehalose	2000-09-21
Koji Sode	TUAT	Modified glucose dehydrogenase	2000-08-15

Table J.2.
Japanese Patents Related to Biosensing in Sites Visited by WTEC Panel (1999–2003).

Researcher	Organization	Patent Title	Year issued
Tadashi Matsunaga	TUAT	Water quality monitoring device	2002-03-27
Tadashi Matsunaga	TUAT	Mold odor substance detector	2001-11-22
Tadashi Matsunaga	TUAT	Aquatic organism fouling-preventing conductive composition, aquatic organism fouling-preventing conductive coating, and a method of preve…	2001-09-26
Tadashi Matsunaga	TUAT	Method of retarding growth of microorganism	2001-09-26
Tadashi Matsunaga	TUAT	Plasmid obtained by cloning eicosapentaenoic acid- biosynthesizing genes and cyanobacterium producing eicosapentaenoic acid	2001-05-29
Tadashi Matsunaga	TUAT	Simultaneous measuring method for multitudinous examination item for diagnosis of diabetes based on chemilumnescence reaction	2001-04-13
Tadashi Matsunaga	TUAT	Electrochemical stain prevention apparatus of submerged structure and process for producing submerged structure used in this apparatus	2000-03-15
Tadashi Matsunaga	TUAT	Electrochemical antifouling device comprising underwater structure and method of producing underwater structure used for the device	2000-03-15
Tadashi Matsunaga	TUAT	Protein-bound magnetic particles and process of producing the same	2000-03-07
Tadashi Matsunaga	TUAT	New DNA sequence and plasmid vector containing the same	1999-10-19
Tadashi Matsunaga	TUAT	Metallic nitride, thermal-sprayed coating thereof and production of member for electrochemical biological control or contamination prevent	1999-09-28

Table J.2.
Japanese Patents Related to Biosensing in Sites Visited by WTEC Panel (1999–2003).

Researcher	Organization	Patent Title	Year issued
Kazuya Kikuchi	University Tokyo	Fluorescent probes for zinc	2002-12-27
Kazuya Kikuchi	University Tokyo	Exposure apparatus and method	2002-12-19
Kazuya Kikuchi	University Tokyo	Measuring method using long life fluorescence of excitation type	2003-01-02
Kazuya Kikuchi	University Tokyo	Connector contact and method of manufacturing the same	2002-12-03
Kazuya Kikuchi	University Tokyo	Fluorescent probes for the quantitation of zinc	2002-11-27
Kazuya Kikuchi	University Tokyo	Ip3 receptor ligands	2002-03-20
Kazuya Kikuchi	University Tokyo	Fluorescent probes for the quantitation of zinc	2002-11-27
Kazuya Kikuchi	University Tokyo	Measuring method using long life fluorescence of excitation type	2003-01-02
Kazuya Kikuchi	University Tokyo	Measuring method using long life fluorescence of excitation type	2003-01-02
Kazuya Kikuchi	University Tokyo	Fluorescent probes for the quantitation of zinc	2002-11-27
Kazuya Kikuchi	University Tokyo	Device for adjusting optical element angle of optical instrument	2001-03-06
Kazuya Kikuchi	University Tokyo	New fluorescent probe which is obtained by modifying both ends of substrate peptide with fluorescent light- emitting compounds and is use	2000-11-21
Yoshio Umezawa	University Tokyo	Probe for analyzing protein-protein interaction and method of analyzing protein-protein interactions with the use of the same	2002-08-07
Yoshio Umezawa	University Tokyo	Probe for visualizing phosphorylation/dephosphorylation of protein and method of detecting and quantifying phosphorylation/dephosphorylation	2002-10-03
Yoshio Umezawa	University Tokyo	Cgmp- visualizing probe and a method of detecting and quantifying of cgmp by using the same	2002-08-28

Table J.2.
Japanese Patents Related to Biosensing in Sites Visited by WTEC Panel (1999–2003).

Researcher	Organization	Patent Title	Year issued
Yoshio Umezawa	University Tokyo	Electrochemical detection method of complementarity to nucleic acid bases	2002-09-19
Yoshio Umezawa	University Tokyo	Cgmp-visualizing probe and method of detecting and quantifying cgmp by using the same	2002-08-28
Yoshio Umezawa	University Tokyo	Probe for analyzing protein-protein interaction and method of analyzing protein-protein interactions with the use of the same	2002-08-07
Yoshio Umezawa	University Tokyo	Visible cgmp probe and method for detecting and determining cgmp therewith	2002-08-28
Yoshio Umezawa	University Tokyo	Probe for analyzing protein-protein interaction and method of analyzing protein-protein interactions with the use of the same	2002-08-07
Yoshio Umezawa	University Tokyo	Cgmp-visualizing probe and method of detecting and quantifying cgmp by using the same	2002-08-28
Yoshio Umezawa	University Tokyo	Method for diagnosing allergy-causing substance	2001-11-16
Yoshio Umezawa	University Tokyo	Method for assaying activity of multiple-resistant protein	2001-08-14
Yoshio Umezawa	University Tokyo	Screening method for agonist	2001-07-06
Yoshio Umezawa	University Tokyo	Light guide plate and its production	2001-02-16
Yoshio Umezawa	University Tokyo	Anion selective electrode	1999-12-10

APPENDIX K. BIBLIOMETRIC STUDY OF WORLD BIOSENSORS RESEARCH, 1997–2002

This comparative bibliometric study of international biosensors research for the period 1997–2002 is the work of Prof. Grant Lewison of the Centre for Information Behaviour and the Evaluation of Research, City University, London, undertaken at the request of the World Technology Evaluation Center. It is intended to inform and complement the WTEC Panel Report on International Research and Development in Biosensing.

SUMMARY OF FINDINGS

1. This study examined world outputs of papers in biosensors research in the Science Citation Index (SCI) from 1997–2002, as retrieved by a "filter" defined in terms of specialist journals and title words by Dr. Jeff Newman of Cranfield University. Output from the United States was compared with that from the European Union + Switzerland (EU+CH), Japan (JP) and China (CN) on a number of indicators, and the papers published by U.S. National Institutes of Health (NIH) grantees in biosensing were separately identified and analyzed.
2. Over the six-year period, there were 4,701 world papers in biosensors research, of which 25% were from the United States, 40% from the European Union and Switzerland, 11% from Japan, and nearly 10% from China. The biosensors subfield has been growing quite rapidly, especially in China, which has almost eight times as many publications as would be expected on the basis of its presence in the biomedical literature. By contrast, the United States only published 63% as many papers as expected, showing that the subfield is not a research priority. However nearly 6% of the U.S. papers were classed as "reviews," an indicator of the reputation of their authors, compared with only 3.5% in Europe and 3% in Japan. The U.S. research team size was a little smaller than that in Europe or Japan, though this is typical in biomedical research.
3. When the papers were classified by the field of their journals and by their citation impact categories, it was clear that the subfield was dominated (45%) by chemistry, especially in China (70%). However, in the United States, over half the papers were in clinical medicine or biomedical research journals, and only one third in chemistry journals. The U.S. papers in all the major fields, and overall, were in the highest citation impact journals; this is a good indicator of

research performance. In comparison, European papers were in lower impact journals, followed by those of Japan and China.

4. U.S. papers from 1997–98 were also highly cited in the year of publication and four subsequent years. They were significantly better cited (mean, nearly 16 cites per paper, cpp) than papers from Europe (10 cpp), Japan (9 cpp) or China (5 cpp), though not better than those of Canada. The papers citing to biosensors research are mainly in chemistry journals and within the same subfield (37% of all citations). There is some evidence that Americans tend to be less aware than Europeans of the biosensors research being conducted in Japan and China.

INTRODUCTION

Study Objectives

The purpose of this bibliometric study was to examine world published output (articles and reviews only from 1997–2002) in peer-reviewed journals covered by the Institute of Scientific Information's Science Citation Index (SCI)©, in the subfield of "biosensors," a subset of the total field of "biosensing." For the purposes of this study, the subfield was defined as follows:

> Biosensors are defined as analytical devices incorporating a biological material (e.g., tissue, microorganisms, organelles, cell receptors, enzymes, antibodies, nucleic acids, natural products, etc.); a biologically-derived material (e.g., recombinant antibodies, engineered proteins, aptamers, etc.); or a biomimic (e.g., synthetic catalysts, combinatorial ligands, imprinted polymers) intimately associated with or integrated within a physicochemical transducer or transducing microsystem, which may be optical, electrochemical, thermometric, piezoelectric, magnetic, or micromechanical.

The papers were identified by means of a "filter" based on specialist journals and title words, and their bibliographic details were downloaded to a spreadsheet for analysis. The following analyses were performed:

- The numbers of papers per year worldwide, and from the United States, Japan, and the European Union + Switzerland
- The sizes of the teams of researchers publishing these papers
- The fraction of papers from the different geographical groups that were classed as "reviews," a mark of the reputation of their authors

- The fields of the journals in which these groups of papers were published
- The potential impact of these groups, as given by the five-year citation impact factors of the journals in which they were published
- The actual impact (numbers of citations over a five-year period beginning with the year of publication) of the papers published in 1997 and 1998
- The trans-field influence of the papers and the nationality of the citing papers, both as ratios of observed-to-expected numbers of citations
- The numbers of papers published from 1997–2002 by biosensing grantees of the U.S. National Institutes of Health (NIH) listed in Appendix D of this WTEC biosensing report (*vs.*), the fields of the journals in which they published, and their potential impact
- The characteristics of the grantees' papers that were also contained in the biosensors file

During the course of the work, it became apparent that a substantial share of world output in the subfield of biosensors came from the People's Republic of China. It was therefore decided to conduct the analysis on five groups of papers: (1) from the United States, (2) from Japan (JP), (3) from the European Union (15 member states) + Switzerland (EU+CH), (4) from China (CN) and (5) from the Rest of the World (RoW). Because of international co-authorship, there was some overlap between the first four of these groups, but the fifth group was distinct.

METHODOLOGY

Definition of the Subfield of Biosensors

The procedure for defining biomedical "filters" has been described in earlier works by this author (Lewison 1996; Lewison 1999). It involves a partnership between a bibliometrician and an expert in the subject. In this study, the subject expertise was provided by Dr. Jeff Newman of Cranfield University. The resulting filter was calibrated against the outputs of 20 leading authors so as to determine the recall of the filter (r), or the fraction of the relevant papers that the filter retrieved. The outputs of the filter, both from the 20 authors and from others, were used to calculate the precision of the filter (p), or the fraction of the papers that the filter retrieved that were relevant. After several rounds of modification of the filter, which was coded BIOSE, the final value of p was 0.88 and that of r was 0.86. These are regarded as good for a somewhat unspecific subject area, being only just below the target values of 0.9.

The bibliographic details downloaded to an Excel spreadsheet were

- Authors' names
- Title of paper
- Document type (article or review)
- Full source (journal, year, volume, issue, pagination)
- Addresses

In the SCI, all the authors' names and all the addresses are given, thus making it possible to identify papers from a country when it does not provide the first address or the address for correspondence.

Papers were downloaded from the SCI in the CD-ROM edition for 1997-September 2003, but only those papers published in the six years 1997–2002 were retained. The final file contained 4,701 papers.

Addresses and Relative Commitments

The papers in the file were filtered using a special macro[4] that could search the whole address field, which sometimes exceeds 255 characters, for the presence of the names of individual countries. These were the United States, the 15 Member States of the European Union, Switzerland, Japan, and some other countries that had been found to be significant contributors to biosensor research from an analysis of the 8,634 addresses listed in the file. In the tables and charts, digraph ISO (International Organization for Standardization) country codes are used, as listed in Table J.1 below.

4 This was kindly provided to the author by Dr. Judit Bar-Ilan of the Hebrew University of Jerusalem.

Table J.1.
ISO Codes and Country Names Used for the Address Analysis of Biosensor Papers

Codes for member states of the European Union are shown in bold.

Code	*Country*	*Code*	*Country*	*Code*	*Country*	*Code*	*Country*
AT	Austria	**DE**	Germany	IN	India	**SE**	Sweden
AU	Australia	**DK**	Denmark	**IT**	Italy	SG	Singapore
BE	Belgium	**ES**	Spain	JP	Japan	SK	Slovakia
BR	Brazil	**FI**	Finland	**LU**	Luxembourg	TR	Turkey
CA	Canada	**FR**	France	KR	South Korea	TW	Taiwan
CH	Switzerland	**GR**	Greece	**NL**	Netherlands	**UK**	United Kingdom
CN	Peoples Rep. China	**IE**	Ireland	**PT**	Portugal	UR	Ukraine
CZ	Czech Republic	IL	Israel	RU	Russia	US	United States

In the SCI, the UK is divided into four separate countries: England, Wales, Scotland, and North Ireland. Care was taken that papers from Denmark Hill (in London) were not included with those of Denmark. Papers from the 15 Member States of the European Union, plus Switzerland, were grouped together for analysis purposes, although they were also attributed to their individual countries. All the counts were integer, that is, a paper with authors in the United States and in Germany would count as unity for each country (and not as 0.5).

The number of papers from each country (and the EU+CH) can be compared with the world total (4,701 papers) to give each country's percentage presence. This value can be compared with the percentage presence of each country in biomedical research overall. This had previously been determined by application of a complex address filter to the SCI (Lewison and Paraje 2004). Thus, the UK published 7.4% of world biosensor papers but 10.0% of biomedical papers, so its relative commitment (RC) to the biosensor subfield was 7.4/10.0 = 0.74. Some other countries, of course, would have an RC greater than unity.

Classification of Journals by Research Level

In order to gain an impression of how basic or applied the work of the different geographical groups was, the journals used for the biosensors were classified by "research level" (RL) on a scale from 1 = clinical or applied to 4 = basic. This scale was originally developed by CHI Research, Inc. (Narin, Pinski, and Gee 1976), as a simple four-category scale based on expert opinion and journal-to-journal citation patterns. However, the system has since been revised; it is now based on the presence of clinical or basic title words in the articles in individual journals (Lewison and Paraje 2004) and gives a value for RL (calc) on a continuous scale from 1 to 4. Examples of journals used for biosensors research in each of six RL groupings are shown in Table J.2.

Table J.2.
Examples of Journals Used for Biosensors Research at Six Different Research Levels, and Percentage of the World Papers in Each Group of Journals

RL calc range	Examples*	% of Set
1.0 – 1.49	*Journal of Clinical Periodontology, Diabetes Care*	0.9
1.5 – 1.99	*Water Science and Technology, Clinica Chimica Acta*	1.9
2.0 – 2.49	*Clinical Chemistry, Water Research, Journal of Biomechanics*	2.7
2.5 – 2.99	*Annals of the New York Acad. of Sciences, Environmental Science & Technology*	3.7
3.0 – 3.49	*Biosensors & Bioelectronics, Electroanalysis, Sensors and Actuators B-Chemical*	44.5
3.5 – 4.0	*Analytical Chemistry, Analytica Chimica Acta, Electroanalysis*	46.2

*Over 90% of all biosensor papers are in journals classed as rather basic (RL calc > 3.0).

Classification of Journals by Potential Impact

The journals were also classified into four groups on the basis of the average number of citations received by papers published in them in the year 1998 and cited in the five years, 1998–2002. This is designated $C_{0\text{-}4}$. For ease of analysis, the journals have been given a potential impact category (PIC) from 1 (low) to 4 (very high) on the basis set out in Table J.3. Overall in biomedicine, typically 10% of papers are in PIC4 journals, 20% in PIC3 ones, 30% in PIC2 ones, and the remaining 40% in PIC1 ones; indeed, the critical values of $C_{0\text{-}4}$ were chosen so that these percentages would be found in most biomedical subfields. Table J.3 shows that, although there is a preponderance of papers in PIC2 journals over ones in PIC1 journals, there are fewer than is normal for biomedicine in the two top categories.

**Table J.3.
Four Journal Potential Impact Categories and Examples of Journals at Each,
with Percentage of the World Papers in Each Group of Journals**

PIC	5-year cite score	Examples	% of set
1	*Below 6*	*Talanta, Analytical Letters, Fresenius J. of Analytical Chemistry*	*28.9*
2	*From 6 to 11*	*Biosensors & Bioelectronics, Electroanalysis, Sensors and Actuators B*	51.1
3	*From 11 to 20*	*Analytical Chemistry, Biochemistry, Chemical Communications*	*13.5*
4	*20 and above*	*Journal of Biological Chemistry, Journal of the Amer. Chem. Society*	5.6

Determination of the Major Fields of Biosensor Papers

The CHI Research, Inc., classification system allocates each journal uniquely to one of about eight major fields, including biomedical research, chemistry, clinical medicine, and physics. The system is a bit arbitrary, but it has been used for some time for the National Science and Engineering Indicators published by the National Science Foundation. It has been used here in order to give an idea of what type of research has been included in the definition of biosensors research. It turns out to be predominantly chemistry (nearly 45%), followed by clinical medicine and biomedical research (about 21% each). However there are some differences between the geographical groups, as will be seen.

Determination of Citation Counts

Citations of papers by other scientific papers, as recorded in the SCI, are often used as a means of evaluating research. For a basic subject such as biosensors, see Table J.2 above, this may well be appropriate, but it is necessary to allow time for the peak of the citation curve to be passed, normally at three years after publication. For this reason, citations have been counted over the same five-year period used for the classification of journals (see Table J.3), so that, in principle, the numbers of citations actually received by a biosensor paper can be directly compared with the average for the journal in which it is published. However this ratio, of observed to expected citations, is NOT a good measure for the evaluation of research. The two scores, $C_{0\text{-}4}$ for the journal and $C_{0\text{-}4}$ for the individual paper, are distinct; however, both are useful indicators.

In this study, citation scores were determined for all world biosensor papers published in the two years 1997 and 1998; they numbered 1,342 in total. Search strategies were created for use with the SCI for the years 1997 to September 2003. Bibliographic details (full source and addresses only) of the citing papers were downloaded to individual files, labeled with codes indicative of the cited paper. A composite file was then created of all the citations to all the papers and citing papers retained only for the five years (1997–2001 or 1998–2002, as appropriate)[5]. These citing papers were then classified, just as the cited papers had been, geographically, by RL, by PIC, and by major field. The previous classifications of the cited papers were also copied across to the file, so that it was possible to create matrices of observed-to-expected citation rates in the different categories in order to test some hypotheses. For example, are U.S. papers cited as frequently by Europeans as one would expect, given the geographical distribution of all the citing papers? Are the citing papers more applied or clinical than the cited ones, which might suggest that the biosensors research is having practical applications?

Papers by NIH Grantees

With the list of grantees of the National Institutes of Health given in Appendix D, it was possible to search the SCI, at least for the last four years (2000–03) when outputs might have been expected, for papers authored by them. In total there are 187 names in Appendix D (a few have more than one grant). The NIH grantees were clearly identified with a research institution in the United States, although some appeared to combine their commercial work (presumably they received grants under the Small Business Innovation Research program) with an academic role. Of the 187 names, 16 did not appear at all during the four years for which they were searched, and a further 29 were ambiguous with clearly different people having the same name, and both (or all) plausibly working on biosensing. Some of the investigators appeared to have come from other countries, or to have gone abroad subsequently; only papers with a U.S. address were retained for these researchers. Altogether, it was possible to identify clearly the outputs of 142 of the grantees, some of whom had unique names and initials, and some of whom could be distinguished from homonyms by the use of a geographical filter.

The papers by these 142 NIH grantees from 1997–2002 were downloaded to a separate file for analysis: it contained 2,671 papers. The papers were categorized by journal in the same way as the ones in the biosensors subfield, and they were also matched across to the biosensors file so that the NIH grantees' outputs within biosensors research could be identified. (These last were in fact a very small minority of the grantees' total output, only 116 papers, or 4.3%.) However the NIH grantees' papers in the biosensor file accounted for 10% of the U.S. total of 1,169 papers.

---

5 About 10% of papers are processed late for the SCI and appear in the following year's CD-ROM. For this reason it was necessary to run the search strategies for one more year in order to collect these "late" citing papers.

Since it was only possible to identify papers by 142/187 = 76% of the grantees, it is necessary to increase this estimated percentage by about one third. Therefore, the estimated contribution of the NIH to U.S. biosensor research is about 13%.

RESULTS

Numbers of Papers from Geographical Areas and Countries

Table J.4 shows the numbers of biosensors papers from 31 leading countries, and the EU and Switzerland together, for 1997–2002. Also given are the numbers of biomedical papers and the ratio of each country's percentage presence in the world, which is its relative commitment (RC) to biosensors research.

Table J.4.
Leading Countries in Biosensors Research, 1997–2002 (in SCI)
BIOSE = biosensor papers, BIOM = biomedical research papers, RC = relative commitment

Code*	Country	BIOSE	% world	BIOM	% world	RC
World		4701	100.00	1586363	100.00	1.00
EU +CH	Europe	1900	40.42			
US	USA	1169	24.87	628837	39.64	0.63
JP	Japan	526	11.19	152714	9.63	1.16
DE	Germany	459	9.76	132192	8.33	1.17
CN	China	449	9.55	19187	1.21	7.90
UK	UK	349	7.42	158092	9.97	0.74
FR	France	273	5.81	99935	6.30	0.92
ES	Spain	206	4.38	41412	2.61	1.68
IT	Italy	202	4.30	70833	4.47	0.96
SE	Sweden	188	4.00	43472	2.74	1.46
RU	Russia	158	3.36	14763	0.93	3.61
CA	Canada	120	2.55	73643	4.64	0.55
IL	Israel	109	2.32	19948	1.26	1.84
KR	S Korea	90	1.91	14467	0.91	2.10
AU	Australia	84	1.79	43036	2.71	0.66
BR	Brazil	78	1.66	16539	1.04	1.59
CH	Switzerland	77	1.64	33677	2.12	0.77
CZ	Czech Rep.	74	1.57	8152	0.51	3.06

Table J.4.
Leading Countries in Biosensors Research, 1997–2002 (in SCI)
BIOSE = biosensor papers, BIOM = biomedical research papers, RC = relative commitment

Code*	Country	BIOSE	% world	BIOM	% world	RC
IN	India	74	1.57	19254	1.21	1.30
IE	Ireland	73	1.55	6559	0.41	3.76
NL	Netherlands	72	1.53	49686	3.13	0.49
DK	Denmark	53	1.13	22386	1.41	0.80
SG	Singapore	50	1.06	3755	0.24	4.49
TW	Taiwan	50	1.06	14199	0.90	1.19
TR	Turkey	47	1.00	12299	0.78	1.29
UR	Ukraine	46	0.98	2017	0.13	7.70
SK	Slovakia	45	0.96	3373	0.21	4.50
GR	Greece	42	0.89	7911	0.50	1.79
AT	Austria	37	0.79	18129	1.14	0.69
BE	Belgium	34	0.72	24989	1.58	0.46
PT	Portugal	29	0.62	5126	0.32	1.91
FI	Finland	26	0.55	19424	1.22	0.45

*There was no output from Luxembourg in biosensors research in 1997–2002.

During the period 1997–2002, world output rose from about 670 papers per year in 1997–98 to almost 950 papers in 2002; this shows that the subfield is growing rapidly—much more so than biomedicine overall, which only rose by 4% over this period. However, this rise was greater in some regions, including the United States, than in others, notably Europe, as can be seen from Figure J.1.

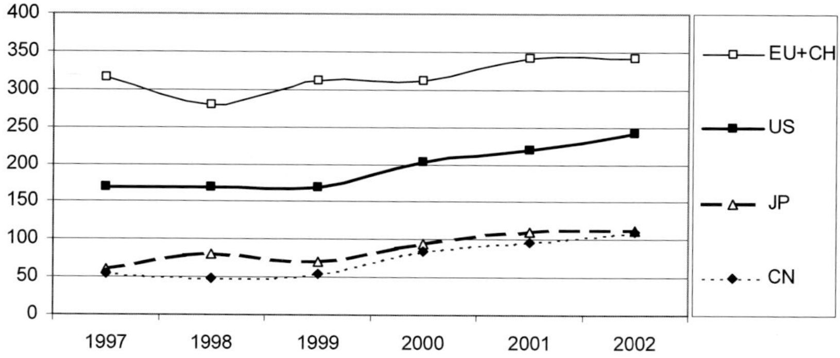

Fig. J.1. Variation in biosensors output from four countries or regions.
EU+CH = Europe, JP = Japan, CN = China.

China's output has more than doubled over the period and was almost equal to that of Japan in 2002. Because of the very high relative commitment to biosensors research in China (see Table J.4) and its high absolute output, it was decided to use it as a geographical comparator in the ensuing analysis, along with Japan and Europe.

Numbers of Authors Per Paper for Geographical Regions

This WTEC report suggests that the U.S. researchers are working in smaller groups than researchers in Europe and Japan. Figure J.2 shows the cumulative distribution of percentages of papers with different numbers of authors for the four geographical regions. It does appear that this suggestion has some basis in the biosensors papers: the median number of U.S. authors is 3.2 compared with 3.5 for the Europeans and 3.7 for the Japanese. (The means are 4.2, 4.3, and 4.5.) There are more than the expected numbers of U.S. biosensors research papers with only two or three authors. However these differences are also observed in biomedicine generally and may be due more to cultural differences in the way authors are included than in actual differences in the size of the research teams.

Percentages of Papers Classed as Reviews

Many reviews are written by invitation, and therefore, the numbers of such documents can be regarded as a mark of appreciation of the standing of the author(s). There were differences in the percentages of such papers between the regions, as shown in Table J.5.

The United States writes a higher than average percentage of reviews, but Canada is, on this indicator, even more highly esteemed. There is quite a big variation between the leading European countries, with Ireland, Spain, Germany, and the UK writing the most reviews.

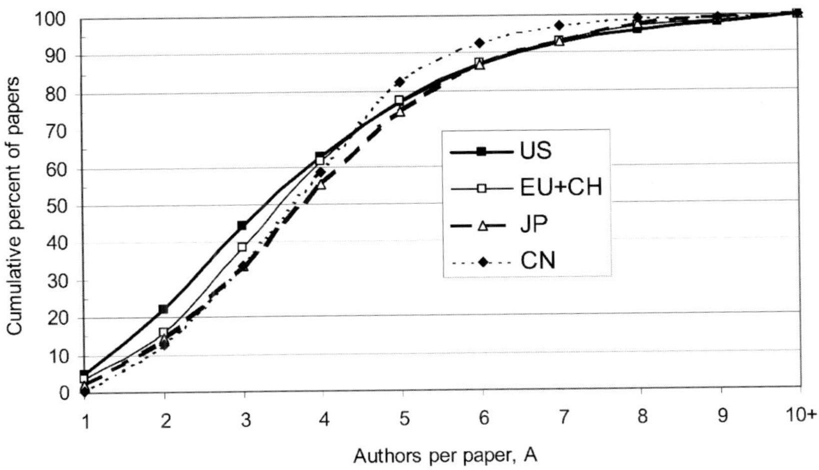

Fig. J.2. Cumulative distribution of numbers of authors on biosensors papers from four geographical regions or countries, 1997–2002.

Table J.5.
Percentages of Biosensor Research Papers Classed by the SCI as "Reviews," 1997–2002

Code	Articles	Reviews	Total	% Rev	Code	Articles	Reviews	Total	% Rev
World	4496	205	4701	4.4	DE	442	17	459	3.7
EU+CH	1833	67	1900	3.5	UK	336	13	349	3.7
US	1100	69	1169	5.9	FR	267	6	273	2.2
JP	510	16	526	3.0	ES	198	8	206	3.9
CN	443	6	449	1.3	IT	199	3	202	1.5
RoW	790	52	842	6.2	SE	183	5	188	2.7
CA	109	11	120	9.2	IE	67	6	73	8.2

Categorization of Papers from the Four Geographical Regions

Figure J.3 shows the distribution of papers from the regions in terms of the major fields of the journals in which they were published. It is apparent that chemistry journals dominate, but particularly so in China (70% of all Chinese papers). In the United States, chemistry only accounts for one third of the research output, and biomedical research journals (which include the multidisciplinary journals such as *Nature* and *Science*) contain almost as much, with nearly 24% of U.S. papers in clinical medicine journals.

This pattern of major fields in turn influences the overall distribution of papers by potential impact category (Figure J.4), because biomedical research and clinical medicine journals tend to be more highly cited than chemistry journals. But despite this caveat, U.S. research in all four of the major fields is in the highest impact journals, as can be seen in Figure J.5. Next comes biosensors research from the EU+CH, then from Japan, then from the Rest of the World, and finally from China.

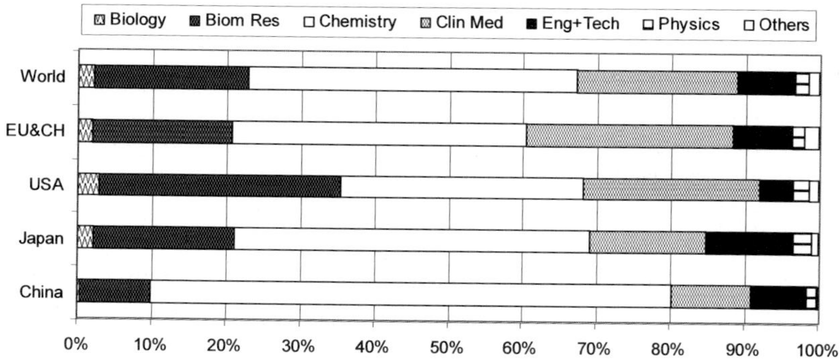

Fig. J.3. Distribution of biosensors papers from four geographical regions by the major field of the journals in which they are published, 1997–2002.

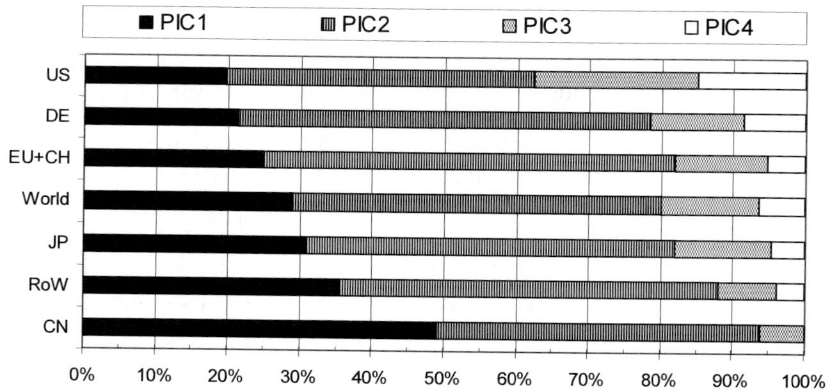

PIC = potential impact category from 1 (low) to 4 (very high)

Fig. J.4. Distribution of biosensors papers from five geographical regions or countries by the potential impact category of the journals in which they are published, 1997–2002. DE = Germany, RoW = Rest of the World (not US, EU+CH, JP, or CN). All differences in PIC distribution between adjacent regions are statistically significant.

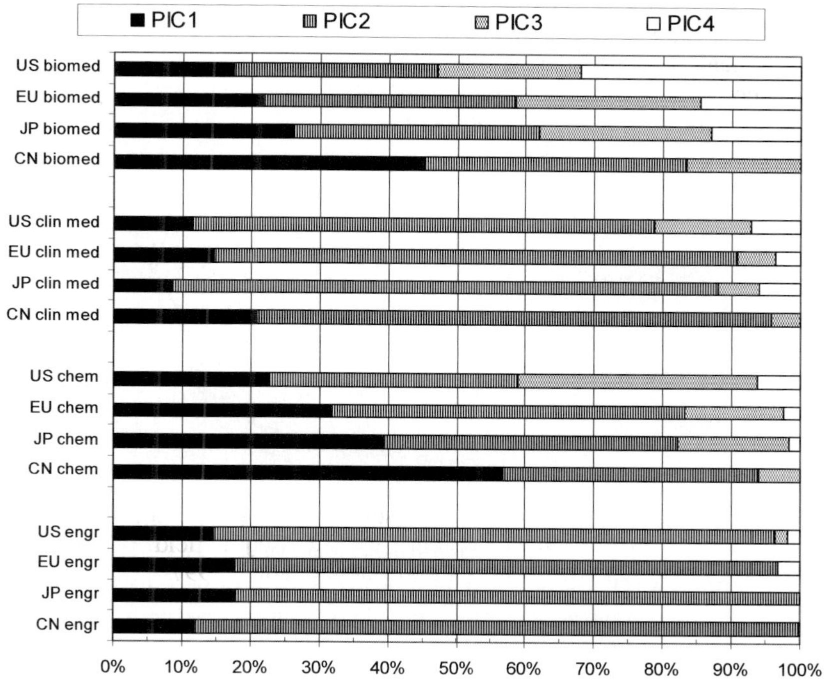

Fig. J.5. Distribution of biosensors papers from four regions or countries by potential impact category of the journals in which they are published, in four major fields: biomedical research, clinical medicine, chemistry and engineering.

Figure J.6 shows the cumulative distribution of papers classified by the research level of the journals, here grouped into ranges of 0.1 in RL. There is some evidence that the research outputs from Europe are somewhat more clinical or applied (median RL = 3.32) compared with ones from the United States and Japan (median RL = 3.44), but the differences between the five geographical regions/countries are rather small.

Papers by NIH Grantees

The papers from three-quarters of the named NIH grantees were downloaded to a separate file for analysis; it was explained in the section above on papers by NIH grantees that the others either had no detectable output, or their output could not be distinguished from that of homonyms. Compared with all U.S. papers, the NIH grantees' papers were slightly more in clinical medicine journals and in engineering and physics journals, see Table J.6, but the differences are not statistically significant.

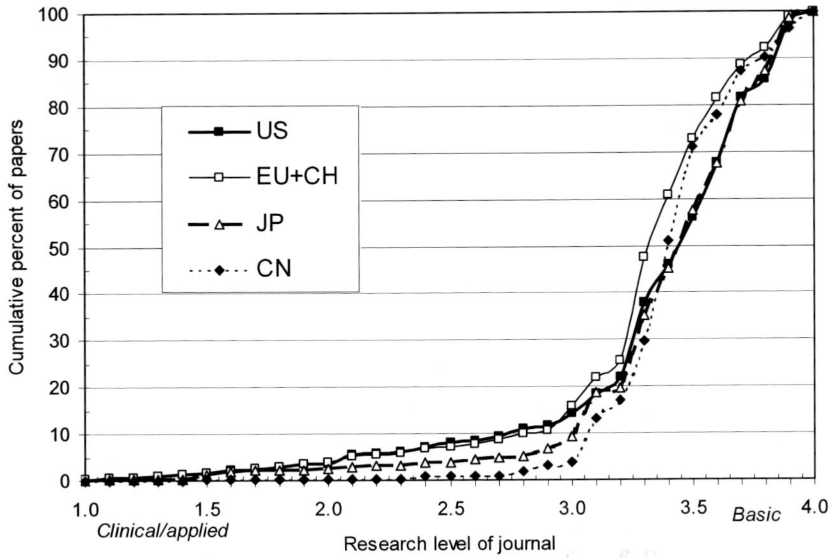

Fig. J.6. Distribution of biosensors papers from four world regions/countries by their journal research level (RL1 = clinical or applied, RL4 = basic research).

Table J.6.
Distribution of US and NIH Grantees' Papers by Major Field of Journal, 1997–2002, % of Totals

Field	All US	NIH
Biology	2.9	1.7
Biomedical research	32.5	31.9
Chemistry	32.8	30.2
Clinical medicine	23.8	25.0
Engineering and technology	4.7	6.9
Physics	2.1	4.3

Figure J.7 shows the distribution of papers by potential impact category of their journals for the NIH grantees that are outside the biosensors subfield, compared with bisensors papers from other U.S. researchers and the NIH grantees.

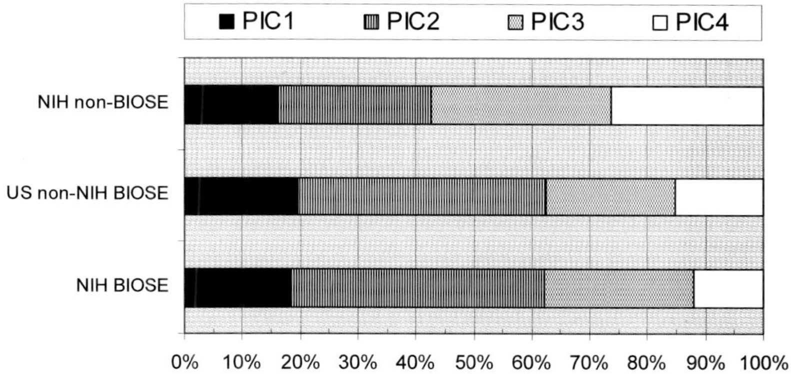

Fig. J.7. Distribution by potential impact category (PIC; 1 = low, 4 = very high) of papers by NIH grantees in biosensing, both within and outside of the biosensor subfield (BIOSE), and papers in biosensing by other U.S. authors.

There was no appreciable difference in PIC distribution between the biosensor papers of the NIH grantees and those of other U.S. researchers in the subfield, but the grantees' papers within the subfield were clearly in lower impact journals than the ones they wrote in other subfields ($p < 0.01\%$). This suggests that biosensors research is published in relatively low-impact journals.

Citation Scores of Biosensor Papers

Table J.7 shows, for the five geographical regions/countries, the numbers of papers published in 1997–1998 that received citation scores over the first five years of their exposure in different groups, called citation categories (citecats).

Table J.7
Citation Score Distribution ($C_{0\text{-}4}$) for 1997–1998 Biosensor Papers from Different Regions

	CC0	CC1	CC2	CC3	CC4	CC5	CC6	Total
Cites:	0	1-5	6-10	11-19	20-39	40-79	80+	
World	176	431	313	250	115	46	11	1342
EU+CH	72	183	156	123	45	14	3	596
USA	25	86	76	65	52	27	5	336
RoW	31	74	56	35	16	4	3	219
Japan	21	49	34	20	10	5	1	140
China	25	45	13	17	1	2	0	103

An "average" citecat value has been calculated. Although, as with the "mean" value of PIC, this is not strictly a correct procedure, it is useful to rank the regions in order in the chart (Figure J.8).

This figure shows that the U.S. papers received more citations than expected: only 7% were uncited (13% for the world), and 25% received 20 or more citations compared with 13% for the world and only 10% for the EU+CH. Despite its relatively large output of papers in biosensors, Chinese research is very poorly cited, with 24% of its papers uncited and only 3% receiving 20 or more citations. The ordering of the geographical groups is rather similar in Figure J.8 to that in Figure J.4, which was based on the potential impact of papers from all six years.

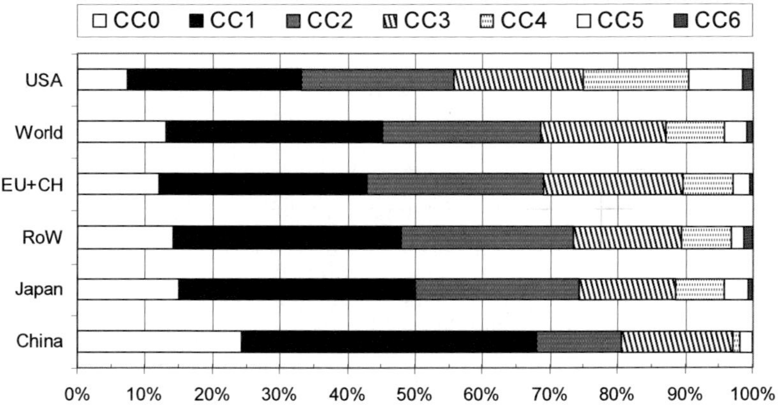

Fig. J.8. Distribution of 1997–1998 biosensors papers by citation category in the five years following publication ($C_{0\text{-}4}$ values).

Characteristics of Papers Citing to the Biosensor Papers

The citing papers were classified by major field, by geographical area, and by research level. Table J.8 shows the numbers of citations received by papers from each of the major fields (different rows) by papers in each major field (different columns). Clearly the subfield is dominated by chemistry, with 45% of the cited papers and as many as 52% of the citing papers in chemistry journals. Of the citing papers within five years of the date of publication of the cited papers, 37% were within the subfield of biosensors.

Table J.8.
Numbers of Citations by Major Field of Citing Journal (columns) to Biosensors Papers in Different Major Fields (rows) in Years 0–4 after Publication for 1997–98 Papers.

Cited field:	Biology	Biom. Res.	Chemistry	Clin. Med.	Earth + Sp.	Engr. +Tech.	Physics	Total
Biology	23	67	40	35	6	5	3	179
Biomed. Res	148	1887	972	470	35	73	72	3657
Chemistry	85	704	4553	546	48	262	86	6284
Clin. Med.	69	581	1363	764	30	182	63	3052
Earth + Sp.	8	17	48	15	40	6	0	134
Engr. + Tech.	13	67	375	79	6	101	40	681
Physics	1	9	46	16	5	12	21	110
Total	347	3332	7397	1925	170	641	285	14097

Table J.9 shows the ratio of observed to expected numbers of citations for each combination of cited and citing fields. The ratios on the diagonal are shown in bold

Table J.9.
Ratio of Observed to Expected Citations by Major Field of Citing and Cited Biosensor Paper, 1997–1998 Publications, Citations in Years 0–4

Cited field:	Biology	Biom. Res.	Chemistry	Clin. Med.	Earth + Sp.	Engr. + Tech.	Physics
Biology	**5.22**	1.58	0.43	1.43	2.78	0.61	0.83
Biomed. Res.	1.64	**2.18**	0.51	0.94	0.79	0.44	0.97
Chemistry	0.55	0.47	**1.38**	0.64	0.63	0.92	0.68
Clin. Med.	0.92	0.81	0.85	**1.83**	0.82	1.31	1.02
Earth + Space	2.43	0.54	0.68	0.82	**24.75**	0.98	0.00
Engr. + Tech.	0.78	0.42	1.05	0.85	0.73	**3.26**	2.91
Physics	0.37	0.35	0.80	1.07	3.77	2.40	**9.44**

Although the ratios on the diagonal are, as expected, all well above unity, there are some unexpected results. Earth and space seems to be a largely self-contained field, with an enormous preponderance of citations from that field (though there are actually more from chemistry journals). Biosensors research in physics journals also influences earth & space, as does research in biology journals.

Figure J.9 shows the mean research level (RL) of the citing journals for each group of cited papers in a set of journals with RL in a range of 0.1, for which there were at least 100 papers. (Most of these were for RL ≥ 3.) As expected, papers in very basic journals (RL > 3.5) tend to be cited by papers in somewhat more applied journals, but for papers in applied journals the reverse is the case—they are cited by more basic journals.

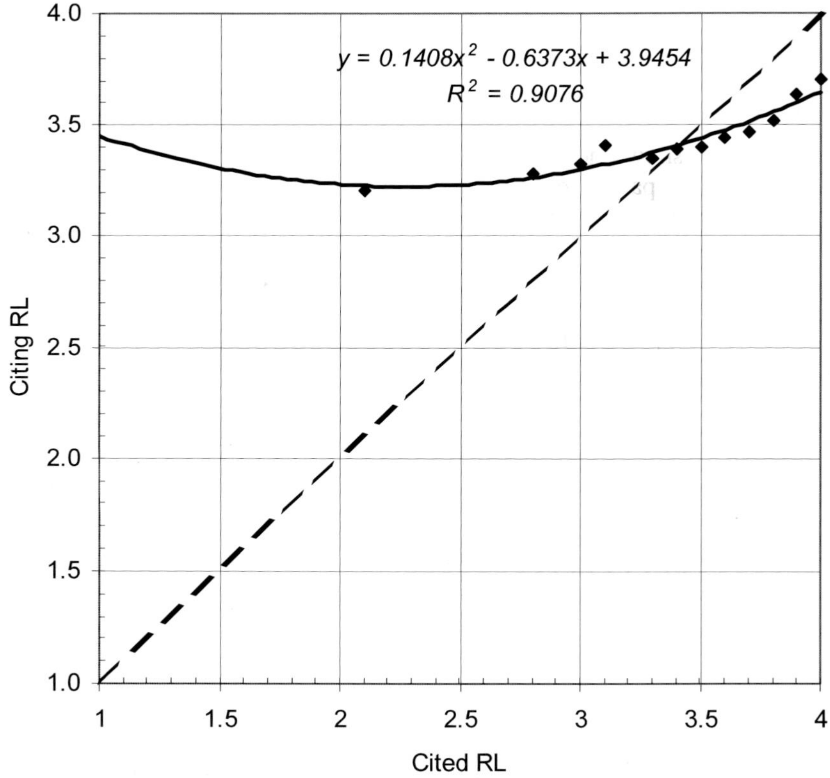

Fig. J.9. Mean research level of citing journals to groups of biosensor papers in different groups of cited journals (classed by research level: 1 = clinical or applied, 4 = basic research).

Most of the citing journals (Figure J.9) have a research level between 3.2 and 3.7. Table J.10 shows the pattern of geographical citations, and is laid out similarly to Table J.8.

Table J.10.
Citations From and To Biosensors Papers from Five Geographical Regions

From\Cited by:	**World**	**USA**	**EU+CH**	**JP**	**CN**	**RoW**
World	16820	5505	6549	1237	1115	3177
USA	6385	3264	1955	345	313	822
EU+CH	6964	1565	3772	334	408	1225
JP	1543	279	438	490	127	285
CN	672	114	187	243	213	153
RoW	2533	669	755	135	146	921

Table J.11 shows how often biosensors papers of the various regions cite biosensors papers of the other regions as a ratio of observed to expected citation rates. Again, the diagonal elements of the matrix are shown in bold, and all are above unity, showing that each region's researchers preferentially cite the papers from their own region. But there are some differences in behavior. U.S. researchers cite Japanese and Chinese research barely half as often as "expected," whereas for European researchers the deficit in citations is only about 28%. This suggests that U.S. researchers are paying less attention to biosensors research from these two countries than the Europeans are, in relation to their total numbers of references. The Japanese are giving a lot of attention to Chinese biosensor research, relatively slightly more even than the Chinese do to their own work.

Table J.11.
Ratio of Observed to Expected Numbers of Citations to Biosensors Papers
From and To Different Geographical Regions

From\Cited by:	**USA**	**EU+CH**	**JP**	**CN**	**RoW**
USA	**1.56**	0.79	0.73	0.74	0.68
EU+CH	0.69	**1.39**	0.65	0.88	0.93
JP	0.55	0.73	**4.32**	1.24	0.98
CN	0.52	0.71	4.92	**4.78**	1.21
RoW	0.81	0.77	0.72	0.87	**1.93**

DISCUSSION AND COMPARISONS

Outputs of Papers

The data in Table J.4 show that U.S. output, although the highest of any single country, is rather low in comparison with its output of biomedical research papers, where it averages almost 40% of world output. In comparison, biosensors research is a relatively weak subfield in the United States, with less than 25% of world output. Nevertheless, its output has been increasing during the last three years, and its share of world output has been rising slightly. Countries with a particularly strong relative commitment to biosensors research are first, China, and then some of the countries of eastern Europe such as Ukraine, Russia, Slovakia, and the Czech Republic. Biosensors research is also fairly strong in some East Asian countries such as Singapore and South Korea, and is above average in Japan and Taiwan.

U.S. biosensors research is characterized by a rather large number of papers in journals classified as "biomedical research," with correspondingly fewer in chemistry and engineering journals. By contrast, China's output is very largely in chemistry journals. European output (the European Union and Switzerland) includes rather more clinical medicine papers than that from other regions of the world.

Impact of Papers

Figures J.4 and J.5 make clear that U.S. biosensors papers are published in relatively high impact journals, both overall and in each of four major fields biomedical research, clinical medicine, chemistry, and engineering. Table J.7 and Figure J.8 show that the United States also has superiority in terms of actual citations received in the year of publication and four subsequent years ($C_{0\text{-}4}$) for 1997–98 papers. The difference in the distributions of citation categories between the U.S. and the European papers is very highly significant (^{2} = 38.8, 5 d/f; $p \ll 0.001\%$). Another measure of esteem of U.S. biosensors research is the percentage of its papers that are classed as "reviews". This was nearly 6% (Table J.5), higher than the average for European papers (3.5%) but lower than the figure for Canada (9%). Canada has an even lower relative commitment to biosensors research than the United States (only 0.55 relative to biomedicine). However its outputs, although small (120 papers in 1997–2002 and 43 in 1997–98), have a potential and actual impact as least as high as those of the United States (mean citecat = 2.49 compared with 2.40 for the United States).

Citing Papers

Analyses of the papers citing to all the biosensors papers published in 1997–98 (n = 1342) indicate that within the five-year window, there were 14,109 citations, or 10.5 per paper. This is not a high average for a biomedical subfield. For U.S. papers, the average score was 15.8 citations; for European papers 9.7 cites, for Japanese papers 9.5 cites, for Chinese papers only 5.4 cites. As expected, papers with authorship from both Japan and the United States, or from Europe and the United States, were more highly cited (mean cites per paper, 18.6 and 17.7 respectively). This apparent effect of international cooperation has been observed previously, but it is largely due to the larger numbers of researchers assembled for international projects. Thus, papers with authors from both the United States and Europe have an average of 6.1 authors per paper compared with 4.3 for all European papers and 4.2 for all U.S. ones; for U.S. papers co-authored with Japanese researchers, the average authorship rises to 7.0.

It appeared that U.S. authors were not citing much to papers from Japan and China (Table J.11), at least in comparison with the Europeans. They were also citing somewhat less to European papers compared with the relative frequency with which Europeans cited to U.S. papers. This, of course, may simply reflect the relatively higher impact of U.S. research observed above. However U.S. research is relatively less cited by Chinese scientists than is European research. This may reflect the smaller U.S. output in chemistry, with which the Chinese are mainly concerned.

REFERENCES

Lewison, G. 1996. The definition of biomedical research subfields with title keywords and application to the analysis of research outputs. *Research Evaluation* 6 (1):25–36.

Lewison, G. 1999. The definition and calibration of biomedical subfields. *Scientometrics* 46 (3):529–537.

Lewison, G., and G. Paraje. 2004. The classification of biomedical journals by research level. *Scientometrics*. In press.

Narin, F., G. Pinski, and H.H. Gee. 1976. Structure of the biomedical literature. *Journal of the American Society of Information Science* 27:25.

Thomson Institute of Scientific Information (ISI®). 2003. Science Citation Index (SCI). www.isinet.com/products/citation/scie/ .

APPENDIX L. GLOSSARY

AFM	atomic force microscopy
AIST	(Japan) National Institute of Advanced Industrial Science and Technology
ANN	artificial neural network (pattern recognition)
AT-cut	piezoelectric quartz crystal blanks cut in a specific direction with respect to the crystal axis
ATRL	Advanced Technology Research Laboratories (of Matsushita)
bioFET	biological field-effect transistor
BOD	biological oxygen demand (biosensors)
BSE	biological systems engineering
bZIP	basic (region) leucine zipper
CALI	chromophore assisted laser inactivation
CCD	charge-coupled device
c-EGFP	cytochrome c-enhanced green fluorescent protein
CFP	cyan fluorescent protein
CHEMFET	chemically modified field effect transistors
CIEF	capillary isoelectric focusing chip
CMOS	complementary metal-oxide semiconductor
COD	chemical oxygen demand (biosensors)
CSEM	Centre Suisse d'Électronique et de Microtechnique
dAT	dialkylthiophene
dNTP	deoxynucleoside triphosphate
DRIE	deep reactive ion etching
ECA	event condition action (distributed active database systems)
ECD	electrochemical detection
EDCs	endocrine disrupting chemicals
EGFP	enhanced green fluorescent protein
ELISA	enzyme-linked immunosorbant assay
EO	electroosmotic
EPFL	École Polytechnique Fédérale de Lausanne
ES	cell differentiation
ESEM	environmental scanning electron microscope
FCLA	fluorescent chemiluminescent agent
FND	ferrocenyl napthalene diimide
FPW	flexural plate wave

FRET	fluorescence resonance energy transfer
FTIR	Fourier transform infrared spectroscopy
GC-MS	gas chromatography-mass spectrometry
GFP	green fluorescent protein
GH	growth hormone
GPC	G protein coupled (receptor)
HF	high frequency
HTS	high-throughput screening
IC	integrated circuit
ICAT	isotope-coded affinity tag
ICPMS	inductively coupled plasma mass spectrometer/spectrometry
IgG	immunoglobulin G
IPG	immobilized pH gradient gel
IR	infrared
ISE	ion-selective electrode
ISFET	ion selective field affect transistor
ISP	integrated sensing and processing
ITO	indium-tin oxide
LAPS	light-assisted potentiometric spectroscopy
LC-MS	liquid chromatography-mass spectrometry
LEED	low energy electron diffraction
LEEPS	low-energy electron point source (microscopy)
LIGA	deep reactive ion etching (a German term, LIthographie, Galvanoformung, und Abformung, that is, X-ray lithography with electrodeposition of metal and sometimes injection molding)
LOD	limit of detection
LSAW	leaky SAW
LSPR	localized plasmon resonance spectroscopy
MALDI	matrix-assisted laser desorption/ionization
MALDI-TOF	matrix-assisted desorption/ionization time-of-flight mass spectrometry
MARS	magnetic acoustic resonant sensor,
MEKC	micellar electrokinetic chromatography
MEMS	microelectromechanical system(s)
MIP	molecularly imprinted polymer
MLR	multiple linear regression
MOEMS	microoptical electromechanical systems

MOMS	microoptical mechanical systems
MOSIS	IC fabrication service set of conventions; see www.mosis.org
MS	mass spectrometry
NEDO	(Japan) New Energy and Industrial Technology Organization
N-EGFP	nucleocapsid-enhanced green fluorescent protein
NEMS	nanoelectromechanical systems
NEST	(Europe) Network of Excellence in Sensing Technology
OCT	optical coherence tomography
PCA	principal component analysis
PCR	polymerase chain reaction
PDMS	poly dimethylsiloxane (polymer patterning)
PEBBLE	probes encapsulated by biologically localized embedding (sensors)
PET	photoinduced electron transfer
pI	isoelectric point
PIC	potential impact category
PLS	partial least squares (or projection to latent structures)
PTFE	polytetrafluoroethylene (polymer)
QCM	quartz (crystal) microbalance (see also QMB)
QMB	quartz (crystal) microbalance (see also QCM)
QMS	quadrupole mass spectrometer
QPLS	quadratic partial least squares
RIANA	RIver ANAlyser (river water analyzer)
RifS	reflectometric interference spectroscopy
RL	(mean) research level (bibliometrics)
ROS	reactive oxygen species (oxidative bursts)
SAW	surface acoustic wave
SCI©	Science Citation Index (bibliometrics)
SECM	scanning electrochemical microscopy
SELEX	segmented large X baryon spectrometer
SERRS	surface enhanced resonance Raman scattering
SERS	surface enhanced Raman scattering (spectroscopy)
SH-APM	shear-horizontal acoustic plate mode
SNOAM	scanning near-field optical/atomic force microscopy
SNOM	scanning near-field optical microscopy
SOM	self-organizing map/mapping (artificial neural networks)
SPR	surface plasmon resonance

ss-DNA	single-stranded DNA
STW	surface transverse wave
TEM	transmission electron microscope/microscopy
TIRF	total internal reflection fluorescence
TSM	thickness-shear mode resonator (popularly known as "quartz crystal microbalance")
UHV-STM	ultra-high vacuum scanning tunneling microscopy
UV	ultraviolet
VOA	variable optical attenuator
XPS	X-ray photoelectron spectroscopy
YFP	yellow fluorescent protein